Flow Dynamics and Tissue Engineering of Blood Vessels

IPEM–IOP Series in Physics and Engineering in Medicine and Biology

About the Series

Series in Physics and Engineering in Medicine and Biology will allow IPEM to enhance its mission to 'advance physics and engineering applied to medicine and biology for the public good.'

Focusing on key areas including, but not limited to:

- clinical engineering
- diagnostic radiology
- informatics and computing
- magnetic resonance imaging
- nuclear medicine
- physiological measurement
- radiation protection
- radiotherapy
- rehabilitation engineering
- ultrasound and non-ionising radiation.

A number of IPEM–IOP titles are published as part of the EUTEMPE Network Series for Medical Physics Experts.

Flow Dynamics and Tissue Engineering of Blood Vessels

Edited by
Arindam Bit
National Institute of Technology, Raipur, Chhattisgarh, India

Jasjit S Suri
AtheroPoint, Roseville, CA, USA

IOP Publishing, Bristol, UK

ISBN 978-0-7503-2088-7 (ebook)
ISBN 978-0-7503-2086-3 (print)
ISBN 978-0-7503-2089-4 (myPrint)
ISBN 978-0-7503-2087-0 (mobi)

DOI 10.1088/978-0-7503-2088-7

Version: 20201101

IOP ebooks

British Library Cataloguing-in-Publication Data: A catalogue record for this book is available from the British Library.

Published by IOP Publishing, wholly owned by The Institute of Physics, London

IOP Publishing, Temple Circus, Temple Way, Bristol, BS1 6HG, UK

US Office: IOP Publishing, Inc., 190 North Independence Mall West, Suite 601, Philadelphia, PA 19106, USA

Arindam Bit wants to dedicate this book to his wife Soma and daughter Anishka for their continuous supports and unbearable understanding while writing every pieces of this book. He also wants to acknowledge all his PhD students and research scholars for their supports with the art and designing of every chapters.

Contents

Preface

Cardiovascular diseases are mounting in numbers among the affected population with an unhealthy life-style. They contribute to 30% mortality rate among non-communicable diseases world-wide (as per the report of WHO 2018). Multiple dimensional strategies are being adapted continuously to address these challenges of cardiovascular disorders. Modern day scientific research aims to address such vital problems with the help of a team of multi-disciplinary research quotients. This book provides a reference to this kind of heterogeneous reader to better understand a medical problem from an engineering approach, and simultaneously utilize engineering tools for clinical articles associated with cardiovascular disorder.

The book is divided into four parts. The first part deals with an introductory explanation of blood vessels by anatomical and physiological aspects, cardiovascular system, coupling of vascular structure with various associated tissues, influence of vascular structure to musculoskeletal structure and vice versa. This part gives an overall introduction about the clinical aspects of blood vessels at various locations in the body to the engineering communities.

The second part deals with the numerical modeling of blood and clinically relevant blood vessels. It describes various mathematical models for realizing blood as a Newtonian and non-Newtonian fluid. Blood flow for various scales of vessels at different Reynolds numbers within the purview of physiological flow rate is also numerically modeled here. Three-dimensional modeling of blood vessels with stenoses is also mentioned in this part. The methodology to handle a patient specific geometrical model for numerical analysis of blood flow in a clinical blood vessel is also explained in this part. A significant feature of this part is an elaboration of the blood flow in a diseased vessel with the help of experimental procedure. A chapter is dedicated to explaining this phenomenon within this part.

The third part describes the protocol for blood vessel fabrication, different materials required for its synthesis, and different technology associated with tissue engineered blood vessel fabrication. A chapter is dedicated to the illustration of different types of biomaterials which can be used for tissue engineering applications, particularly for soft tissue engineering. It highlights silk fibroin as a biomaterial for soft tissue application. Another chapter of this part illustrates various fabrication techniques for the production of blood vessels, emphasizing the formation of different layers of the vessels to house different types of cells on it. The third chapter of this part deals with the bioprinting technique exclusively for the development of blood vessels. Different types of bioinks are described, with their physical and chemical features for optimizing 3D bioprinting techniques for the development of the blood vessel. The fourth chapter of this part illustrates various kinds of bioreactors being employed for the production and culture of cells in a 3D annular construct of blood vessel.

The fourth part of the book represents the controlling aspects to be considered while designing and developing artificial blood vessels. It consists of three chapters. The first chapter deals with the biological and physiological controlling aspects of blood vessels of different circulatory systems. The second chapter illustrates control engineering of circulatory systems with different feedback, linear, non-linear, and pumping systems. The final chapter explains the controlling aspects and control system modeling of a cardiovascular unit with different sub-units, at different clinical conditions.

Thus, this books defines complete engineering and bio-engineering of blood vessels. It can serve as a reference book, textbook, and comprehensive guide for all researchers working in the field of vascular engineering. This book also opens up the opportunities for a minor discipline in vascular engineering. From such a perspective, this book will provide the founder of such a centre with comprehensive guidelines to build an inter-disciplinary research team to successfully address the cardiovascular engineering challenges.

Arindam Bit, PhD
Jasjit Suri, PhD

Editor biographies

Arindam Bit

Arindam Bit received his PhD in Biofluidic Engineering at Jadavpur University, India in 2016. He is working as an Assistant Professor in the Department of Biomedical Engineering at National Institute of Technology, Raipur, India. He has more than 30 research publications in peer-reviewed journals. He served as guest editor to *Journal of Healthcare Engineering*. He is a reviewer to many journals such as *Journal of Bionanoscience*, *Journal of Healthcare Engineering*, *Journal of Psycholinguistic Research*, *Journal of Cognitive Neurodynamics*, *Journal of Scientific Report-Nature*, *Journal of Medical & Biological Engineering & Computing*, *Progress in Computational Fluid Dynamics: An International Journal*, *Acta Biomechanics and Bioengineering*, and *Journal of Computer Methods in Medicine*. His primary research interest includes biofluid mechanics, tissue engineering, microfluidic systems for tissue engineering and point-of-care device development, organoid development, and cognitive neuroscience.

Jasjit S Suri

Jasjit S Suri has spent over 25 years in the industry as an innovator, scientist, a visionary, and is an internationally known world leader in the field of biomedical engineering and its management. Dr Suri has published ~700+ technical articles, which inlcudes (a) over 100 journal papers/book chapters in artificial intelligence (AI) alone, (b) 200+ NIH high impact factor publications listed in PubMed.gov, (c) 100+ journals in vascular technology alone, (c) ~50+ books, (d) 50+ US/European Patents and Trademarks, all this, leading to 16 000 citations and H-index of 60. Dr Suri was awarded the *Director General's Gold Medal* in 1980, *Fellow of American Institute of Medical and Biological Engineering* by National Academy of Sciences, Washington DC (2004) for his outstanding contributions in imaging, *Life Time Achievement Award* from Marquis for his outstanding contributions in healthcare imaging (2018) and *Fellow of American Insitute of Ultrasound in Medicine* (2019) for outstanding contributions to AI application in ultrasound.

Contributors

Arindam Bit
National Institute of Technology
Raipur, India

Jasjit S Suri
Point of Care Division
Global Biomedical Technologies Inc., United States

Karabi Ganguly
JIS College of Engineering
Kolkata, India

Khemraj Deshmukh
National Institute of Technology
Raipur, India

Sharda Gupta
National Institute of Technology
Raipur, India

Swati Sikdar
JIS College of Engineering
Kolkata, India

Achalla Ranjani
National Institute of Technology
Raipur, India

Bharti Mishra
National Institute of Technology
Raipur, India

Sonali Gore
Indian Institute of Technology Gandhinagar, India

IOP Publishing

Chapter 1

Anatomy and physiology of blood vessels

Arindam Bit, Jasjit S Suri and Achalla Ranjani

A transport system is an inherent feature of an organ system. In human beings, the vascular system is one representation of a transport system. It innervates in most of the organ systems. Innervation of blood vessels in organs is associated with exchange of gases, nutrients, and maintaining homeostasis of organs. Based on transport of gases, blood vessels are classified as oxygenated (arteries) and de-oxygenated (veins). Nomenclature of a blood vessel is done based on the organ (or organ system) they are innervating. This chapter reveals the structural and functional details of various categories of blood vessel.

1.1 Introduction

Blood vessels transport blood throughout the body. They are one of the most important tissues as they are present extensively in each and every organ of the human body. Blood vessels are of mainly three kinds: arteries, capillaries and veins. Arteries take blood through the heart to other parts of the body while veins supply blood to the heart. Capillaries are the smallest blood vessels, acting as the site for exchange for oxygen, waste products and nutrients in between tissues and blood.

The heart is the major organ among the whole blood vascular system and is made of tube-like muscle. The blood is supplied to all the parts and organs of the body by contraction through a sequence of tubes called arteries. These arteries will finally end in minute vessels called arterioles that later open into microscopic and tiny vessels known as capillaries. The blood flows through the capillaries and collects finally in larger blood vessels called veins. The blood returns to the heart by veins. The blood flow with the help of the heart and blood vessels is considered as blood circulation. The human heart through which the blood is circulated throughout the body is divided into two halves by septa, i.e. right half and left half. Each half is again classified into cavities where the upper cavity is known as the atrium and the lower one as the ventricle. The right half of the heart contains impure blood also called venous blood and the left half contains pure blood or arterial blood. Pure blood is transported to a large artery called the aorta from the left ventricle. From there, blood will be distributed to various body parts through the capillaries. The impure blood is collected by veins that supply the blood back to the right atrium. From the right atrium, the impure blood is sent to the right ventricle and then supplied by pulmonary arteries to the lungs. Again, the blood is transported to the left atrium by means pulmonary veins. Blood is carried to the left ventricle from the left atrium and the process is continuous.

In this chapter, we will be discussing the vascular part of the circulatory system which includes various blood vessels that transport blood throughout the human body and circulatory systems.

1.2 Structure of blood vessel

All the blood vessels have the same features but will vary slightly in structure. Arteries have walls thicker than veins as they receive blood from the heart, which creates great pressure. Each and every blood vessel has a hollow passage through which the blood flows that is known as the lumen. Arteries have narrower lumina than those of veins, which helps in maintaining blood pressure throughout the system. Arteries are more rounded in structure than veins due to the thicker walls and small lumina. Arteries and veins have large walls that are made of living cells such as elastic and collagenous fibers. These cells need nourishment and removal of waste which cannot be fulfilled by blood present in the lumen. Moreover, the walls of the vessels are too thick to pass through the cells. To overcome this problem, large blood vessels have smaller blood vessels inside the walls called vasa vasorum which means vessels of the vessel. These smaller blood vessels provide proper exchange of nutrients. All the blood vessels have three different layers called tunica. From inner side to outer side, these layers are tunica intima, tunica media and tunica externa.

1.2.1 Tunica intima

Tunica intima, also known as tunica interna, is made of layers of epithelial and connective tissue. Tunica intima is lined by endothelium that is simple squamous epithelium continuously throughout the vascular system. The major reason for a blood clot is damage to the endothelium by which blood is exposed to the collagenous fibers present beneath. Basal lamina, also called basement membrane, is present next to the endothelium, binding it to connective tissue. This membrane maintains flexibility providing strength and is permeable, which enables material to pass through. The outer layer is thin containing aerolar connective tissue with elastic fibers that provide flexibility and have collagenous fibers which help in providing additional strength.

Thick layers of elastic fibers called internal elastic membranes are present in large arteries forming a boundary with tunica media. This elastic membrane will provide structure that allows the stretching of a vessel. Small openings are present, which allows the exchange of substances among the layers. This internal membrane is absent in veins. However, most of the veins, especially in lower limbs, have valves that are developed by thickened endothelium and connective tissue that extends into the lumina.

1.2.2 Tunica media

Tunica media is the layer present in the middle of the vessel wall and is thickest among all the layers in arteries. Tunica media in arteries is compared to veins. It contains layers of smooth muscles that are given support by connective tissue, which is mainly composed of elastic fibers mostly organized in circular sheets. Layers of longitudinal muscle are present towards the outer portion of the tunic. The diameter of the vessel lumen decreases with contraction and increases with the relaxation of the muscles. In arteries, vasoconstriction, that is smooth muscle contraction in tunica media, decreases blood flow, which narrows the lumen and increases blood flow. In the same way, vasodilation relaxes the smooth muscle making the lumen widen and there will be a decrease in blood pressure. Both the vasoconstriction and vasodilation are maintained by vascular nerves called nervi vasorum with the meaning 'nerves of the vessel'. These nerves are located within the blood vessel walls. They are generally made of sympathetic fibers among which some induce vasodilation and others vasoconstriction. This induction entirely depends on the neurotransmitters and their receptors present on the target cell. Hormones and chemicals also regulate blood vessels.

The smooth muscle layers present in tunica media are made up of a network of collagenous fibers that helps in binding tunica media to other tunica. Apart from collagenous fibers, there are also many elastic fibers appearing as wavy lines. In larger arteries, the elastic membrane that is present externally, which is also known as external elastic lamina, acts in separating the tunica media from the tunica externa.

1.2.3 Tunica externa

The tunica externa is also called tunica adventitia and is the topmost tunic. It is made of a sheath of connective tissue which is comprised of collagenous fibers. A small amount of elastic fibers are also present in tunica externa. Tunica externa present in veins is composed of smooth muscle fibers and is considered as the thickest among the tunica in veins. In the case of some large arteries, it is usually thicker than tunica media. The outermost layers of tunica externa have connective tissue as a support that is present outside the vessels. These connective tissues will help in holding the vessels in the right position so that any movement cannot disrupt the blood flow (structure and function of blood vessels).

1.3 Types of blood vessels

1.3.1 Arteries

An artery can be defined as a blood vessel carrying blood to remaining parts of the body away from the heart. Arteries are usually composed of thick walls so as to resist high blood pressure ejected from the heart [1].

1.3.2 Pulmonary artery

Pulmonary arteries have the function of transporting the blood that is deoxygenated to the lungs. Similar to all the other arteries, the pulmonary artery carries blood to various organs of the body away from the heart. The pulmonary artery starts at the tip of the right ventricle in the heart which is also called the pulmonary trunk as it is short and wider in shape. When leaving the heart, it divides in two ways, left and right pulmonary artery. The left pulmonary artery passes by the sac that is present covering the heart called the pericardium and is the shorter. The right pulmonary artery is large and passes through the upper chest and enters the lungs. The right pulmonary artery lies within the pericardium for more than three quarters of its length and runs parallel to the right back of the aorta and superior vena cava.

1.3.3 Coronary artery

The coronary arteries helps in supplying the blood that is rich in oxygen to the heart muscle which is in need of oxygenated blood to be supplied and carry away the oxygen-depleted blood. They are present wrapped around the heart and are divided into tiny branches to carry blood to the heart muscle.

Two types of coronary arteries are present, the left main coronary artery (LMCA) and the right coronary artery (RCA). The LMCA transports blood to the left side of the heart muscle, i.e. left ventriculum and left atrium. It is further distributed into branches, the left anterior artery which arises into the left coronary artery supplying blood to the left side of the heart while the circumflex artery diverges towards the left coronary artery encircling the muscle of the heart. It carries blood to the exterior side and the backside of the heart. The right coronary artery functions in supplying blood to the right ventricle, right artrium, sinoartrial (SA) and atrioventricular (AV) nodes. It also supplies blood to the septum of the heart. It is further divided into small branches, i.e. acute marginal artery and right posterior descending artery.

1.3.4 Systemic artery

The systemic artery is the artery of the system circulation of the cardiovascular system. These arteries transfer oxygenated blood far from the heart to various body parts and transport back deoxygenated blood to the heart. The systemic artery is of two types, i.e. muscular and elastic depending upon the amounts of muscle and elastic tissues in the tunica media. In general, the larger arteries are usually elastic and the smaller ones are muscular. Systemic arteries transport blood to arterioles, and from there to capillaries at which the exchange of nutrients and gases occurs. Figure 1.1 shows a flow diagram representing the systemic circulatory system.

1.3.5 Hepatic artery

The common hepatic artery starts from the celiac artery of the liver running towards the right, forming a co-curve from the aorta which is concave. It supplies blood to various organs such as the liver, pylorus, pancreas and duodenum, which are rich in oxygen. The common hepatic artery is further divided into two branches, proper

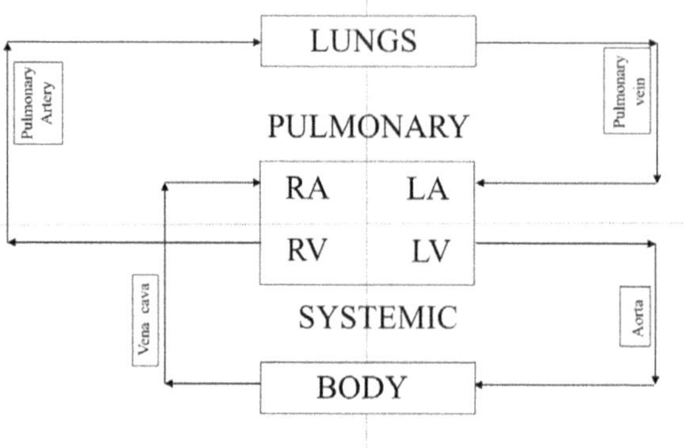

Figure 1.1. Flow diagram of the systemic circulatory network.

hepatic artery and gastroduodenal artery. The proper hepatic artery is the last section of the common hepatic artery that arises in the direction of the liver in the hepatoduodenal ligament. The proper hepatic artery is again divided into branches known as right hepatic artery and left hepatic artery. Here the right hepatic artery starts from the superior mesenteric that runs within the space between the pancreas and vena cava. The left hepatic artery starts in the left gastric artery. It divides into small branches that lead to the oesophagus and stomach. The gastroduodenal artery will be passing behind the duodenum dividing into the right gastropiploic artery and superior pancreaticoduodenal artery. These arteries help in supplying blood to the pancreas and duodenum. Figure 1.2 illustrates various segments of the hepatic arterial system.

1.3.6 Carotid artery

The carotid arteries are the vessels present in the neck that supply blood to the brain, face and neck. Two types of ceratoid arteries are present, i.e. right carotid artery and left ceratoid artery. The left one starts in the arch of the aorta, while the right artery is in the brachiocephalic trunk that is the biggest branch formed from the aortic arch. Every ceratoid artery is again classified into external and internal carotid artery. The internal carotid artery passes through the ceratoid canal present in temporal bone in the cranial cavity. It further gives rise to the ophthalmic branch in the eyeball and further divides into anterior and middle cerebral arteries. The internal carotid along with vertebral arteries provide the primary supply for the brain. In the case of the external carotid artery, it passes through the upper portion of the side of the neck and also behind the lower part of the parotid gland. Here it divides into branches that are, (a) the superior carotid artery that supplies to the thyroid gland and larynx, (b) the lingual artery to the sublingual gland and lung, (c) the facial artery to the palate, tonsil, face and submaxillary gland, (d) the occipital branch supplying the sterno-mastoid muscle and then the back of the scalp, (e) the posterior auricular artery to the back of the ear and also to the adjacent part of the scalp, (f) the superficial temporal branch to the scalp in the front portion of the ear and the transverse facial branch to the back portion of the face, (g) the maxillary branch which gives branches to the

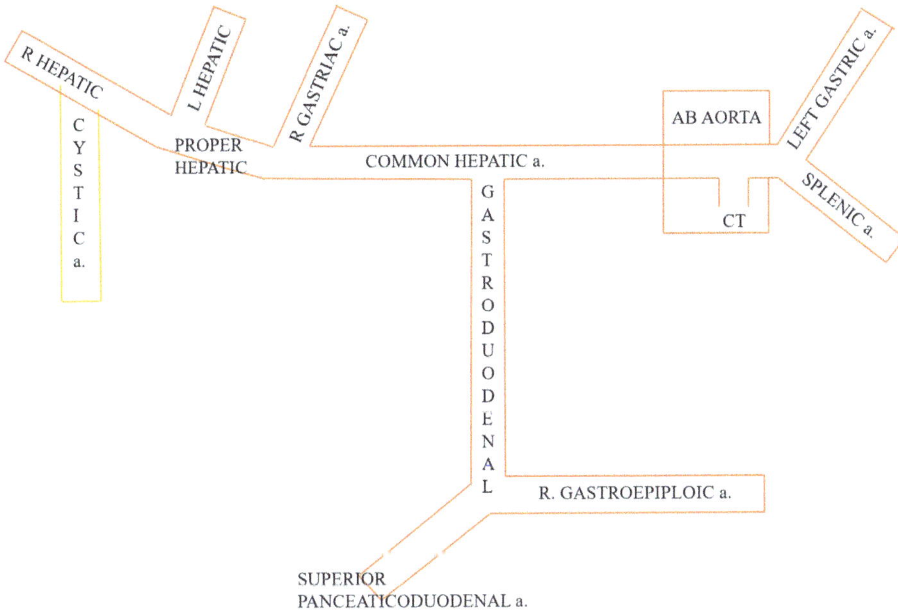

Figure 1.2. Structural representation of the hepatic arterial system.

muscles of mastication, meningeal matters to dura matter, dental branches to teeth, and respective branches to palate, nose, tympanum and (h) the ascending pharyngeal branches supplying tonsils, palate, pharynx and duramatter.

1.3.7 Retinal artery

The retinal artery also called the central retinal artery runs inferior to the optic nerve branching off from the ophthalmic artery. It goes through the optic nerve that is near to the eyeball which further sends branches to the internal area of the retina which is the only way of blood supply to the majority of the portion. The central retina along with its accompanying vein passes through the optic nerve entering the eye bulb through the porous opticus. Further on the artery will be divided into lower and upper branches. These branches will further divide into nasal, lateral and temporal branches. These branches will run between the hyaloid membrane and the nervous layer and further pass forward dividing dichotomously. A minute capillary plexus will arise from these branches which do not extend beyond the inner nuclear layer. These branches of the central retinal artery are terminal arteries.

1.3.8 Splenic artery

The splenic artery is one among the three main celiac artery branches. The celiac artery arises from the largest artery in the abdomen that is the abdominal artery. The splenic artery supplying oxygenated blood to the spleen also has several branches which supply blood to the pancreas and stomach. The splenic artery's largest branch is the left gastroepiplotic artery that runs towards the interior of the stomach through a large membrane called the greater omentum that hangs down from the stomach [2].

1.3.9 Capillaries

A capillary can be defined as a small microscopic channel supplying blood directly to the tissues by the process of perfusion. In capillaries, exchange of various gases and materials takes place within the surrounding blood and cells along with the interstitial fluid. The capillary lumen has a diameter ranging from 5–10 micrometers; small in size which can barely allow an erythrocyte to pass through it. Flow of blood by capillaries is known as microcirculation.

The capillary wall is formed of a layer made up of endothelial cells covered by a basement membrane that is made of smooth muscle fibers. Various endothelial cells will be bordering in a large capillary, while in a small capillary a single layer will wrap around it.

1.3.10 Fenestrated capillaries

A fenestrated capillary is a capillary that has fenestrations (pores) along with the firm junctions in the lining of the endothelial. These pores enable the capillary to be accessible to larger molecules. The quantity of fenestrations and their accessibility varies according to their location. Fenestrated capillaries are seen commonly in the small intestine, whicht is a major location for absorption of nutrients and also in the kidneys that filter the blood. Also, they are found in the choroid plexus of the brain and some endocrine structures such as the hypothalamus, thyroid, pituitary, and pineal glands.

1.3.11 Sinusoidal capillaries

A sinusoid capillary, also called sinusoid, is an uncommon category of capillary. These are flat in structure having large basement membranes and intercellular gaps that are incomplete along with intercellular clefts and pores. The openings are large enough that they enable the flow of large molecules that include cells and plasma proteins. The flow of blood through the sinusoidal capillaries is very slow so that it allows larger time for transport of various gases, nutrients and other waste products. Sinusoids are present in liver, bone marrow, lymph nodes, spleen and various endocrinal glands that include adrenal and pituitary glands. These organs are unable to perform their functions without these capillaries.

1.3.12 Continuous capillaries

This is the most commonly seen capillary present in mostly all vascularized tissues. Continuous capillaries are lined completely by endothelial cells with firm junctions among them. These junctions are generally impermeable, allowing only water and ions to pass through them. Sometimes, these tight junctions are incomplete in capillaries which leaves intercellular clefts allowing interchange of water and some tiny molecules between blood plasma and interstitial fluid. Materials which are permeable and pass through cells are products that are metabolic such as water, glucose, hormones, gases and various leukocytes [3]. They are not related with the brain and are rich in the transport vesicles that lead to endocytosis or exocytosis. Continuous capillaries that are present in the brain are a section of the blood–brain barrier. Here in the brain, they have tight and firm junctions with intercellular clefts, along with a basement membrane and also astrocyte extensions which called feet. The structures together will act in preventing the movement of almost all substances.

1.3.13 Veins

A vein can be defined as a blood vessel that is elastic in nature carrying blood to heart the that is collected from different body parts. Veins are one of the components of the cardiovascular system that provides nutrients by blood circulation to cells of the body. The venous system is considered a low pressure system which depends on muscle contractions for returning blood to the heart.

1.3.14 Pulmonary vein

Pulmonary veins return blood to the atrium of the heart from the lungs. Four pulmonary veins are present in which two work in draining each of the lungs. The four types of pulmonary veins are:

- right superior pulmonary vein—removes blood from the right middle and upper lobes of the lung;
- right inferior vein—removes blood from the lower right lobe of the lungs;
- left superior vein—drains blood from the left upper lobe of the lungs; and
- left inferior vein—drains off blood from the left lower lobe of the lungs.

All these veins together form a capillary network above the air sac walls. Here they are extended with the network of capillary of the pulmonary artery, joining and finally forming a single vessel for each lobule of the lungs. These uniting vessels result in the formation of an individual trunk for each lobe, among which three are for the right lung and the remaining two are for the left lung [4]. The vein that arises from the right lung's middle lobe unites generally with the vein that comes from the upper lobe and finally forming two separate trunks from each lung. These trunks will create perforations in the pericardium layer, which is fibrous, and then open individually into the back and upper part of the left atrium. Near the base of the lung, the superior pulmonary vein lies in the front just down the pulmonary artery while the inferior pulmonary vein lies at the base part of the lung's hilus that is posterior to the uppermost vein. The right pulmonary vein goes at the back of the right atrium and the superior vena cava while the left pulmonary vein goes ahead of the thoracic aorta [2].

1.3.15 Systemic veins

The systemic veins are of three categories: (a) the heart veins, (b) the veins of the uppermost parts such as head, neck and the thorax that terminate in the superior vena cava and (c) the veins of the body parts lying in the lower side such as abdomen, and pelvis that end in inferior vena cava [2].

1.3.16 The heart veins

The veins present in the heart collect blood from the heart muscle together jointly forming a large vessel called the coronary sinus. The coronary sinus can be defined as a broad venous channel which is approximately 2.25 cm in length and present in the posterior portion of the coronary sulcus that is surrounded mostly by muscular fibers of the left atrium. Finally, it terminates in the right atrium in the middle of the orifice of the inferior vena cava and artrioventricular opening. The opening of the coronary sinus is protected by a valve that is semilunar and is called the valve of coronary sinus.

The tributaries of the coronary vein are the great vein, the middle vein, small cardiac veins, the posterior vein of the left ventricle, and the oblique vein of the left atrium.

The great cardiac vein, also called the left coronary vein, starts near the top of the heart and passes down through the anterior longitudinal sulcus up to the lower part of the ventricle. In the coronary sulcus, it turns to the left reaching the backside of the heart opening into the left of the coronary sinus.

The small cardiac vein or right coronary vein passes in the coronary sulcus that lies in the middle of the right atrium and ventricle and finally terminates in the coronary sinus at the right terminus. It gets blood from behind the right atrium and right ventricle.

The middle cardiac vein starts at the top of the heart, runs through the posterior longitudinal sulcus and opens into the coronary sinus at the extreme right.

The posterior vein of the left ventricle passes through the area that is diaphragmatic of the left ventricle and then to the coronary sinus and finally ends in the great cardiac vein.

The oblique vein of the left atrium is a tiny blood vessel that passes diagonally behind the left atrium ending in the coronatery sinus situated near the extreme left [2].

1.3.17 The veins of the head and neck

These veins are classified into three categories:

- the veins present in the exterior part of the head and face;
- the veins in the neck;
- the veins of the brain, the diploic veins, and the venous sinuses of the duramatter.

1.3.18 The veins present in the exterior part of the head and face

The veins that are present on the surface of the head and face are: frontal vein, supraorbital vein, angular vein, anterior facial vein, Occipital vein, superficial temporal vein, internal maxillary vein, posterior facial vein and posterior auricular vein.

The **frontal vein** starts in the forehead in a network of veins communicating with the frontal tributaries present in the superficial temporal vein. The vein divides forming a trunk running downwards near to the forehead's middle line that is parallelly located to the vein of the other side. These veins are further divided by an oblique branch known as the nasal arch, at the tip of the nose. It receives smaller veins from the back of the nose. Near the tip of the nose, the veins divide uniting the supraorbital vein at the middle angle of the orbital forming angular vein.

The **supraorbital vein** starts in the forehead communicating with the superficial temporal vein's frontal branch. It travels downwards, that is, on the surface of the frontalis muscle joining the frontal vein of the orbit's medial angle forming the angular vein.

The **angular vein** that is formed by the joining of the frontal and supra orbital vein passes downwards beside the tip of the nose to the lower margin of the orbit forming the anterior facial vein.

The **anterior facial vein** that starts beside the tip of the nose is the extension of the angular vein. It is present next to the external maxillary artery. It joins along with the posterior facial vein and forms the common facial vein crossing the external carotid artery entering the internal jugular vein at a junction near the lower hyoid bone. The facial veins do not have valves and their walls are also not particularly soft compared to the majority of superficial veins. Branches of the anterior facial vein are getting into a branch of the deep facial vein with considerable size from the pterygoid venous plexus. Also, it is united by the superior and inferior palpebral, labial, buccinator and masseteric veins.

The **occipital vein** starts at the backside of the vertex of the skull in a network. It starts as a single blood vessel from the plexus and finally joins the deep cervical and the vertebral veins. Sometimes, it comes after the occipital artery ending in the internal jugular vein.

The **superficial temporal vein** starts at the vertex and beside the skull in a network communicating with the frontal supraorbital veins, a similar vein of the other end and also with the posterior auricular and occipital veins. Frontal and parietal branches originate from this network and unite above the zygomatic arch forming trunk of the vein. This trunk will be further joined by the middle temporal vein. Then it crosses the posterior root of the zygomatic arch entering the parotid gland. It then joins with the internal maxillary vein giving rise to the posterior facial vein.

The **internal maxillary vein** is a short trunk accompanying the internal maxillary arteries first part. This vein is formed by joining of veins of the pterygoid network and runs backwards in the middle of the mandible, neck and sphenomandibular ligament. It unites with the temporal vein forming the posterior facial vein.

The **posterior facial vein is** organized by joining of both superficial temporal and internal maxillary veins. It further diverges into two branches, the anterior branch passing forward which joins with the anterior facial vein forming the common facial vein and *a posterior* branch. It is further united by the posterior auricular vein giving rise to the external jugular vein.

The **posterior auricular vein** starts beside the head in the network. It passes along with the branches of the occipital and superficial temporal veins. It passes down at the back of the auricular joining the posterior facial vein's posterior division forming the external jugular vein.

1.3.19 The veins in the neck

The veins that are present in the neck function in returning blood to various body parts from the head and face. The veins in the neck are: external jugular vein, anterior jugular vein, posterior external jugular vein, internal jugular vein and vertebral vein.

The **external jugular vein** gets a major amount of the blood from the outer region of the cranium and internal parts of the face. This vein is constructed by the union of the dorsal division of the posterior facial vein with the posterior auricular vein. The external jugular vein varies in size from other veins. Occasionally it is double in size as it is inversely proportional to the other veins in the neck. It has four valves in two pairs, where the lower set is present at the start of the subclavian vein and the upper pair are present at the top of the clavicle.

The tributaries of the external jugular vein lead to receiving of the occipital vein occasionally and the posterior external jugular vein. It also gets the transverse cervical vein, transverse scapular vein and anterior jugular vein near its termination.

The **anterior jugular vein** starts at the hyoid bone by joining of various superficial veins of the submaxillary region. It passes downwards in the middle of the median line and antecedent border of the sternocelinomastoideous and lower part of the neck and finally opens into the ending of the external jugular veins, or in a few cases opens in the subclavian vein. Its major branches are the small thyroid vein and laryngeal veins. Valves are absent in this vein.

The **posterior external jugular vein** brings back the blood from the skin and also from superficial muscles present in the back and upper portion of the neck that is present in between the splenius and tarzenius. It starts in the occipital region and opens in the external jugular vein.

The **internal jugular vein** assembles blood from the upper portion of the face, brain and from neck. It starts in the dorsal part of the jugular foramen that is the tip of the skull. It is connected with the transverse sinus. At the starting point it is dilated, called the superior bulb. It passes down the side of the neck vertically. Initially it will be lying obliquely to the internal carotid artery and further to the common carotid artery. It is united at the tip of the neck forming an innominate vein. There is a second dilation just a little above its termination called the inferior bulb. The right internal jugular vein is situated at the small gap from the common carotid artery at the bottom of the neck. It goes through the sub clavian artery's primary part. The left internal jugular vein intervenes with the common carotid artery. The right vein is larger than the left vein.

The major branches of the internal jugular vein are: lingual veins, pharyngeal veins, inferior petrosal sinus, superior thyroid vein, middle thyroid vein, common facial and occipital veins.

The vertebral vein is formed from various tributaries in the sub occipital triangle.

1.3.20 The veins of the brain

The veins present in the brain are thin without valves and muscular tissue is absent. They pass through the archanoid membrane and meningial layer of dura mater and finally open cranial venous sinuses. These veins are diverged into two pairs, cerebral and cerebellar veins. The cerebral veins are further classified into external and internal veins as they take away the blood from the outer and internal parts of the brain hemisphere. The external veins are further divided into superior, middle and inferior cerebral veins. The internal cerebral veins are further divided into the terminal vein and choraoid veins. The cerebellar veins are present in the external part of the cerebellum that are divided into two pairs, superior and inferior. Superior cerebellar veins go through the super vermis and open in straight sinus and internal cerebral veins. Here the inferior cerebellar veins are larger in size and open in transverse, accipital and superior petrosal sinuses.

1.3.21 Opthalmic vein

The ophthalmic veins are without veins and are divided into superior ophthalmic vein and inferior opthalmic vein. The superior ophthalmic vein starts at the internal orbital angle in a nasofrontal vein that is in contact with the angular vein. It gets tributaries to the vessel branches. It goes through the top of the rectus lateralias and then passes through the middle portion of the superior orbital fissure. Finally, it terminates in the cavernous sinus. The inferior ophthalmic vein starts in the venous network of the front part and middle wall of the orbit. It gets veins from the obliquus inferior, rectus inferior, eyelids and lacrimal sac. It runs backwards below the orbit and classifies into two divisions. One travels through the inferior orbital fissure joining the ptyregoid venous pluxes and the other passes through the superior orbital fissure entering the cranium. It finally ends in the cavernous sinus through an individual opening or jointly with the superior ophthalmic vein [2, 5].

1.3.22 Hepatic vein

The hepatic veins starts at the termination of the hepatic artery and poratal vein in the liver. They are divided further into two categories, i.e. upper and lower category. The upper category has three large veins that pass towards the posterior side of the liver and open into the inferior vena cava. The lower part is small in size, and starts from caudate and right lobes of the liver. The hepatic veins are destitute of valves and run singly with direct contact in the hepatic tissue [2, 6].

1.3.23 Splenic vein

The splenic vein is also called the lienal vein. It is made of 5–6 huge branches of blood vessels that return blood away from the spleen. They jointly form a vessel passing from left to right that grooves the back and upper part of the pancreas lying below the lineal artery. It finally terminates in the back side of the neck and pancreas joining at a right angle with the superior mesentric forming the portal vein. The splenic vein gets blood from tributaries that are the left gastroepiploic vein, short gastric veins, pancreatic veins and inferior mesenteric veins [2].

1.4 Circulatory networks

1.4.1 Pulmonary circulation

This is one of the systems of circulation where the de-oxygeneated blood carried from heart to lungs is shunted to make it saturated again with oxygen before the blood is distributed into the systemic circulation. Deoxygenated blood collected from lower body parts will be entering the heart over the inferior vena cava while deoxygenated blood is transported by the superior vena cava from the upper portion of the body. Both the superior and inferior vena cava pump blood into the right atrium. Blood passes into the right ventricle via the tricuspid valve and before delivering the blood to the lungs, it flows via the pulmonary valve into the pulmonary artery. Once entering the lungs, blood is distributed to various pulmonary capillaries, where carbon dioxide is released and it fills with oxygen. After the blood is completely saturated with oxygen, it is supplied to the left atrium through the pulmonary vein that pumps blood into the left ventricle via the mitral valve. The left ventricle with high contraction releases oxygen rich blood into the aorta by means of the arotic valve. Here the systemic circulation begins.

1.4.2 Coronary circulation

Coronary circulation can be defined as a part of the systemic circulatory system. Here, blood is supplied to the heart tissues providing drainage from the heart tissue. The human heart has two coronary arteries arising from the aorta beyond the semilunar valves. Arotic pressure that increases during diastole above the valves, pumps blood into coronary arteries and later pumps into the heart's musculature. The deoxygenated blood will be returned to the heart chambers through coronary veins. Mostly they will merge leading to the formation of the coronary venous sinus that drains blood into the right atrium. Generally 70%–75% of the available oxygen is extracted from the blood by the heart in coronary circulation. It is larger than the amount of extracted oxygen from other organs through other circulations.

1.4.3 Systemic circulation

Systemic circulation can be defined as blood vessels where oxygenated blood is transported to tissues and then deoxygenated blood is returned from tissues to the body. Pumping of blood is done through aorta from the left ventricle and it flows to arterioles through arterial branches through capillaries. Once it attains equilibrium with the tissue fluid, blood is drained into the veins through venules. The blood returns through venae cavae to the right atrium. Pressure formed in the arterial system, which is the result of heart action and distention of blood helps in maintaining systemic blood flow. This systemic pathway contains various parallel circuits, each having its arteriolar resistance determining blood flow of pressure and overall flow independently.

1.5 Physiology of blood flow

1.5.1 Blood flow between capillaries and tissues

The blood flow via capillaries are majorly regulated by hydrostatic pressure and oncontic pressure. The blood flowing through the capillary applies pressure on the capillary wall due to the pressure of blood flowing from the arteriole. Blood pressure (BP) develops hydrostatic pressure that helps in the flow of the blood from capillary pores to the interstitial compartment. The hydrostatic pressure developed in the capillary will be higher at the end of the arterial side and lower at the venous side.

Oncontic pressure is another force influencing the blood flow. The major principle being this pressure is osmosis. The plasma proteins in the blood which cannot pass through the walls of the capillary will develop the osmotic pressure. This pressure pulls the blood from the surrounding tissue (high concentration) into the capillary (low concentration). This whole pressure is called oncontic pressure.

Another major factor that plays a vital part in the blood pressure is the design of the capillaries that differs depending upon the location in the body and affects the permeability.

1.5.2 Regulation of blood pressure

The major factors influencing the blood pressure are:

- total blood volume in the body;
- the amount of pumped out blood from the heart in a minute, called the cardiac output;
- peripheral vascular resistance (PVR), that is, resistance of the blood flow in the arterial system.

Some of the short term and long term physiological mechanisms regulating the blood pressure (BP) are the following.

1.5.3 Baroreceptor response

The vasomotor present in the brain's medulla oblongata, transfers signals through sympathetic nerve fibers to vascular smooth muscle. This finally results in vaso-dilation or vasoconstriction simultaneously effecting BP and flow of blood to the tissues. Changes in the BP are detected by baroreceptors such as mechanical pressure sensors present in the arterial wall of the arotic arch carotid sinus.

1.5.4 Chemoreceptor response

This is the similar phenomenon that takes place through chemical induced reflex by chemoreceptors. They are found in specialized cells in the aortic arch and common carotid arteries. These chemoreceptors will detect the oxygen changes, pH and levels of carbon dioxide.

1.5.5 Rennin–angiotensin–aldosterine activation system

This is a hormonal system in the kidneys and adrenal glands known as the RAAS system. It plays a vital role in long term regulation of BP. The RAAS triggers the sympathetic nervous system that functions in regulating BP and plasma sodium. The RAAS starts with breakdown of angiotensinogen which is a plasma protein that is produced by the liver through rennin (an enzyme which is produced by the kidneys). When BP is low, rennin is released and triggers a series of enzymatic reactions. Angiotensinogen produces angiotensin I which is an inactive peptide. An enzyme produced by lungs, ACE (Angiostenin Converting Enzyme), is involved in converting angiotensin I to angiotensin II, which is a potent vasoconstrictor triggering the increase in BP.

1.5.6 Autoregulation of blood flow

Autoregulation refers to adjusting the blood flow automatically by altering the diameter of the arterioles. Some organs and tissues have this ability to auto regulate the blood flow. Autoregulation is crucial in some organs such as the kidneys, heart and brain. in these organs, when the perfusion pressure drops, the organ reduces the vascular resistance by local vasodilation, which leads to blood flow increase.

1.6 Conclusion

Blood that will be pumped through the heart passes through a number of blood vessels such as arteries, arterioles, capillaries, venules and veins before pumping back to the heart. Arteries have thick walls as the arterial system is considered a high-pressure system, whereas the venous system is of low-pressure. So veins have wider lumina with thin walls. Capillaries have only one layer of tunica intima that is formed of simple squamous epithelium and a small amount of connective tissue. Tunica media is thicker with smooth muscle and connective tissue while tunica externa is made of a layer of connective tissue. Physiology of blood flow through blood vessels through capillaries is mainly regulated by oncontic and hydrostatic pressure while the blood pressure will be controlled by varying short term and long term physiological regulation.

References

[1] Gray H and Goss C M 1974 Anatomy of the human body *Am. J. Phys. Med. Rehab.* **53** 293

[2] Klimek-Piotrowska W, Hołda M K, Piątek K, Koziej M and Hołda J 2016 Normal distal pulmonary vein anatomy *PeerJ* **4** e1579

[3] Hurst J W, Logue R B, Schlant R C and Wenger N K (ed) 1978 *The Heart: Arteries and Veins* vol 986 (New York: McGraw-Hill)
[4] Townsley M I 2012 Structure and composition of pulmonary arteries, capillaries, and veins *Compr. Physiol.* **2** 675–709
[5] Schumacher M and Tubbs R S 2020 The emissary veins *Anatomy, Imaging and Surgery of the Intracranial Dural Venous Sinuses* pp 175–83 (Amsterdam: Elsevier)
[6] Blumgart L H and Hann L E 2012 Surgical and radiologic anatomy of the liver, biliary tract, and pancreas *Blumgart's Surgery of the Liver, Pancreas and Biliary Tract* (Philadelphia, PA: WB Saunders) pp 31–57

IOP Publishing

Chapter 2

Neurovascular structure and function

Arindam Bit, Jasjit S Suri and Bharti Mishra

In the previous chapter we have understood the basic configuration of blood vessels, their structure and functions. It has been observed that transitory motion of blood through the conduits of vessels maintains homeostasis throughout the body. Such phenomena are supported by regulation of blood pressure on the wall of the vessels by a feedback mechanism. It builds coordination between the circulatory system and many other controlling systems. One such coordination results in the development of a physical phenomenon called neuro-vascular coupling. This chapter illustrates the anatomical and physiological aspects of various such neurovascular ventures present in the human body. A later section of this chapter deals with the various clinical features of neurovascular disorders in patho-physiological conditions.

2.1 Introduction

The neurovascular unit (NVU) facilitates the integrated flow of blood through vasculatures. Arteries innervating to the brain, arising from sub-arachnoid space are composed of a basement membrane, smooth muscle cells (SMC), endothelial cells (ECs), pia mater, the peri-vascular space and astrocyte end feet. SMC of vessels and pia mater lamination begin to lose integrity as they continue deeper into the brain and, between the endothelial cells and astrocyte end feet, they gain pericytes. Also, the other components of the NVU are connected with the astrocyte contact, neuronal and cerebral vasculature where the function of the entire unit is influenced by them.

Mural cells are nested with the basement membrane which encapsulates blood vessels and turbinates into thin lining along capillaries, post-capillary venules and pre-capillary arterioles (Agulhon *et al* 2008). The structure and function of pericytes vary with configurational setting in the vasculature. Here, a variety of functions including upkeep of vascular permeability, angiogenesis, formulation of blood vessel, filtration of cellular toxic agents, supervision of cerebral blood flow, as well as restriction about entry to the immune cell are taking place (Abbott *et al* 2010,

doi:10.1088/978-0-7503-2088-7ch2

Abo-Ramadan *et al* 2009, Abouheif *et al* 2011). Pericytes communicate directly with cerebral endothelial cells (Akoudad *et al* 2016, Ali *et al* 2014). Further, during the advancing process of NUV, pericytes activate polarization of astroglial end-feet around vessels. Degeneration of pericytes in diseased condition leads to increase in vessel permeability

Vasculature is in physical contact with the neuronal processes in the brain and their coalition arbitrates a surge in CBF (cerebral blood flow) during enhancement in diabolic demands belonging to neuronal activities at that location, this is termed as functional hyperemia. Provincial regulation of CBF in preview of neurons comes out as a convoluted function of: (1) augmentation phase of the brain, (2) segments of brain and the populace of neural stroma supplied through the blood–brain barrier (BBB), (3) extent and magnitude of neuronal activity, (4) existence and features of glial cells which serve as resident arbitrators of neuronal stimuli, and (5) effects of brain laceration or degeneration which lead to alteration of other factors. For the development of the BBB, association of cerebral vasculature with neuronal processes is important. Vascular patterning is driven by vascular endothelial growth factor (VEGF) signaling during advancement (Allen *et al* 2010). And post-developmentally the maintenance of the vascular network is done by neuronal activity (Armulik *et al* 2010), where the chief arbitrators of cerebrovascular permeability are astrocytes.

The bodily, alchemical, and diabolic support cells of the central nervous system turn out to be astrocytes and these are dispersed all over the brain. These are amalgamated, star-shaped and highly outspread, and their structure and functions depend on their location and with the cell populace with which they combine. To modulate neuronal function, astrocytes transfer processes between a few neurons to several millions of synapses (Attwell *et al* 2010). Astrocytes are centrally mounted between endothelial cells and neurons in the connection with the NVU, which directs them to respond vigorously to neuronal metabolism and synaptic activity which helps in maintaining CBF.

These are thin, homogeneous sheets of specialized, self-assembled extracellular matrix (ECM). The basement membrane is formed by secreted proteins and mounted between astrocytes and pericytes and between ECs (entrohinal cortex) and pericytes. Iso-form of ECM protein (like fibronectin, laminins, collagens, vitronectin, etc) are found embedded in pivotal proteins of basement membranes. It also houses enzymes associated with ECM degeneration and processing, proteins binding to ECM, and soluble factors which are growth factors and cytokines. ECM and support proteins are fundamental for effective functioning of NVU because they transfer the incitement state of receptors on cellular components.

These glial cells are elementary immune cells located throughout the brain and spinal cord and attribute to 10%–15% of all cells found within the brain. In early advancement process, these yolk sac derived myeloid precursors embed the brain (Baker *et al* 2013, Barber *et al* 2000). Here, these precursors are upgraded to enhanced plastic cells with mobile features.

Microglia become stimulated under certain clinical conditions and this activation state is contemplated conversed as M1 or M2 forms, accompanied by the cells branching towards both pro-inflammatory or anti-inflammatory executions, accordingly. Unique strategies were also adapted for modulating microglial polarization as a conceivable therapeutic outlook (Baumbach *et al* 2013, Baumbach *et al* 2003). In the neurovascular unit there reside perivascular microglia which originate from central nervous system (CNS) resident microglia (Bell *et al* 2010, Belvindrah *et al* 2007). To facilitate angiogenesis, the microglial cells associated with vasculature interact with tipped cell while sprouting vessels. Perivascular macrophages also act as a first line of protection inhibiting attacks of pathogens within the adult brain (Boileau *et al* 2010).

As a response to physiological stimulus some interactions happen which are also called intracellular interactions, between glial, vascular and neural components of the NVU which facilitate the activity based regulation of CBF regulation, vascular permeability, and neuroimmune responses.

The time and space domain simultaneous phenomena of increased blood flow to CNS tissue is known as functional hyperemia which happens in response to neural activities (Bozic *et al* 2013). The astrocytes can step in as a relay between endothelial cells and neurons as they conduct effective transfer attributed to blood substrate, in particular essential gases with nutrients to the diabolic oxidative part related to the brain such as regional cluster regarding neurons and their connected astrocytes and towards SMCs. In astrocytes, arachidonic acid (AA) is produced due to the elevated $[Ca^{2+}]_i$ which activates phospholipase A2 (PLA2) and this AA accumulates at the astrocyte end feet to the constrictive ingredient of vascular walls and here this is transformed towards the essential attributes: 20-HETE, that triggers vascular narrowing or vasoconstriction. The accumulation of AA then converts to vasoactive metabolites such as prostaglandin (PG) and epoxy-eicosatrienoic acid (EET), which are discharged to evoke vasodilation (Chan *et al* 2013).

An increase in $[Ca^{2+}]$ ion in the astrocyte end feet also activates elevated-conductance calcium-gated potassium channels and incites K^+ influx from astrocyte to vessels, producing vasodilation, localized signaling of potassium channel integrated with neuronal activity for the vasodilating brain. Elevated $[Ca^{2+}]$ ions trigger neuronic nitric oxide synthesis and produce nitric oxide (NO) in neurons, and that inhibits vasoconstriction due to its presence on cyclic guanosine monophosphate in an arterial smooth muscle cell (Charidimou *et al* 2016).

2.2 Pathology in neurovascular units

Several diseases can impair brain functions by causing abnormality in cellular communication within the NVU. Two of such diseases are Alzheimer's disease (AD) along with CNS neoplasms.

Alzheimer's disease: Neurovascular dysfunction causes cognitive deterioration in patients with AD. It is said to be manifested due to impaired clearance of amyloid β (Aβ) peptide proteins that are toxic from the structure when the trans membrane recipient amyloid precursor protein (APP) remains split by β-secretase and γ-secretases indirection of discharge dispersible Aβ. Alteration inside genetics which encrypts the APP substrate otherwise encrypts proteins and is associated with formulation of Aβ consistently synthesizing predominantly passed down models of AD (Charles *et al* 2010).

Neoplastic disease: One more cerebrovascular abnormality possibly initiates in CNS neoplasms, incorporating the two primary and secondary tumors. In both cases, neoplastic cells can survive in the advancing stage of the illness in the perivascular regions of the NVU; after that they can alter, desolate and generate the latest vasculature with abnormal NVU configuration along with executive operation. Investigation of the NVU in neoplastic abnormalities will foster new therapeutic methodologies (Chazal *et al* 2000).

2.3 Medial neurovascular structures

In the case of posterior tibial tendon insufficiency, medial displacement osteotomy of the calcaneus is a procedure for improving the condition. Calcaneal osteotomy is an extra articular, joint spearing procedure that is used in the correction of cavovarus and planovolgus foot deformity and also for more and definitive re-configuring of the mechanical force on the foot. Also, an anatomic study was done to obtain correlation between medial neurovascular anatomy and osteotomy. It provided information regarding ensuring structural integrity during the procedure (Chen *et al* 2014).

A cadaveric study with 22 fresh frozen cadavers of below the knee specimens was made, and dissection was performed medially to examine the medial plantar nerve (MPN), lateral plantar nerve (LPN) and prosterior tibial artery (PTA). Also, it has been seen that around four neurovascular assemblies intersect one another, and most of them are branches of the LPN. It has been observed that medial neurovascular structures may be in crisis while performing a calcaneal osteotomy. Two structures were found crossing the osteotomy site at various positions and most of these were LPN and PTA (Cherry *et al* 2014).

2.4 Cerebral vascular disease in ischemic stroke

Cerebrovascular disease is seen as a dominant area among health concerns worldwide. The abnormalities of the cerebral circulation are responsible for hemorrhagic stroke, dementias and different types of neurological deterioration. Ischemic stroke acts mainly as an emanation about carotid and cerebrovascular abnormality, including both large and small-vessel distortion (SVD). Irrespective of gender ischemic thump originated due to thrombosis in cerebral as well as carotid arteries, hemorrhagic strokes are often found. Micro vascular alterations across small-vessel distortion yield micro spots regarding ischemia and micro hemorrhages. After institution, it takes many years to progress for both large and small vascular diseases. At the time of aging along with the subsistence of vascular threats, the rate of advancement of large and small ischemic happenings and micro bleeds increases. It has been observed that epigenetic factors like unhealthy diet practices can impact the rate of advancement of vascular abnormalities in the brain. It is also found that the changes in function and architecture of the vasculature promote hypoperfusion by affecting CBF and adaptive vascular responses. The regulation of CBF is controlled by many mechanisms, and the determination of baseline CBF and underline amendments in profusion is done in reaction to particular conditions which are altered by vascular disease and ischemic events (Chillon *et al* 2001).

Neurovascular coupling relates with the correlation around regional neural action along with succeeding alterations in CBF. fMRI data predicts alteration in hemodynamics during neurovascular coupling counter, which involves a chain of tuned cells that have neuronal stimulation with unified allowance of astrocytes and ECs. Optimum blood flow in the cerebral network is maintained by cerebral autoregulation. Due to the highest metabolic stress, the brain plays a significant role for the sustainability of life. Of the many components association, myogenic function of resistance vessels is highly significant. These components subsidize to autoregulation within vasculature squeezes when intravascular pressure is aggravated, and dilate in the presence of decreased intravascular pressure. In diseased conditions, these mechanisms can change: such as, during chronic hypertension the autoregulatory curve shifts to higher levels of pressure, which shows the regulation of CBF through perfusion pressure. Hence, the brain can be at a danger of decreased CBF and prone to internal injury during reduction in perfusion pressure, which appears along systemic hypotension of upstream vessels during neuro-degeneration (Cipolla *et al* 2009).

The multicellular interaction, BBB, activation of glia, infiltrations of immune cells and neuronal deaths include the injury and repair of neurovascular structure. The above are the neuronal components of NVU, the non-neuronal components such as molecular and cellular components of vascular injury post-stroke also add to it (Cipolla *et al* 2008, 2014).

2.5 Vascular risk factors

As the age increases the vascular events such as strokes may also increase markedly. One of the leading dangers for vascular abnormalities is chronic hypertension, and one of the prominent threats for stroke is ischemic and hemorrhagic stroke inadvertent about the region (Cuéllar *et al* 2015).

As vascular disease is any condition that affects the network of blood vessels, small vascular disease and large vascular disease are the cerebrovascular diseases, in which the small vessels are those at the surface of the brain and within the parenchyma which includes small arteries, arterioles, capillary and venules (Zonta *et al* 2003). The vascular resistance distribution in the brain is very peculiar, as major vascular opposition limits within arterioles and arteries upstream and downstream within circulation of parenchyma. In the pial circulation the attribution points are small arterioles. Myogenic mechanism is the mechanism in which arteries and arterioles react to an accession or depression of blood pressure to keep the flow of blood inside the blood vessel. The middle cerebral artery act as the extent of myogenic tone (Fonseca *et al* 2014).

During the disease the alteration in vascular structures and mechanics differs in large vessels versus small vessels so an amalgamation between these two is needed. During neurovascular coupling (NVC) pial and parenchymal arterioles dilate because in the brain the resistance of large arteries is high and also resident microvascular pressure reduces during NVC due to an absence of upstream vasodilation (Davoodi *et al* 2009).

Cerebrovascular diseases such as stroke onset and neurovascular injuries are firmly related with endothelial based aberrations. Alteration in endothelium encourages oxidative stress, increased vascular tone, inflammation, atherosclerosis along with embolus, and compromised blood brain barrier integrity. These aberrations are collectively termed as endothelial dysfunction; these dysfunctions add to NVC (neurovascular coupling) and in the long term they also add to cellular injury and degradation (Ciuceis *et al* 2014).

2.6 Neurovascular mechanics

Vascular resistance and local hemodynamics are majorly influenced by even slight alterations in the mechanical or structural state of the vasculature. It also affects the extent of injury due to initiation of ischemia and the collateral dependent CBF's effectiveness. Due to constrictions in vascular lumen and degradation of vessels, the vascular resistance increases. Degradation of cerebral arterioles are a function of time and disease advancement in human and preclinical models. Alterations in vascular structure can have a significant impact on vasodilator capacity, dormant CBF (hypoperfusion), and vasodilator response (Silva *et al* 2016).

During hypertension the cross-sectional area of the vessel wall is enhanced, termed hypertrophy, as a modifying response to decrease the wall stress, which is common and visible in nearly all types of hypertension. In the cerebral arteries thickening of the wall is present, especially in the starting phase of abnormality. The vascular resistance is impacted by hyperplasia along the vessel wall and also incites a decrease within lumen width existent in maximally dilated vessels (Drizenko *et al* 2004).

The structural alterations in the vasculature, particularly in lumen diameter of dilated vessels, occurs because of alterations in wall tautness, also known as vascular remodeling. Vascular hypertrophy and inward remodeling happens all through hypertension and this shows relocation of the vessel wall in which lumen diameter is decreased, even in conditions of mostly dilated vessels (Endres *et al* 2011).

Inward remodeling is maximally observed in resistance vessels, and the vessels' resistance occurring from this phenomena guards peripheral vessel and the BBB from elevated upstream pressure. In pial and parenchymal arterioles vascular remodeling is existent. Also, it is apparent that infused remodeling happens in micro cerebral arteries in humans having hypertension (Faber *et al* 2011).

Loss of vessels in capillaries and arterioles is noted as vascular rarefaction. This reduction in vessels can occur due to some abnormality and also as a result of genetic components. It impacts angiogenesis or vessel destruction. In chronic hypertension, loss of arterioles and capillaries in the brain and decrease in micro vascular density are evident (Fan *et al* 2015).

In microcirculation, the connections between arteries or arterioles are known as auxiliary vessels. This amount regarding damage of the cell along with degeneration post-blockage of sourced arteries upstream is actuated by the structural and functional condition of collateral vessels in patients. A continuous loss in the size and diameter of the collateral vessel is observed due to hypertension in pial circulation. Also, collateral vessels' resistance is enhanced significantly due to alteration in vascular schematic associated with increased size of infract in models of ischemia. Collateral vessels are very important for proper regulation of CBF, loss of autoregulation and decrease in perfusion pressure; they can also influence management of vascular based stroke (Faraci 2011).

Proper conduction of CBF and the activities of the BBB is affected by ischemia induced injury. Also the preservation of vascular function after a stroke is very important. After an ischemic injury the interactions between neurons and the vasculature is hampered. Neurovascular connection is damaged by the two global and focal ischemia. Also, the damage of NVC followed by changes of focal ischemia with or without reperfusion can be stopped by inducing ischemic tolerance through lipo-polysaccharide. Also, after global I/R glucose exploitation and somatosensory initiation is decreased (Feigin *et al* 2016).

2.7 Pathology of microvascular components in the NVU

As the elemental composition of the BBB is constructed by endothelial cells and pericytes, so the impairment in the BBB is one of the most debilitating ramifications of strokes. This impairment allows infiltration of intravenous enzymes, fluids along with immune cells towards the extra-cellular area which results in vaso-active edema. But the complete profile of BBB damage post-ischemia remains ambiguous. The structural changes in endothelial cells along with pericytes are formulating factors fundamental to primary BBB leakage. The basic building blocks of the BBB are cerebral endothelial cells as they serve crucially in handling the exchange of ions, cells and molecules between the CNS and its perimeter. The endothelium experiences structural changes as the first stride to unlock the BBB post an ischemic injury (Ferkel *et al* 2001, Fillingham *et al* 2017).

Cerebral I/R triggers vascular alteration as there is an increase in paracellular permeability; it also results in the separation of endothelial cells from vessel walls and reduction of endothelial paracellular junctions. These are intracellular junctions in which confluence of nearby endothelial cells is combined; these junctions can be said to be cultivated protein complexes (ocludin and cloudin) including tight junctions (TJs) and adherence junctions. Also, adhering proteins (zonula occluden 1,2,3) serve as a primary element of TJs. The active interactivity between actin cytoskeleton and junction protein are crucial for proper function of junction integrity and endothelial permeability (Filosa *et al* 2006, Flammerc *et al* 2010).

Actin filaments of f-actin cells, which are scattered throughout endothelial cells, aid in maintaining endothelial shape. Many signaling instances are critical in proper functioning of actin dynamics like phosphorylation of myosin light chains (MLCs), which is peculiar to contraction of actin along with myosin, also for dis-arrangements about the endothelial cell junction. The actin filaments are polymerized as a feedback to certain stressors and this happens along with enhanced cytoskeletal tensions, which results in impaired junction sealing efficiency and contracted cell morphology (Fogal *et al* 2012, Golanó *et al* 2006).

Also, it is evident that cytoskeleton re-allocations and configurational modifications in endothelial cells are a means for early BBB disturbance post-stoke. BBB disruption is a reason of parenchymal cell injury. Hence to prevent succeeding unfavorable incidents post-stroke, preserving endothelial cell structure is very crucial (Goldmann *et al* 2016, Gordon *et al* 2008).

It has been observed that aging degrades endothelium-based control of vascular tone, hence aging has certain effects on the NVU and vascular component, as it reduces their regenerative capacity as well. The connection between pericytes and capillaries is also affected by ageing, as it also affects the microglial functions by leaving them less mobile and less efficient in the CNS and less effective in the repair of the NVU. Other elements of the NVU such as neurons, astrocytes, and oligodendrocytes can be influenced through aging as it makes the system vulnerable to strokes and ischemic brain injury. The change in the BBB is exacerbated because of hypertension and diabetes metilus—it causes instability in plasma glucose level. It also causes variations in BBB transport, damages TJ integrity and increases oxidative stress. Hence comorbidities must be prevented to save the NVU damage after ischemia (Greene *et al* 2001, Gruionu *et al* 2000).

2.8 Neurogenesis and neurovascular homeostasis

NVUs of the CNS can be said to be multicellular complexes comprising neural cells, blood vessels and a setting of extra cellular matrix proteins. The function of the extra cellular matrix is to form continuity between cascades of propagated signaling events and neurovascular units. It further helps in integrating CNS development and physiology (Hall *et al* 2006).

Ablation of ECM receptors in neural progenitor cells results in developmental of serious ECM pathologies. An integrin function in the adult brain was investigated by examining mice containing a homozygous-null β8 gene mutation. It was observed given adult β8–/– mice show a substantial anomaly within homeostasis of the NVU, containing an enhanced population of intracerebral blood vessels alongside profound perivascular astrogliosis. Within the region throughout neurogenesis, neural progenitor cells accumulate over blood vessels within neurovascular apertures. Here, β8 integrin plays a significant role for normalizing NPC proliferation. Cultural analysis of NPCs shows a strong correlation between survival defects and growth. It diminishes integrin mediated activated proteins in ECM structure. Thus, it states the significance of integrin proteins in regulation of neurovascular units in the brain. The vertebrate CNS consists of a detailed mesh of vasculature that peculiarly consorts with neuronal and glial ingredients to form structural complexes (Hamid *et al* 2008, Heskova *et al* 2009).

The cellar membrane which consists of growth factors and ECM proteins becomes the medium in which neural and vascular cells interact with each other. Also, for the effective formulation and function of the vasculature within the CNS, absolute supervision of cell–ECM interaction within neurovascular structure is required (Hicks *et al* 2010). In the case of many ECM protiens, integrins are heterodimeric receptors. Here, most of the integrins are asserted in post-natal astrocytes and N^+PCs (neural progenitor cell).

There are a few studies that have suggested that integrin is important for normal function and control of neurovascular development of avß8 in the brain, despite the fact that its functional role in adult neurovascular physiology avß8 is not very clear. It has been reported that elimination of integrin genes for progressing neural ancestry along with astrocytes formulates intra-cerebral hemorrhage in embryonic and neonatal blood vessels (Zhu *et al* 2002). The avß8 integrin-led control of neurovascular growth relies upon stimulation of dormant forms of integrin that are discharged through cells as extracellular matrix-interacted networks comprising much less bioactivity (Hillman 2014, Hu *et al* 2017).

Integrin adheres to the hibernating shapes of TGFß3 TGFß1, arbitrates their bio activation and impels canonical receptor prompting that TGFß1 shows phenotypes similar to one seen in gene knockouts. However, originally, β8–/– neonates were discovered based upon intracerebral hemorrhage phenotype and all genotypes have been validated on event of disengaging to augment β8 gene sequences with the use of genomic DNA separate from ear snips. Almost 90% of β8–/– neonates survived even after the first post-natal week. Other findings were a reduction in olfactory bulb size, improper neurosphere proliferation, minimized TGFβ1 activation in β8–/– neurospheres and effects on the blood vessel regulation in the adult and embryonic brain (Hughes *et al* 2006, Hynes 2002, Iadecola 2004, 2013).

2.9 Neurovascular structure at coronal segment

Several studies were conducted to see the impacts of anatomical and structural variations, such as dehiscence and protrusion (bulging out) of the paranasal sinus (air-filled sacs on either side of the nasal cavity that filter air breathed through the nose and also lighten the skull) and the related neurovascular structures. There are four paired sinuses in the head, and the most posterior and least accessible of these is the sphenoid sinus. The sphenoid sinus acts as geared through multiple neurovascular architecture including inner carotid artery, optic canal, vidian canal along with maxillary nerve (Joutel *et al* 2014, Jung *et al* 2013, Kanekiyo *et al* 2013).

For study about anatomic variations in the neurovascular structure, coronal CT scans were analyzed and evaluated and as a result as many as 66% of the cases were found to have heterogeneity within neurovascular confines regarding the sphenoid sinus, which is most conventional. Variations were those involving the vidian canal and were around 42%, also variations including the bony canal about the inner carotid artery (ICA) were 33%; variations involving the maxillary nerve were 21%, along with variations involving optic nerve of 9% (Kaplan *et al* 2005).

The sphenoid sinus is covered with extremely sensitive structures, including the optic nerve along with the ICA. The trans-spheroid transferase to contact pituitary adenomas turns out to be most frequent, as well as a dissimilar anatomical structure based on the sphenoid sinus that may adversely influence the transferase to the sella through the nose. As we know, variations of sphenoid sinus are not associated with any significant disease but the chances of impairment of normal drainage pathways by them is high. This produces hindrances for endoscopic access to peripheral zones, and therefore, focus on occult disease and enhances the risk of investigational crisis. Hence, an increased risk of mishaps during endoscopy or any other critical examination before surgery is present and it also enhances the probability of harming the optic nerve, thereby, adding to the complications of intranasal sinus surgery. Both the ICA and optic nerve are very important cranial nerves and the former is liable to the threat of wounding or bruising when there are alterations in the sphenoid sinus (Kaur *et al* 2011, Kim *et al* 2015, Korczyn 2015).

A study about idiopathic tarsal tunnel syndrome (TTS) includes study on depression or squeezing of the posterior tibial nerve, which promotes symptoms propagation along the downstream path of the nerve from the ankle to the foot. Idiopathic TTS comprises the instance where compressive lesion is difficult to identify. A neurovascular point of depression may arise whilst neurons along with arterio-venous complexes scuttle along within the tarsal tunnel, compressing the nerves in a confined tunnel. Patients with TTS experience one or more of symptoms such as tingling, burning, and pain sensation similar to electrical impulses (Kruse *et al* 2015, Kunz *et al* 2007).

There are treatment options such as rest, ice packs, oral medication, immobilization, injection therapy, physical therapy, orthotic devices, shoes and bracing, and in extreme cases surgery is advised. But the surgical outcome of idiopathic TTS is reported unsuccessful compared to that imputable to ganglion, tarsal association, or tumor. Therefore, revised surgical intervention for treatment management of idiopathic TTS needs to be considered. Newly revised procedure should have primary consideration of enhancing the efficiency of neuro-vascular compression phenomena in such patients (Lacoste *et al* 2014, Lee *et al* 2015).

In one research, patients having idiopathic TTS were surgically handled and most of them were complaining about severe pain in the foot-sole. It shows sensory disruption in the sole due to scarcity in association of the calcaneal part corresponding to the posterior tibial nerve. Patients undergoing neurovascular decompression will exhibit high resolution of reduction in symptoms in their feet. Simultaneously, no injury contagion can be found after subsequent surgical procedures. Hence, it can be concluded that neurovascular compression syndrome drives idiopathic TTS (McCarty *et al* 2005, Merlini *et al* 2016).

2.10 Neurovascular structure and pathology near the foot

A study was conducted to investigate and understand the anatomic markers in neurovascular structures in the ankle for arthroscopic atlas. Ankle arthroscopy has been used to address a variety of ankle pathologies such as impingement, synovitis, intra-articular dissociation, osteochondral lesions, loose bodies, etc. It is a minimally invasive procedure with minimized morbidities. It was important that rigor mortis must not have set in, and the passive extent of movement on ankle joint had to be free from detectable defects. The ankle had been regulated in a plantigrade configuration, zero-degree reverse along with neutral configuration. After that, four ankle openings were located by 23-gauge needles which were introduced via the joint capsule. The point of the openings from reciprocal neurovascular architecture can be calculated through a thickness gauge (Modrick *et al* 2009, Moore *et al* 2015, Mozaffarian *et al* 2016).

Such a portal consisted of AL point to extension about superficial peroneal nerves. The lateral nerve serves as an interposed dorsal cutaneous division, whereas the median dorsal part executes at the dorsal end. It also consists of AM opening towards saphenous vein along with the nerve and its extensions. It also assembles the PM point to the neurovascular bundle at the posterior tibial region. To reduce intra-observer error, an average of three measurements was taken. The resulting effect was that the inter-nodal distances also reduce between the great saphenous vein and the nerve, which can be considered to be an anatomical hazard (Mu *et al* 2008, Nissen *et al* 2007).

2.11 Neurovascular pathology at the shoulder joint

Thoracic arteries with neurovascular junctions are sometimes entrapped between the collar bone and first rib. This leads to compression of the neurovascular segment. Such phenomena lead to development of different entrapment syndromes within the area about the posterior cervical triangle along with the shoulder (Nossa *et al* 2007).

The nerve trunk is organized in three levels, of which each unit is surmounted with connective tissue encapsulation. The nerve fiber exists as the elementary unit along with the smallest level of organization and conducting element of the nerve trunk along with a phenomenon related to the neuron. All fibers are covered with concentric connective tissue known as endoneurium which is delicate but definitive. Nerve fibers are bundled to form the perineurium. At the end this bundle forms a nerve trunk which is encapsulated recursively by the epineurium, which is the heaviest layer of connective tissue (O'Donnell *et al* 2016, Ögüt *et al* 2004).

Branches of the epineurium engraved inside the body of the nerve also separate the funiculi. The conjunctive tissue at every stage is divided flexibly and asymmetrically also contains the primary elements of all connective tissues which can be cells, fibers, and ground elements. It also has an extensive arterial network which is freely anastomosing. Also, these arterial networks freely anastomose with each other. This result in removal of a nerve and this can from its base for ample length without affecting its blood supply (Pantoni *et al* 2014, Petrovic *et al* 2016).

Entrapment neuropathies (loss of mobility of the peripheral nerve) are abrasions of individual peripheral nerves, which result from an injury at an accessible anatomic location. Nerve compression usually happens at the aperture region of neural length in fibrous tissue or at the osseofibrous canal. The reasons and mechanisms for damage lead to effective compression, angulation, and even compromise vascular structure (Piquilloud *et al* 2011, Pires *et al* 2013).

It can be said that as movement occurs in the tissue, structures articulate over one another and cause friction. It is believed that while examining the symptoms resulting from the entrapment neuropathy: two factors are to be taken into account, one of which is local and happens at the point of entrapment, and the other is more peripheral and belongs to the regions and structures rooted by the injured nerve bundles. The depression or friction causes a reaction in the bundles, associating with the connective tissue near the point of entrapment, which makes the nerve more vulnerable to further inflammation (Polesello *et al* 2014, Pooja *et al* 2014).

The degree of one such impact is closely associated with the acuteness and continuation of the entrapment pressure. The primary strain of quantal amount accelerates a localized inflammatory response which is generic for any nerve degeneration and inflammatory responses (Proctor *et al* 2005).

There can be various causes of some entrapment syndromes such as a sprain, fracture, any alterations in tasks associated with new postures, an alteration in an exercise routine, a phenomenal degenerative course or advancing variations in the cervical rib. Many entrapment syndromes appear suddenly and have no apparent reason (Rapsomaniki *et al* 2014).

Thoracic-outlet-syndrome distorts the distance between the collarbone and first rib of the thoracic outlet. It is a frequent, generic diagnosis in which patients exhibit symptoms characteristic of entrapment (pain, numbness, coldness and weakness) about the brachial plexus along with the subclavian-axillary vessels. It specifically inculpates a major area of the plexus, starting far away to the intervertebral foramina, also spreading alongside the coracoid process and to the infusion of pectoralis muscle. An integration of subclavian-axillary vessels can also be seen as they innervate across the first rib from the thorax as well as pursue the brachial plexus. Several customized analyses are implemented on these portions of the plexus and vessels, and these are combined to the span of thoracic outlet syndrome (TOS). A few of such ailments are anterior scalene muscle, costoclavicular, cervical rib, pectoralis minor muscle, superior thoracic outlet, costodorsal outlet, and hyper abduction. The symptoms belonging to TOS harvest against association of the ventral rami of C8 along with T1 and ulnar nerve (Rezzouk *et al* 2002, Rizzoni *et al* 2009, 2015).

The brachial plexus belongs to the mesh of nerves, and is formed by the pre-existing ramus of the four bottommost cervical nerves along with the initial thoracic nerve (C5, C6, C7, C8, T1). This plexus starts from the spinal cord, innervates over the cervicoaxillary canal over the initial rib, and terminates on the arm. The brachial plexus is prone to intrinsic and extrinsic depression and perioperative destruction. Depressive (compressive) neuropathies occur due to direct pressure to the nerve. Entrapment neuropathy illustrates a compressive neuropathy rooted by the patient's body (Rodrigues *et al* 2010, Rosenberg *et al* 2016).

The suprascapular nerve slants around the rearmost cervical triangle as well as terminating at the lateral side of the scapula after originating from the adjacent part of the superior trunk around the brachial plexus. Later it cuts across the superior boundary of the scapula by transmitting through the scapular notch. The supra-scapular nerve compression is due to its depression as it transmits over the suprascapular notch. This entrapment neuropathy originates frequently in the form of severe, deep, aching pain which mitigates from the uppermost portion of the scapula to the lowermost ipsilateral shoulder (Yang *et al* 2007). Thus, the nerve innervates through a tiny fibro-osseous opening. In the suprascapular cavity, the nerve bifurcates towards the supraspinatus muscle. Then it further advances towards the neck and the scapula, facilitating extensions for infraspinatus muscle and shoulder joint. The dorsal scapula nerve is an extension of the ventral ramus of the cervical spinal nerve. It proceeds along the mid-scalene myocardium and then plunges deeper into the levator scapulae and rhomboid muscles. Consistently, it stimulates both these muscles. The entrapment of the nerve is due to its passage through these muscles. Symptoms of entrapment comprise pain, dispersed pain in the shoulder region, and inflammation of scalene muscle (Rothwell *et al* 2005, Ruhrberg *et al* 2013).

The extended thoracic nerve appears to be assembled through subsections of ventral rami of fifth–seventh cervical nerves. The subsections against C5 to C6 proceed along mid-scalene muscle, and the nerve is intended posteriorly with respect to the brachial plexus. Further, it plunges through the anterolateral conditions around the thoracic wall, proceeding cursorily to the serratus anterior muscle. There is restriction in the movement of this nerve as this anchored adjacently to the mid-scalene muscle and joined remotely to jaggedness of the serratus anterior muscle which it supplies. Mostly injury appears due to extensive elongation of the nerve such as when the distance between head and shoulder increases accidentally. It results in an aching pain in the posterior area of the shoulder (Salminen *et al* 2011, Shen *et al* 2005).

Adjacent to the shoulder, there are points where crucial terminal nerves of the brachial plexus are susceptible to entanglement. It can be said that by elevation of axillary and radial nerves, nothing is formulated until it approaches the shoulder level. Definitely some of the significant motors of shoulder are innervated with the help of axillary nerve only. It can be said that axillary nerve dysfunction (AND) or neuropathy of the axillary nerve is a phenomena in the shoulder area which is decided by loss of movement or sensation. AND is also caused by extreme stress or damage to the axillary nerve serving the deltoid muscles of the shoulder (Sitler *et al* 2002, Strbian *et al* 2008, Sun *et al* 2006).

2.12 Neurovasculature at the hip joint and meniscus

This study was done to see the neurovascular structures which are at risk during hip arthroscopy, which is a minutely invasive surgical procedure performed to address several abnormalities relating to the hip joint. It has certain benefits such as: enhanced stability conservation at the joint, minimized avascular necrosis risk at the head of the femur, and infections in the post-operative time period. Also, it facilitates rapid rehabilitation (Takano *et al* 2006).

Arthroscopic meniscus repair is a procedure to regain torn knee cartilage. The torn meniscus is repaired by minimally invasive techniques. This was a cadaveric study. Here all-inside repair technique is used because of its ease of insertion and lessening of operative time. However, some danger to the significant adjacent neurovascular structures can still be possible, which can cause significant problems to patients, hence the proximity of these neurovascular architectures must be considered very carefully (Tamaoki *et al* 2019, Trainer *et al* 2008).

We know that acromio clavicular joint (ACJ) injuries are frequently caused, as they represent 9%–12% of all the injuries of the shoulder and around 50% of all sports-associated shoulder injuries. Typically the ACJ reconstruction must have a minimally invasive technique and an estimation of the safety of this procedure is essential (Tureyen *et al* 2011).

This research was done to investigate neurovascular and neuromuscular dispensation through the triangular space of the shoulder, which is also the medial space made through the long head of the triceps, teres minor and teres major. It was a cadaveric study for which 35 samples from adult Korean cadavers were taken in which 6 were females and 12 were males (aged 42–102 years). However, in this study it was examined whether or not any nerve was presented, and the sequence in which the artery plunged through the muscle from the triangular space was assessed (Ucerler *et al* 2007, Vcltkamp *et al* 2002).

It was found that the circumflex scapular artery passes from triangular space. Some authors previously studied these arteries but not in the preview about triangular space. Here it was studied if the artery going over the triangular space provided blood to the encapsulating muscle. In the study there were four specimens in which arterial distribution in all encapsulating muscles was observed. It is clinically very significant as the conditions included discomfort and pain in the posterior shoulder and weakness in abducted external rotation. This study was very significant (Wang *et al* 2012, Wardlaw *et al* 2013, Wolf *et al* 2001).

2.13 Conclusion

An integrated form of anatomical and patho-physiological features associated with neurovascular junction was presented in this chapter. It has been observed that NVUs were present in proximal regions of musculoskeletal units. Due to their presence in the form of innervation or muscle collateral coverage, these units are often subjected to compression (or depression) and tension, thereby leading to development of severe pathological circumstances. It has also been observed that during several musculoskeletal operations, a pre-operative simulation process is mandatory in order to avoid destruction to the associated NVUs.

References

Abbott N J, Patabendige A A, Dolman D E, Yusof S R and Begley D J 2010 Structure and function of the blood–brain barrier *Neurobiol. Dis.* **37** 13–25

Abo-Ramadan U, Durukan A, Pitkonen M, Marinkovic I, Tatlisumak E, Pedrono E, Soinne L, Strbian D and Tatlisumak T 2009 Post-ischemic leakiness of the blood–brain barrier: a quantitative and systematic assessment by Patlak plots *Exp. Neurol.* **219** 328–33

Abouheif M M, Shibuya H, Niimoto T, Kongcharoensombat W, Deie M, Adachi N and Ochi M 2011 Determination of the safe penetration depth during all-inside meniscal repair of the posterior part of the lateral meniscus using the FasT-Fix suture repair system *Knee Surg. Sports Traumatol. Arthrosc.* **19** 1868–75

Agulhon C, Petravicz J, McMullen A B, Sweger E J, Minton S K, Taves S R, Casper K B, Fiacco T A and McCarthy K D 2008 What is the role of astrocyte calcium in neurophysiology? *Neuron* **59** 932–46

Akoudad S, Wolters F J, Viswanathan A, de Bruijn R F, van der Lugt A, Hofman A, Koudstaal P J, Arfan Ikram M and Vernooij M W 2016 Cerebral microbleeds are associated with cognitive decline and dementia: the Rotterdam study *JAMA Neurol.* **73** 934

Ali M, Sundaraj K, Ahmad R B, Ahamed N U, Islam M and Sundaraj S 2014 Evaluation of repetitive isometric contractions on the heads of triceps brachii muscle during grip force exercise *Technol. Health Care* **22** 617–25

Allen C, Srivastava K and Bayraktutan U 2010 Small GTPase RhoA and its effector rho kinase mediate oxygen glucose deprivation-evoked *in vitro* cerebral barrier dysfunction *Stroke* **41** 2056–63

Armulik A and et al 2010 Pericytes regulate the blood–brain barrier *Nature* **468** 557–61

Attwell D, Buchan A M, Charpak S, Lauritzen M, MacVicar B A and Newman E A 2010 Glial and neuronal control of brain blood flow *Nature* **468** 232–43

Baker W B, Sun Z, Hiraki T, Putt M E, Durduran T, Reivich M and Greenberg J H 2013 Neurovascular coupling varies with level of global cerebral ischemia in a rat model *J. Cerebral Blood Flow Metab.* **33** 97–105

Barber P A, Demchuk A M, Zhang J and Buchan A MASPECTS Study Group 2000 Validity and reliability of a quantitative computed tomography score in predicting outcome of hyperacute stroke before thrombolytic therapy *Lancet* **355** 1670–4

Baumbach G L 2013 Nox_2 deficiency prevents hypertension-induced vascular dysfunction and hypertrophy in cerebral arterioles *Int. J. Hypertension* **2013** 793630

Baumbach G L, Sigmund C D and Faraci F M 2003 Cerebral arteriolar structure in mice overexpressing human renin and angiotensinogen *Hypertension* **41** 50–5

Bell R D, Winkler E A, Sagare A P, Singh I, LaRue B, Deane R and Zlokovic B V 2010 Pericytes control key neurovascular functions and neuronal phenotype in the adult brain and during brain aging *Neuron* **68** 409–27

Belvindrah R, Hankel S, Walker J, Patton B L and Müller U 2007 β1 integrins control the formation of cell chains in the adult rostral migratory stream *J. Neurosci.* **27** 2704–17

Boileau P, Old J, Gastaud O, Brassart N and Roussanne Y 2010 All-arthroscopic Weaver-Dunn-Chuinard procedure with double-button fixation for chronic acromioclavicular joint dislocation *Arthroscopy: J. Arthrosc. Rel. Surg.* **26** 149–60

Bozic K J, Chan V, Valone F H III, Feeley B T and Vail T P 2013 Trends in hip arthroscopy utilization in the United States *J. Arthroplasty* **28** 140–3

Chan S L and Baumbach G L 2013 Deficiency of Nox_2 prevents angiotensin II-induced inward remodeling in cerebral arterioles *Front. Physiol.* **4** 133

Charidimou A, Pantoni L and Love S 2016 The concept of sporadic cerebral small vessel disease: a road map on key definitions and current concepts *Int. J. Stroke* **11** 6–18

Charles N A and Holland E C 2010 The perivascular niche microenvironment in brain tumor progression *Cell Cycle* **9** 3084–93

Chazal G, Durbec P, Jankovski A, Rougon G and Cremer H 2000 Consequences of neural cell adhesion molecule deficiency on cell migration in the rostral migratory stream of the mouse *J. Neurosci.* **20** 1446–57

Chen B R, Kozberg M G, Bouchard M B, Shaik M A and Hillman E M 2014 A critical role for the vascular endothelium in functional neurovascular coupling in the brain *J. Am. Heart Assoc.* **3** e000787

Cherry J D, Olschowka J A and O'Banion M K 2014 Neuroinflammation and M2 microglia: the good, the bad, and the inflamed *J. Neuroinflammation* **11** 98

Chillon J M and Baumbach G L 2001 Effects of an angiotensin-converting enzyme inhibitor and a β-blocker on cerebral arteriolar dilatation in hypertensive rats *Hypertension* **37** 1388–93

Cipolla M J and Bullinger L V 2008 Reactivity of brain parenchymal arterioles after ischemia and reperfusion *Microcirculation* **15** 495–501

Cipolla M J, Chan S L, Sweet J, Tavares M J, Gokina N and Brayden J E 2014 Postischemic reperfusion causes smooth muscle calcium sensitization and vasoconstriction of parenchymal arterioles *Stroke* **45** 2425–30

Cipolla M J, Smith J, Kohlmeyer M M and Godfrey J A 2009 SKCa and IKCa Channels, myogenic tone, and vasodilator responses in middle cerebral arteries and parenchymal arterioles: effect of ischemia and reperfusion *Stroke* **40** 1451–7

Cuéllar A, Cuéllar R, Cuéllar A, Garcia-Alonso I and Ruiz-Ibán M A 2015 The effect of knee flexion angle on the neurovascular safety of all-inside lateral meniscus repair: a cadaveric study *Arthrosc.: J. Arthrosc. Rel. Surg.* **31** 2138–44

da Fonseca A C C, Matias D, Garcia C, Amaral R, Geraldo L H, Freitas C and Lima F R S 2014 The impact of microglial activation on blood-brain barrier in brain diseases *Front. Cell. Neurosci.* **8** 362

Davoodi M, Saki N, Saki G and Rahim F 2009 Anatomical variations of neurovascular structures adjacent sphenoid sinus by using CT scan *Pak. J. Biol. Sci.* **12** 522

De Ciuceis C, Cornali C, Porteri E, Mardighian D, Pinardi C, Fontanella M M and Rosei E A 2014 Cerebral small-resistance artery structure and cerebral blood flow in normotensive subjects and hypertensive patients *Neuroradiology* **56** 1103–11

De Silva T M and Faraci F M 2016 Microvascular dysfunction and cognitive impairment *Cell. Mol. Neurobiol.* **36** 241–58

De Silva T M, Kinzenbaw D A, Modrick M L, Reinhardt L D and Faraci F M 2016 Heterogeneous impact of ROCK2 on carotid and cerebrovascular function *Hypertension* **68** 809–17

Drizenko A, Demondion X, Luyckx F, Mestdagh H and Cassagnaud X 2004 The communicating branches between the sural and superficial peroneal nerves in the foot: a review of 55 cases *Surg. Radiol. Anat.* **26** 447–52

Endres M, Heuschmann P U, Laufs U and Hakim A M 2011 Primary prevention of stroke: blood pressure, lipids, and heart failure *Eur. Heart J.* **32** 545–52

Faber J E, Zhang H, Lassance-Soares R M, Prabhakar P, Najafi A H, Burnett M S and Epstein S E 2011 Aging causes collateral rarefaction and increased severity of ischemic injury in multiple tissues *Arterioscler. Thromb. Vasc. Biol.* **31** 1748–56

Fan Y, Lan L, Zheng L, Ji X, Lin J, Zeng J and Sun J 2015 Spontaneous white matter lesion in brain of stroke-prone renovascular hypertensive rats: a study from MRI, pathology and behavior *Metabol. Brain Dis.* **30** 1479–86

Faraci F M 2011 Protecting against vascular disease in brain *Am. J. Physiol.-Heart.Circ. Physiol.* **300** H1566–82

Faraci F M and Heistad D D 1990 Regulation of large cerebral arteries and cerebral microvascular pressure *Circ. Res.* **66** 8–17

Feigin V L, Roth G A, Naghavi M, Parmar P, Krishnamurthi R, Chugh S and Estep K 2016 Global burden of stroke and risk factors in 188 countries, during 1990–2013: a systematic analysis for the Global Burden of Disease Study 2013 *Lancet Neurol.* **15** 913–24

Ferkel R D, Small H N and Gittins J E 2001 Complications in foot and ankle arthroscopy *Clin. Ortho. Rel. Res.* **391** 89–104

Fillingham Y A, Riboh J C, Erickson B J, Bach B R Jr and Yanke A B 2017 Inside-out versus all-inside repair of isolated meniscal tears: an updated systematic review *Am. J. Sports Med.* **45** 234–42

Filosa J A, Bonev A D, Straub S V, Meredith A L, Wilkerson M K, Aldrich R W and Nelson M T 2006 Local potassium signaling couples neuronal activity to vasodilation in the brain *Nat. Neurosci.* **9** 1397–403

Flammerc A J and Luschera T F 2010 Three decades of endothelium research *Swiss Med. Wkly* **140** w 13122

Fogal B and Pober J S 2012 Vascular endothelial cells as immunological targets in atheroscleroisis *Inflamm. Atheroscler.* pp 87–114 (Vienna: Springer)

Golanó P, Vega J, Pérez-Carro L and Götzens V 2006 Ankle anatomy for the arthroscopist. Part I: the portals *Foot Ankle Clin.* **11** 253–73

Goldmann T, Wieghofer P, Jordão M J C, Prutek F, Hagemeyer N, Frenzel K and Locatelli G 2016 Origin, fate and dynamics of macrophages at central nervous system interfaces *Nat. Immunol.* **17** 797

Gordon G R, Choi H B, Rungta R L, Ellis-Davies G C and MacVicar B A 2008 Brain metabolism dictates the polarity of astrocyte control over arterioles *Nature* **456** 745–9

Greene D L, Thompson M C, Gesink D S and Graves S C 2001 Anatomic study of the medial neurovascular structures in relation to calcaneal osteotomy *Foot Ankle Int.* **22** 569–71

Gruionu G, Constantinescu G M and Laughlin M H 2000 An anatomical study of the arteries feeding the triceps brachii muscle of swine *Anat. Histol. Embryol.* **29** 31–6

Hall P E, Lathia J D, Miller N G, Caldwell M A and Ffrench-Constant C 2006 Integrins are markers of human neural stem cells *Stem Cells* **24** 2078–84

Hamid O, El Fiky L, Hassan O, Kotb A and El Fiky S 2008 Anatomic variations of the sphenoid sinus and their impact on trans-sphenoid pituitary surgery *Skull base* **18** 009–15

Heskova G, Mellova Y, Holomanova A, Vybohova D, Kunertova L, Marcekova M and Mello M 2009 Assessment of the relation of the optic nerve to the posterior ethmoid and sphenoid sinuses by computed tomography *Biomed. Pap. Med. Fac. Univ. Palacky Olomouc Czech Repub.* **153** 149–52

Hicks K, O'Neil R G, Dubinsky W S and Brown R C 2010 TRPC-mediated actin-myosin contraction is critical for BBB disruption following hypoxic stress *Am. J. Physiol.-Heart Circ. Physiol.* **298** C1583–93

Hillman E M 2014 Coupling mechanism and significance of the BOLD signal: a status report *Ann. Rev. Neurosci.* **37** 161–81

Hu X, De Silva T M, Chen J and Faraci F M 2017 Cerebral vascular disease and neurovascular injury in ischemic stroke *Circ. Res.* **120** 449–71

Hughes S, Gardiner T, Hu P, Baxter L, Rosinova E and Chan-Ling T 2006 Altered pericyte–endothelial relations in the rat retina during aging: implications for vessel stability *Neurobiol. Aging* **27** 1838–47

Hynes R O 2002 Integrins: bidirectional, allosteric signaling machines *cell* **110** 673–87

Iadecola C 2004 Neurovascular regulation in the normal brain and in Alzheimer's disease *Nat. Rev. Neurosci.* **5** 347–60

Iadecola C 2013 The pathobiology of vascular dementia *Neuron* **80** 844–66

Joutel A and Faraci F M 2014 Cerebral small vessel disease: insights and opportunities from mouse models of collagen IV–related small vessel disease and cerebral autosomal dominant arteriopathy with subcortical infarcts and leukoencephalopathy *Stroke* **45** 1215–21

Jung S, Gilgen M, Slotboom J, El-Koussy M, Zubler C, Kiefer C and Mordasini P 2013 Factors that determine penumbral tissue loss in acute ischaemic stroke *Brain* **136** 3554–60

Kanekiyo T, Cirrito J R, Liu C C, Shinohara M, Li J, Schuler D R and Bu G 2013 Neuronal clearance of amyloid-β by endocytic receptor LRP1 *J. Neurosci.* **33** 19276–83

Kaplan L D, Flanigan D C, Norwig J, Jost P and Bradley J 2005 Prevalence and variance of shoulder injuries in elite collegiate football players *Am. J. Sports Med.* **33** 1142–6

Kaur J, Tuor U I, Zhao Z and Barber P A 2011 Quantitative MRI reveals the elderly ischemic brain is susceptible to increased early blood—brain barrier permeability following tissue plasminogen activator related to claudin 5 and occludin disassembly *J. Cereb. Blood Flow Metabol.* **31** 1874–85

Kim J Y, Kim N and Yenari M A 2015 Mechanisms and potential therapeutic applications of microglial activation after brain injury *CNS Neurosci. Therap.* **21** 309–19

Korczyn A D 2015 Vascular Parkinsonism—characteristics, pathogenesis and treatment *Nat. Rev. Neurol.* **11** 319

Kruse L M, Yamaguchi K, Keener J D and Chamberlain A M 2015 Clinical outcomes after decompression of the nerve to the teres minor in patients with idiopathic isolated teres minor fatty atrophy *J.f Shoulder Elbow Surg.* **24** 628–33

Kunz A, Park L, Abe T, Gallo E F, Anrather J, Zhou P and Iadecola C 2007 Neurovascular protection by ischemic tolerance: role of nitric oxide and reactive oxygen species *J. Neurosci.* **27** 7083–93

Lacoste B, Comin C H, Ben-Zvi A, Kaeser P S, Xu X, Costa L D F and Gu C 2014 Sensory-related neural activity regulates the structure of vascular networks in the cerebral cortex *Neuron* **83** 1117–30

Lee S, Wu Y, Shi X Q and Zhang J 2015 Characteristics of spinal microglia in aged and obese mice: potential contributions to impaired sensory behavior *Immun. Ageing* **12** 22

McCarty J H, Lacy-Hulbert A, Charest A, Bronson R T, Crowley D, Housman D and Hynes R O 2005 Selective ablation of αv integrins in the central nervous system leads to cerebral hemorrhage, seizures, axonal degeneration and premature death *Development* **132** 165–76

Merlini M, Wanner D and Nitsch R M 2016 Tau pathology-dependent remodelling of cerebral arteries precedes Alzheimer's disease-related microvascular cerebral amyloid angiopathy *Acta Neuropathol.* **131** 737–52

Modrick M L, Didion S P, Lynch C M, Dayal S, Lentz S R and Faraci F M 2009 Role of hydrogen peroxide and the impact of glutathione peroxidase-1 in regulation of cerebral vascular tone *J. Cereb. Blood Flow Metabol.* **29** 1130–7

Moore S M, Zhang H, Maeda N, Doerschuk C M and Faber J E 2015 Cardiovascular risk factors cause premature rarefaction of the collateral circulation and greater ischemic tissue injury *Angiogenesis* **18** 265–81

Mozaffarian D, Benjamin E J, Go A S, Arnett D K, Blaha M J, Cushman M and Howard V J 2016 Executive summary: heart disease and stroke statistics—2016 update: a report from the American Heart Association *Circulation* **133** 447–54

Mu Z, Yang Z, Yu D, Zhao Z and Munger J S 2008 TGFβ1 and TGFβ3 are partially redundant effectors in brain vascular morphogenesis *Mechan. Dev.* **125** 508–16

Nissen C W and Chatterjee A 2007 Type III acromioclavicular separation: results of a recent survey on its management *Am. J. Ortho.-Belle Mead* **36** 89

Nossa J M, Fuerte Diaz M, Rueda Escallon G and Pesantez R 2007 Artroscopia de cadera: indicaciones y resultados *Rev. Colomb. Ortop. Traumatol.* 246–56

O'Donnell M J, Chin S L, Rangarajan S, Xavier D, Liu L, Zhang H and Lopez-Jaramillo P 2016 Global and regional effects of potentially modifiable risk factors associated with acute stroke in 32 countries (INTERSTROKE): a case-control study *Lancet* **388** 761–75

Ögüt T, Akgün I, Kesmezacar H, Türker T, Üzün I, Demirci S and Akkin S M 2004 Navigation for ankle arthroscopy: anatomical study of the anterolateral portal with reference to the superficial peroneal nerve *Surg. Radiol. Anat.* **26** 268–74

Pantoni L and Gorelick P B 2014 *Cerebral Small Vessel Disease.* (Cambridge: Cambridge University Press)

Petrovic-Djergovic D, Goonewardena S N and Pinsky D J 2016 Inflammatory disequilibrium in stroke *Circ. Res.* **119** 142–58

Piquilloud G, Villani F and Casoli V 2011 The medial head of the triceps brachii. Anatomy and blood supply of a new muscular free flap: the medial triceps free flap *Surg. Radiol. Anat.* **33** 415–20

Pires P W, Dams Ramos C M, Matin N and Dorrance A M 2013 The effects of hypertension on the cerebral circulation *Am. J. Physiol.-Heart Circ. Physiol.* **304** H1598–614

Polesello G C, Fernandes A E O, de Oliveira L P, Linhares J P T and Queiroz M C 2014 Medial hip arthroscopy portals: an anatomic study *Arthrosc.: J. Arthrosc. Rel. Surg.* **30** 55–9

Pooja Naik L C, Sajja R K, Naik P and Cucullo L 2014 Diabetes mellitus and blood-brain barrier dysfunction: an overview *J. Pharmacovigil* **02** 125

Proctor J M, Zang K, Wang D, Wang R and Reichardt L F 2005 Vascular development of the brain requires β8 integrin expression in the neuroepithelium *J. Neurosci.* **25** 9940–8

Rapsomaniki E, Timmis A, George J, Pujades-Rodriguez M, Shah A D, Denaxas S and Williams B 2014 Blood pressure and incidence of twelve cardiovascular diseases: lifetime risks, healthy life-years lost, and age-specific associations in 1· 25 million people *Lancet* **383** 1899–911

Rezzouk J, Durandeau A, Vital J M and Fabre T 2002 Long head of the triceps brachii in axillary nerve injury: anatomy and clinical aspects *Rev. Chir. Ortho. Rep. L'app.Mot.* **88** 561–4

Rizzoni D, De Ciuceis C, Porteri E, Paiardi S, Boari G E, Mortini P and Rizzardi N 2009 Altered structure of small cerebral arteries in patients with essential hypertension *J. Hypertens.* **27** 838–45

Rizzoni D, De Ciuceis C, Salvetti M, Paini A, Rossini C, Agabiti-Rosei C and Muiesan M L 2015 Interactions between macro-and micro-circulation: are they relevant? *High Blood Press. Cardiovasc. Prevent.* **22** 119–28

Rodrigues S F and Granger D N 2010 Role of blood cells in ischaemia–reperfusion induced endothelial barrier failure *Cardiovasc. Res.* **87** 291–9

Rosenberg R N, Lambracht-Washington D, Yu G and Xia W 2016 Genomics of Alzheimer disease: a review *JAMA Neurol.* **73** 867–74

Rothwell P M, Coull A J, Silver L E, Fairhead J F, Giles M F, Lovelock C E and Binney L E 2005 Population-based study of event-rate, incidence, case fatality, and mortality for all acute vascular events in all arterial territories (Oxford Vascular Study) *Lancet* **366** 1773–83

Ruhrberg C and Bautch V L 2013 Neurovascular development and links to disease *Cell. Mol. Life Sci.* **70** 1675–84

Salminen A, Ojala J, Kaarniranta K, Haapasalo A, Hiltunen M and Soininen H 2011 Astrocytes in the aging brain express characteristics of senescence-associated secretory phenotype *Eur. J. Neurosci.* **34** 3–11

Shen Q, Ren H, Cheng H, Fisher M and Duong T Q 2005 Functional, perfusion and diffusion MRI of acute focal ischemic brain injury *J. Cerebr. Blood Flow Metabol.* **25** 1265–79

Sitler D F, Amendola A, Bailey C S, Thain L M and Spouge A 2002 Posterior ankle arthroscopy: an anatomic study *JBJS* **84** 763–69

Strbian D, Durukan A, Pitkonen M, Marinkovic I, Tatlisumak E, Pedrono E and Tatlisumak T 2008 The blood–brain barrier is continuously open for several weeks following transient focal cerebral ischemia *Neuroscience* **153** 175–81

Sun H, Breslin J W, Zhu J, Yuan S Y and Wu M H 2006 Rho and ROCK signaling in VEGF-induced microvascular endothelial hyperpermeability *Microcirculation* **13** 237–47

Takano T, Tian G F, Peng W, Lou N, Libionka W, Han X and Nedergaard M 2006 Astrocyte-mediated control of cerebral blood flow *Nat. Neurosci.* **9** 260–7

Tamaoki M J, Lenza M, Matsunaga F T, Belloti J C, Matsumoto M H and Faloppa F 2019 Surgical versus conservative interventions for treating acromioclavicular dislocation of the shoulder in adults *Cochrane Datab. System. Rev.*

Trainer G, Arciero R A and Mazzocca A D 2008 Practical management of grade III acromioclavicular separations *Clin. J. Sport Med.* **18** 162–6

Tureyen K, Bowen K, Liang J, Dempsey R J and Vemuganti R 2011 Exacerbated brain damage, edema and inflammation in type-2 diabetic mice subjected to focal ischemia *J. Neurochem.* **116** 499–507

Ucerler H, Ikiz Z A A and Uygur M 2007 A cadaver study on preserving peroneal nerves during ankle arthroscopy *Foot Ankle Int.* **28** 1172–8

Veltkamp R, Rajapakse N, Robins G, Puskar M, Shimizu K and Busija D 2002 Transient focal ischemia increases endothelial nitric oxide synthase in cerebral blood vessels *Stroke* **33** 2704–10

Wang J, Sun L, Si Y F and Li B M 2012 Overexpression of actin-depolymerizing factor blocks oxidized low-density lipoprotein-induced mouse brain microvascular endothelial cell barrier dysfunction *Mol. Cell. Biochem.* **371** 1–8

Wardlaw J M, Smith C and Dichgans M 2013 Mechanisms of sporadic cerebral small vessel disease: insights from neuroimaging *Lancet Neurol.* **12** 483–97

Wolf E M and Pennington W T 2001 Arthroscopic reconstruction for acromioclavicular joint dislocation *Arthrosc.: J. Arthrosc. Rel. Surg.* **17** 558–63

Yang Z, Mu Z, Dabovic B, Jurukovski V, Yu D, Sung J and Munger J S 2007 Absence of integrin-mediated TGFβ1 activation *in vivo* recapitulates the phenotype of TGFβ1-null mice *J. Cell Biol.* **176** 787–93

Zhu J, Motejlek K, Wang D, Zang K, Schmidt A and Reichardt L F 2002 β8 integrins are required for vascular morphogenesis in mouse embryos *Development* **129** 2891–903

Zonta M, Angulo M C, Gobbo S, Rosengarten B, Hossmann K A, Pozzan T and Carmignoto G 2003 Neuron-to-astrocyte signaling is central to the dynamic control of brain microcirculation *Nat. Neurosci.* **6** 43–50

Chapter 3

3D bioprinting in tissue engineering and regenerative medicine

Arindam Bit, Jasjit S Suri and Sharda Gupta

This chapter is primarily focused on bioprinting technology for biomedical applications. Bioprinting can be utilized for fabrication of a wide range of tissue, based on which this chapter describes in detail its application in tissue regeneration. Further, the difficulties and potential in developing a construct for tissue regeneration are discussed herein. In this book chapter, the significance of 3D bioprinting in tissue restoration will be discussed in depth.

3.1 Introduction

The conventional therapies for diseased tissue substitutes are organ transplantation (allografts, autografts and xenografts) and medical implants. The ultimate aim regarding tissue engineering is to fabricate serviceable tissue for the replacement of impaired tissue. Tissue engineering has been used in clinical applications to overcome the constraints of organ shortage. Tissue engineering for regeneration of tissue can be broadly divided into two approaches: top-down approach and bottom-up approach. In the top-down approach, scaffold is prepared; cells are seeded within it and matured inside a bioreactor to regenerate the desired tissue. However, this approach is unable to mimic the microarchitecture of tissue. Traditional 3D printing is one of the top-down approaches of tissue engineering. Limitations of these techniques are non-uniform distribution of cells, low vascularization and limited diffusion, whereas the bottom-up approach overcomes the limitations of the top-down approach. In this method, cellularized scaffold is prepared which mimics the extracellular matrix. This approach has advantages such as high cell density and superior diffusion.

The 3D printing technique, evolved from the additive manufacturing technique, works with the elementary principle of layer-by-layer assembly with biomaterial to prepare the required geometry. Many traditional 3D printing techniques like

doi:10.1088/978-0-7503-2088-7ch3

selective laser sintering (SLS), stereolithography (SLA) along with fused deposition modelling (FDM) have been in the field of tissue engineering along with recuperative medicine, as shown in figure 3.1. Using this classical 3D printer, solid scaffolds are fabricated using thermoplastics or resins as biomaterial. This type of system either polymerizes liquid resin, hot filament while passing through the nozzle or sinters material of powder form. The principle of conventional tissue engineering is to seed isolated cells on the pre-formed solid scaffolds, placing the construct inside the bioreactor and then implanting it into patients. The 3D printing techniques are capable of meeting the requirements of hollow organs, but still there are some limitations, such as non-uniform seeding of cells into constructs, placement of different types of cells in defined positions, inability to reconstruct complex 3D organs. If the cell seeding is non-uniform through the scaffold, then incomplete healing may occur. Moreover, high temperature and use of toxic solvent can decrease cell survival. Also, increased time span for cell expansion may delay the treatment.

Hence, improvement in pre-existing 3D technology to tissue engineering leads to evolution of 3D bioprinting. Harsh processing conditions such as exposure to UV light, high temperature and organic solvents make 3D printing incompatible for the incorporation of cells while printing. Bioprinting allows printing of tissue constructs using hydrogel material at ambient conditions allowing incorporation of cells with biomaterials as ink. Current 3D bioprinting technique overcomes the shortcomings of 3D printing technique by organizing cells, macromolecules, extracellular components and biomaterials in a defined pattern in three or two dimensions to mimic functional tissue. The biomaterials available for 3D bioprinting are polymers (PCL,

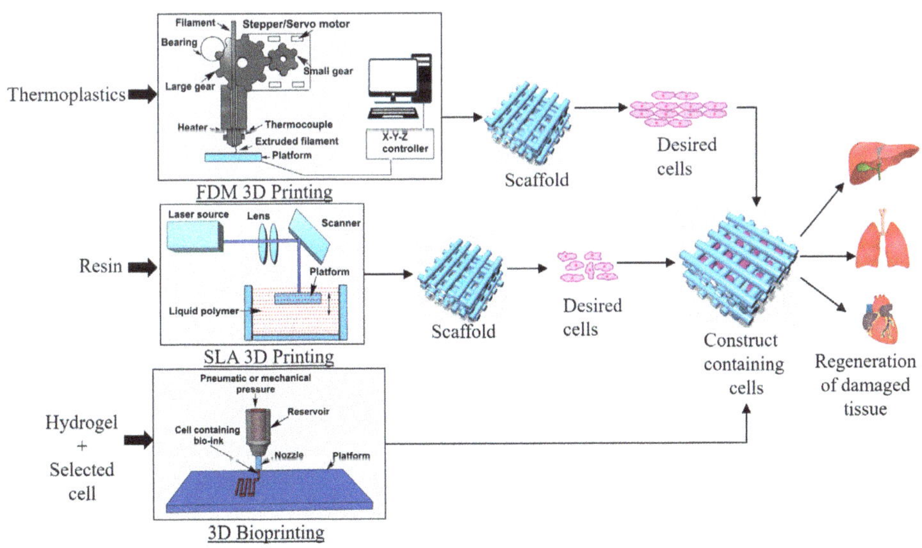

Figure 3.1. Figure showing different types of 3D printing techniques.

PLGA), extracellular matrix material (gelatin, collagen), proteins (silk) and hydrogels. The unique advantage of designing a scaffold with 3D bioprinting alleviates the problem occurring due to limited donors by using cell-infused bioink to print tissue constructs. Bioink is the most important component of 3D bioprinting. It consists of biomaterials, cells, growth factors and bioactive components. The advantages of bioprinting technology are positioning of cells at accurate position, *in situ* printing of cells, and ability to print high cell density tissues and vascularized tissues.

To date, many tissue constructs have been prepared using *in vitro* bioprinting. However, this printing approach is nowadays replaced by *in situ* and *in vivo* bioprinting. In this approach, denovo organs are fabricated directly at the damaged site of the patient. The applicability of *in vitro* tissue printed outside is to design functional tissue. The 3D Bioprinting technique helps to create tissue engineered constructs to heal the morbidity of the damaged site leading to the development of complex organs. A 3D model of an organ can be fabricated from DICOM data obtained from imaging techniques. A suitable image recognition algorithm is used to extract the desired structure and convert it into standard tessellation language (.STL) format. From this .STL file, a 3D model can be created using 3D bioprinter. 3D bioprinting mechanisms are used in wide range of application areas, as depicted in figure 3.2.

The fabrication of a functional organ of clinical dimensions is very challenging. Some of the difficulties are: addition of different cell types to prepare complex organ geometry, integration of vascular network and structure and functionality. Bioprinting is the simultaneous printing of proteins, living cells, DNA, drug particles, biomaterials, growth factor within prescribed layer-by-layer manner for the manufacturing of organs, tissues, organ-on-chip model or tissue construct. The bioink solution deposited on the support material can be polymerized through various crosslinking mechanisms (ionic, photopolymerization or enzymatic). Dependent upon working fundamentals, the bioprinting technology can be divided

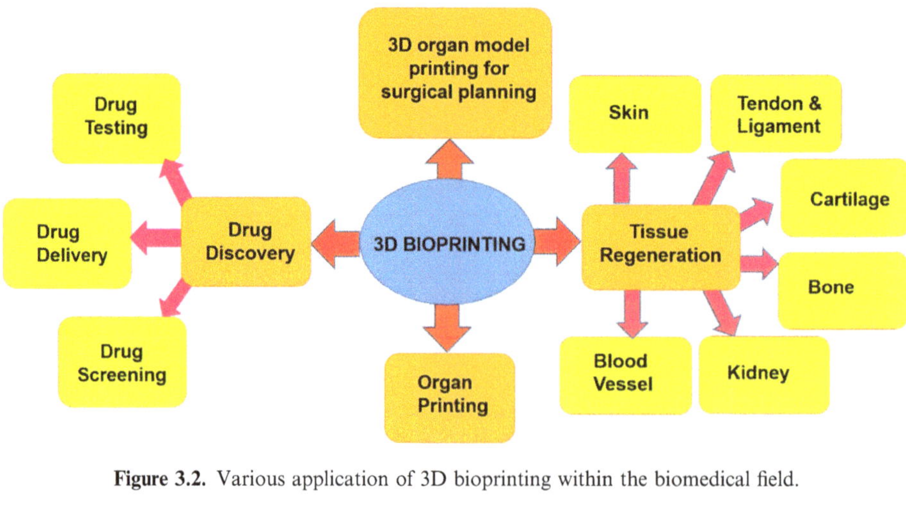

Figure 3.2. Various application of 3D bioprinting within the biomedical field.

into the following types: inkjet based bioprinting, laser based bioprinting and extrusion based bioprinting.

3.2 Types of 3D bioprinting

On the basis of the material delivery system, the 3D bioprinting technique can be classified as contact bioprinting and non-contact bioprinting (Mironov *et al* 2006):

- contact bioprinting: there is a contact between substrate and biomaterial delivery tool and substrate. Ex-extrusion based bioprinting;
- non-contact bioprinting: there is no contact between substrate and biomaterial delivery tool and substrate. Ex-inkjet based bioprinting.

Three main type of 3D Bioprinting techniques are used widely nowadays which are described in the following sections.

3.2.1 Extrusion-based bioprinting

In this 3D bioprinting technology, materials are dispensed by force (either pneumatic or mechanical) through a nozzle. It is one of the least expensive methods and is relatively slow compared to laser-based and inkjet-based technologies. In this bioprinting method, the temperature of the nozzle and the working platform can be controlled, allowing printing of a large variety of cell encapsulated biomaterials with high cell densities and viabilities. For resoluteness of extrusion, bioprinting is said to be directly proportional to the needle's diameter; material viscidity along with effect of shear stress while distributing is taken into consideration with this type of bioprinting. The biomaterial is melted through deposition and fused after extrusion (Gupta *et al* 2018).

3.2.2 Inkjet bioprinting

This bioprinting technology offers high-resolution patterning of a small amount of biomaterial to form a substrate. The basic printing process of an inkjet printer is: firstly, formation of a drop directed to a specific location; secondly, interaction of a drop with the substrate. Based on formation of drops of ink, inkjet printing can be further categorized as: (1) continuous inkjet printing (CIJ); (2) drop-on demand (DOD) printing . The CIJ printing mechanism operates at higher frequency, whereas DOD works at lower frequency. The droplet size formed by CIJ is twice that of the orifice size. The small volume droplets formed by DOD have greater resolution.

In the continuous inkjet printing approach with application of external pressure, a force is applied on the ink allowing streams of droplets to come out of the orifice continuously. As the ink in this approach is electrically conductive, the drops when formed are charged and are deposited on the desired place by magnetic or electric field. The extra droplets are captured in the reservoir and recirculated, as shown in figure 3.3.

In contrast, in DOD printing, a pressure pulse is applied into the ink and the drop is generated only when needed, as illustrated in figure 3.4.

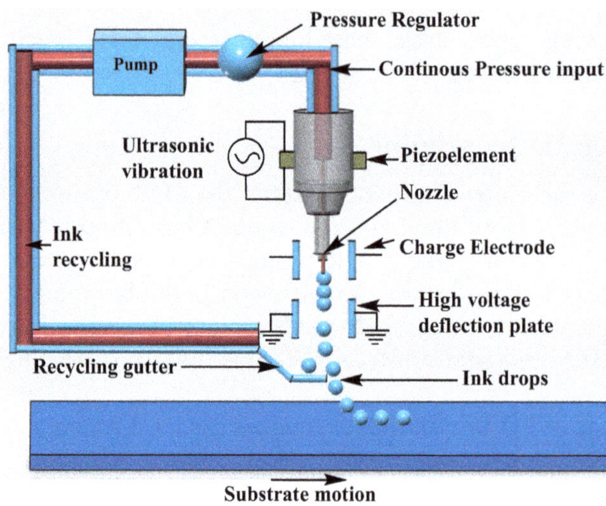

Figure 3.3. Representation of continuous inkjet 3D bioprinting technique.

During the printing process, the biomaterial to be printed in liquid form is jetted continuously onto the printer platform through the actuating units, forming ink droplets on the platform. Before the next layer of droplets is added in the platform, the ink droplet solidifies into the pre-defined geometry. Nevertheless, this technology has some disadvantages, such as: the material to be printed must have low viscosity; stress generated during bioprinting can affect cell viability; due to printer head clogging, only limited biomaterials can be used (Gupta *et al* 2018).

3.2.3 Laser based bioprinting

The principle elements of laser based bioprinting are:

- pulsed laser source: this pulse actuates metal film vaporization and hence forms a droplet, which deposits on the receiving substrate;
- focusing system;
- target to act as a support for printing material (ribbon): a thin layer of bioink spreads on a thin laser absorbing layer. For different types of bioinks and cells, different types of ribbon are needed;
- receiving substrate.

The laser from a pulsed laser source passes through the focusing system (lens) and falls on ribbon. This technique has additional advantages, such as: preventing cell clogging (as the nozzle is independent); having control on cell density; high deposition speed; control on cell organization and non-contact printing. A detailed description of the working principle of a bioprinter is explained elsewhere (Gupta *et al* 2018)

These bioprinters have the ability to print tenable cells of various types towards spatially determined 3D arrays, which mimics the structure of tissue or organ. Following the development of bioprinters, it became popular to print patient specific tissue constructs to supplement the activity of the natural tissue. Some of them are discussed in the following sections.

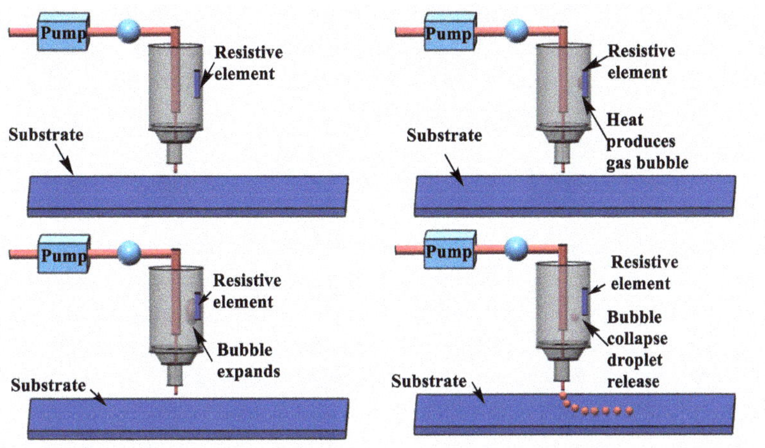

(a)

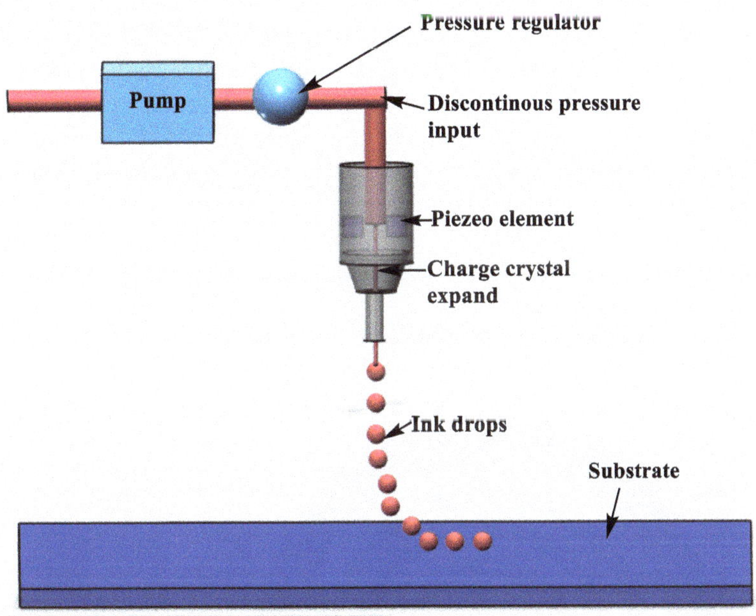

(b)

Figure 3.4. Figure showing DOD 3D bioprinting technique using (a) resistive element and (b) piezo-element.

3.3 Hard tissue engineering

3D bioprinting technique has the potential of regeneration of hard tissue such as bone and cartilage. A mineralized extracellular matrix is required for hard tissue engineering to mimic the tissue environment (Sun *et al* 2018). Hard tissues contain unique cell types, for example, bone consists of osteoblast cells with calcified extracellular matrix (ECM) and cartilage consists of chondrocyte cells. The biological functions performed by hard tissues are metabolism and hematopoiesis. The hard tissues require strong mechanical support hence the biomaterial selected should provide strength to the construct developed.

3.3.1 Bone

Bone is a type of connective tissue with calcified ECM. There are three main type of cell: osteoblasts, osteocytes and osteoclasts, as illustrated in the 3D model of bone in figure 3.5. The main challenges associated with bone bioprinting are selection of biomaterial (it must be biocompatible, promote uniform distribution of cells, be printable) and functional vasculature. All these challenges can be addressed by employing a suitable bioprinting technique.

Keriquel and colleagues have demonstrated *in situ* laser-assisted bioprinting of mesenchymal stromal cells for *in vivo* regeneration of bone tissue with high resolution and precision (Keriquel *et al* 2017).

Byambaa and co-authors fabricated a large bone tissue construct containing perfusable vascular lumen with functional vasculature by utilizing an extrusion-based direct-writing bioprinting method utilizing gelatin methacryloyl (GelMA) hydrogel on less methacryloyl replacement for bioink (Byambaa *et al* 2017).

Titanium due to its low antigenicity and good durability can be used as a biological implant material. However, when it encounters tissues, load transmits into the implants and it tends to loosen with time. Hence, a functional connection, i.e. osseointegration of a titanium implant is required. McBeth *et al* improved this osseointegration of a titanium implant by printing a gelatin methacryloyl (GelMA) scaffold via an extrusion-based 3D bioprinter. This scaffold either can be grafted or printed directly upon the titanium implant. This scaffold formed promotes calcium

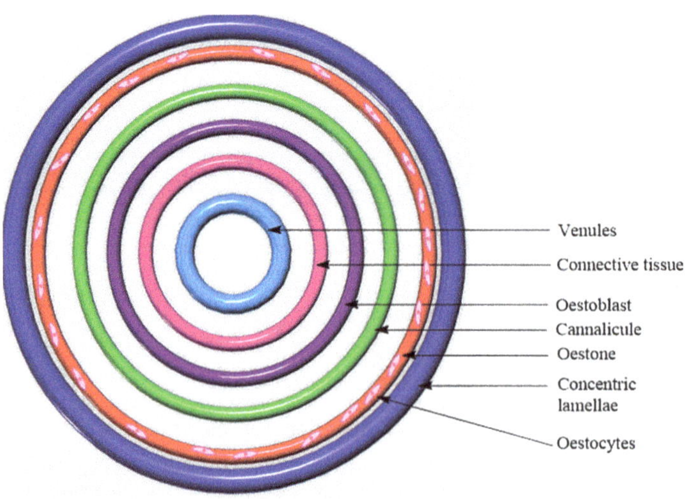

Figure 3.5. Schematic diagram of 3-D model of trabecular bone.

mineral discharge through osteoblasts and standard human primary osteoblasts without any exogenous osteogenic factors (McBeth *et al* 2017).

Hydroxyapatite is the main component of natural bone and poses osteoconductivity, which helps to create a bone-like microenvironment. The addition of hydroxyapatite in hydrogel improves osteogenic differentiation and cell proliferation (Demirtaş *et al* 2017). Wang *et al* described impairment of bone regeneration due to high levels of tumor necrosis element (TNF)-a. They utilized the antagonist outcome of Atsttrin and combined it with alginate (Alg)/hydroxyapatite (nHAp) to repair complex calvarial bone defects and improved tumor necrosis factor (TNF/TNFR) signals (Wang *et al* 2015). Bendtsen and co-authors prepared a novel hydrogel using hydroxyapatite (HA), alginate and polyvinyl alcohol (PVA) hydrogel, and further mouse calvaria 3T3-E1 (MC3T3) cells were added into it. Later, a porous scaffold formed, promoting enhanced uniform healing of the bone defect (Bendtsen *et al* 2017). Even complex craniofacial bone defects can be reconstructed by bioprinting of stromal vascular fraction (SVF) derived cells within polycaprolactone/hydroxyapatite (PCL/HAp) hydrogel under short-term hypoxia condition. The autogenous vascularized bone printing proves the ability of 3D bioprinting in tissue regeneration applications (Kuss *et al* 2017).

Hyaluronic acid (HA) is glycosaminoglycan carrying a high molecular weight, one of the polymers present in the extracellular matrix for giving mechanical support to ECM. Poldervaart *et al* made HA into a 3D printable scaffold by modification with methacrylate groups and further addition of photo initiators. The resultant methacrylated HA (MeHA) gel formed is resistant to degradation, has good biocompatibility, mechanical stiffness, allows primary cell survival, supports osteogenic differentiation and has low cytocompatibility. Hence, the required characteristics of bone tissue engineering are fulfilled (Poldervaart *et al* 2017).

A comparison was studied between chitosan and alginate as bioink when printed through an extrusion-based bioprinter in terms of mechanical support, mineralization and osteogenic differentiation, and it was shown that chitosan proved to be superior when compared to alginate (Demirtaş *et al* 2017).

Cui *et al* also prepared a vascularized bone biphasic assembly using FDM and SLA 3Dbioprinter (dual 3D bioprinting approach). To obtain biomimetic vascularized construct, the polylactide (PLA) fibers along with cell-encumbered gelatin methacrylate (GelMA) hydrogels and addition of hMSCs and human umbilical vein endothelial cells (HUVECs) were used (Cui *et al* 2016).

Bioactive glasses have the potential to provide angiogenesis, increased mineralisation of bone related cells and oesteogenesis. Wang *et al* studied the influence of bioglass upon biomineralization of bone related SaOS-2 cells. Towards this alginate/gelatin, hydrogel is taken and supplemented with bioglass. In addition, they found that incorporation of bioglass increases the SaOS-2 cells' potency to mineralize and proliferate (Wang *et al* 2014). Gao *et al* also conducted research to evaluate the osteogenesis property of bioactive glass. For this they co-printed hMSCs strung up in poly(ethylene glycol) dimethacrylate (PEDGMA) along with bioactive glass and hydroxyapatite. However, they found that hydroxyapatite promotes oestogenesis as compared to bioactive glass (Gao *et al* 2014). Murphy and co-authors used two syringe-based extrusion bioprinting systems in which one syringe consists of PCL polymer, bioactive glass and chloroform and the other contains human adipose stem cells (ASCs) suspended in Matrigel. Firstly, the syringe containing bioactive glass is allowed to print, on the top of which another layer of bioink is printed. This 3D bioprinted polycaprolactone/bioactive borate glass scaffold has the prospect to be utilized for bone tissue engineering with enhanced angiogenesis (Murphy *et al* 2017).

Gao *et al* contrived hMSCS incorporated PEG-GelMA hydrogel scaffold using an inkjet bioprinter having simultaneous photopolymerisation capacity. This scaffold shows improved mechanical strength required for cartilage and bone tissue engineering (Gao *et al* 2015). In further studies, Gao and co-authors combined acrylated peptides and acrylated poly-(ethylene glycol) (PEG) to form a hydrogel into which human mesenchymal stem cells (hMSCs) were added for preparing bioink for inkjet bioprinting to print bone and cartilage. The scaffold exhibited good biocompatibility with enhanced chondrogenic and osteogenic differentiation (Gao *et al* 2015).

3.3.2 Cartilage

Cartilage is a connective tissue, which lacks a vasculature system and a nervous system. It contains special types of cell called chondrocytes, as shown in figure 3.6. Due to absence of blood vessels, diffusion of nutrition takes place through a matrix, hence it has slow self-healing capability. Cartilage is a type of hard tissue but is elastic and flexible in nature. Cartilage is broadly classified into three types: fibrous, hyaline (or articular) and elastic cartilage. For the fabrication of artificial cartilage, bioink containing chondrogenic cells is preferred.

In cartilage-based tissue engineering, materials such as polycaprolactone (PCL), alginate and methacrylate-containing materials are blended with suitable hydrogel and cells to be used as bioink compatible with cartilage cells to promote chondrogenesis. Bioink selection is one of the challenges of bioprinting. To support the chondrocytes phenotype, many researches on alginate-blended bioink have been

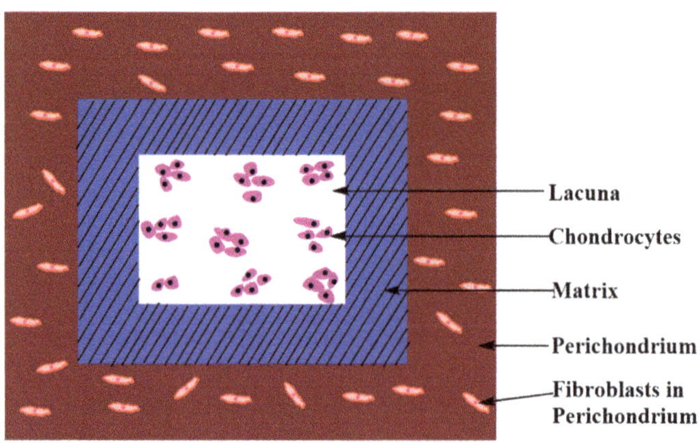

Figure 3.6. Diagram showing a 3-D model of cartilage.

carried out. To regenerate cartilage by the use of a multi-head deposition system (MHDS), Kundu *et al* prepared a PCL framework into which chondrocyte cell-constituted alginate hydrogel was dispensed to prepare a hybrid scaffold to prepare cartilaginous ECM for cartilage regeneration (Kundu *et al* 2015). Markstedt *et al* merged the shear thinning attributes about nano fibrillated cellulose (NFC) with alginate to bioprint this hydrogel after addition of chondrocytes cells to fabricate cartilage tissue (Markstedt *et al* 2015). Muller *et al* introduced a sulfated version of alginate to bind growth factors (fibroblast, transforming and hepatocyte growth factor). The gel disc made of alginate sulfate-nano cellulose material allows cell spreading and proliferation (Müller *et al* 2017). Apelgren and team fabricated *in vivo* human cartilage by combining nano cellulose and alginate with human chondrocytes and hMSCs using an extrusion printer (Apelgren *et al* 2017). Abbadess and co-authors prepared polyethylene glycol together with in part methacrylated poly (*N*-(2-hydroxypropyl) methacrylamide mono/dilactate) triblock copolymers as hydrogel to fabricate a cartilage repair construct. This combination of polymer supported cartilage formation and provided good mechanical property (Abbadessa *et al* 2016).

Mouser and co-authors bioprinted chondrocytes-laden gelatin-methacryloyl (GelMA) blended with gellan gum to evaluate its suitability for use as bioink for cartilage fabrication. In addition, they found that this hydrogel supports cartilage matrix production and at high yield stress, viscosity is decreased, resulting in high printability of hydrogel (Mouser *et al* 2016).

In another study, Cui and co-authors suspended human chondrocytes into poly (ethylene glycol) diacrylate (PEGDA) to print neo-cartilage by delivering the bioink at a precise location using a thermal inkjet printing method. This neo-cartilage formed exhibits outstanding collagen (type II) and glycosaminoglycan (GAG) production (Cui *et al* 2016).

3.4 Soft tissue engineering

Soft tissue engineering is challenging as it aims to treat defect with high volume-to-surface ratio. The prevailing therapeutic treatment by surgery or autologous tissue transfer introduces further complication of morbidity at the donor site and even scar formation at the injury site.

3.4.1 Vascular tissue

Any tissue construct of either large or small tissues or an organ requires vasculature to allow blood flow for its functioning. The incorporation of a vessel-like structure is of critical importance to develop functional tissue suitable for tissue regeneration. Extracellular matrix-based biomaterials (ex-gelatin, collagen) have limited mechanical strength to withstand the weight of hollow tubes but are biocompatible and allow cell proliferation. Hence, they are blended with suitable biomaterial having mechanical stability to be used as bioinks to fabricate highly organized and intact vessels. The vascular tissue is made of the following main layers: tunica interna, tunica media along with tunica external, as shown in figure 3.7. The innermost layer (tunica intima) comprises endothelial cells and provides resistance to infections and non-thrombogenicity. The next layer is the basement layer containing collagen type IV, laminin. The middle tunica media layer contains collagen types I and III and smooth muscle cells. These smooth muscle cells are reliable towards the dilation and contraction of vessels. The composition of the outermost layer named tunica adventitia is fibroblasts. There is a coordination between all the layers to perform their function such as perfusion. Depending on the size of vessels, vasculature can be classified as macro vasculature and microvasculature. Macro vasculature contains arteries and veins, whereas microvasculature consists of arterioles, capillaries and venules (Yuan and Rigor 2010). The interconnected vascular networks within tissue constructs supporting transportation of nutrients, oxygen and waste products has many applications in organ repair and transplantation. Conventional microfabrication techniques are unable to create such perfusable hollow vascular tubes leading to formation of a necrotic region in bulk tissues. Jia *et al* biofabricated the vascular network through patterned cells and biomaterial deposition using an advanced extrusion system with cell feedback bioink containing gelatin methacryloyl (GelMA), 4-arm poly(ethylene glycol)-tetra-acrylate (PEGTA) along with sodium alginate (Jia *et al* 2016). Kolesky and co-authors constructed thick vascularized tissue (⩾1 cm) using fugitive ink (pluronic and thrombin), silicone ink, cell-charged ink (such as gelatins and fibrinogen), also cells as bioink to print the construct within 3D perfusion chips (Kolesky *et al* 2016).

Construction of small sized vessels of diameter <6 mm (periphery artery, coronary artery) is very challenging. Lee *et al* bioprinted a small diameter vessel by adding bone marrow-originated mesenchymal stem cells (bMSCs) within sodium alginate for a second layer on top of diagonal cross-stripped PCL scaffold layer. Then a third layer of helical form PCL scaffold is placed on the top of second layer. The research is conducted for studying the influence of pulsatile flow upon bMSCs, and it is observed that cells differentiated into endothelial-like cells (Lee *et al* 2018).

3.4.2 Skin

Bioprinting technique is feasible to print artificial skin by combining two or more types of skin cells (figure 3.8). In the case of extensive burn injury, it is difficult to opt

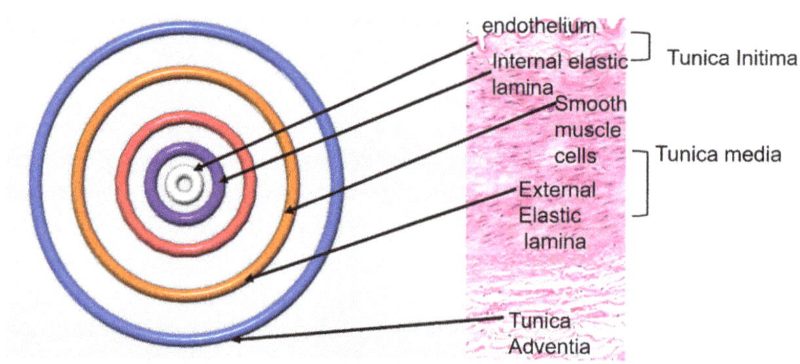

Figure 3.7. Schematic diagram of 3-D model of blood vessels.

for conventional treatment techniques such as autografts, allografts. This treatment option has many disadvantages such as limited size, very long preparation time and contracture.

Kyle *et al* proposed on-site *in situ* skin repair by printing keratinocytes and fibroblasts directly on the wound model using inkjet bioprinting. These constructs leads to production of skin identical to normal murine skin (Binder *et al* 2010). Skardal *et al* printed collagen-fibrin gel containing amniotic fluid-derived stem cells (AFS) over the wound of mice for skin regeneration. The incorporation of AFS cells modulates the inflammatory response and the prepared construct exhibits increased neovascularization and maturation of blood vessels (Skardal *et al* 2012).

3.5 Tissue engineering for application in specific organs

The only treatment for end-stage organ injury/failure is replacement of the organ. There is an unmet demand for organ donors. This gap of donor and organs can be covered by fabricating artificial organs using 3D bioprinting. This section explains about the status of organ bioprinting. The potential of organ bioprinting is to convert bioprinted cell-laden hydrogel into a vascularized organ-like structure.

3.5.1 Liver

The liver is the largest and unique organ responsible for blood detoxification, and also it is crucial for metabolism of drugs. Some of its important functions are: bile secretion, detoxification and homeostasis. Anatomically, the hepatocytes are polyhedral cells constituting 80% liver volume, as depicted in figure 3.9. The liver function decreases with damage of hepatocytes leading to liver failure. The end stage liver disease, chronic liver injury or tumors of liver origin leads to liver malfunction. The advancement of stem cell based research allowed differentiation of stem cells into hepatocytes leading to development of artificial liver either by liver support

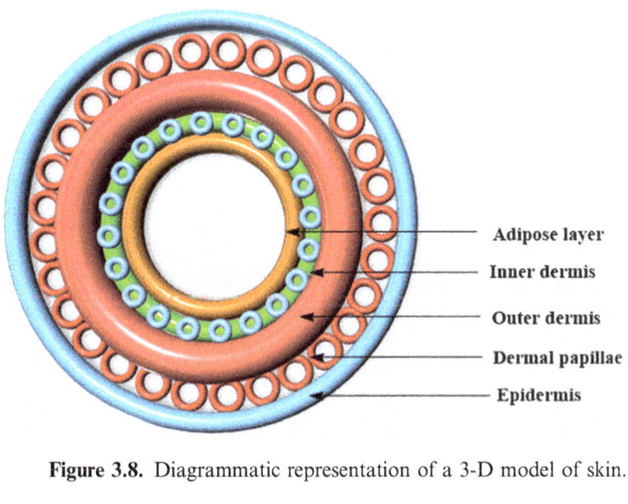

Figure 3.8. Diagrammatic representation of a 3-D model of skin.

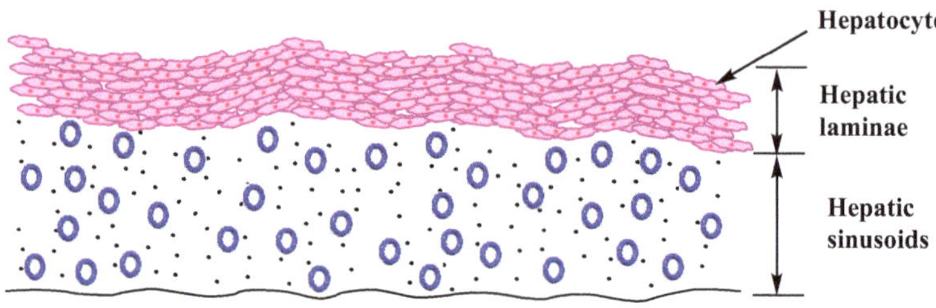

Figure 3.9. Schematic diagram of a 3-D model of the liver.

systems, 3D bioprinting or by seeding cells in a decellularized scaffold. Faulkner-Jones *et al* addressed the bioprinting of human influenced pluoripotent stem cells (hiPSCs) for the very first time using a valve-implanted printing technique, and established that post-printing hiPSCs differentiates through to hepatocyte cell types (HLCS) (Faulkner-Jones *et al* 2015). Ma *et al* employed a DLP oriented 3D bioprinter to print an *in vitro* 3D liver model mimicking several features of liver by patterning hiPSC-derived hepatic primogenitor cells (hiPSC-HPCs) within the GelMAas first layer. And subsequently printing supporting cells such as human umbilical vein endothelial cells along with adipose derived stem cells within GelMA and GMHA as a second layer (Ma *et al* 2016).

The bioprinting approach for liver model fabrication includes fabrication of 3D architecture combined with a bioreactor to prepare an organ-on-chip platform. During clinical trial of new drugs, the liver is most susceptible to toxic chemicals present in drugs. As the liver is responsible for clearance of those toxic agents it is susceptible to damage i.e. hepatotoxicity. Hence, nowadays prediction of the toxicity of new drugs on tissues can be done quickly using liver-on-chip or mini-liver (microfluidic chip containing liver cells) instead of performing *in vivo* studies. Bhise and co-authors fabricated and demonstrated the use of a liver-on-chip platform, as a means to assess the drug toxicity. They developed a perfusable bioreactor with printer to bioprint human hepato-carcinoma cells (HepG2/C3A) spheroids. The printed spheroid was mixed with GelMA along with 2-hydroxy-4′-(2-hydroxyethoxy)-2-methylpropiophenone solution to again bioprint this solution onto a glass slide to prepare hepatic construct (Bhise *et al* 2016).

To assess drug induced liver injury (DILI) many researchers have created a liver tissue model to perform drug diffusion and test toxicity. Nguyen and co-authors have bioprinted human hepatic stellate cells (HSCs) along with human umbilical vein endothelial cells (HUVECs), and cryopreserved primary human hepatocytes containing bioink within plates to develop 3D liver tissue (Nguyen *et al* 2016). Zhong *et al* prepared bioink through joining L02 cells (human hepatocytes cell line HL-7702) with collagen and chitosan solution buffered with morpholinoethane sulfonic acid solution (MES) and NaOH hydrogel to print a 3D scaffold in favor of liver tissue engineering (Zhong *et al* 2016). In related studies, Massa *et al* utilized a sacrificial bioprinting technique by printing agarose fiber into HepG2/C3A cell-laden GelMa hydrogel poured into a PMMA mold. The mold was then exposed to UV light for crosslinking polymer solution and agarose fiber then removed manually to form a hollow microchannel in the construct. In addition, HUVECs are added to prepare a vascularized liver tissue model (Massa *et al* 2017).

3.5.2 Kidney

The kidney derived from mesoderm is the main organ of the body responsible for excretion of waste, xenobiotics and water homeostasis. The nephron is the functional unit of the kidney containing the glomerulus (for filtration of blood) and renal tubule (for flow of glomerular filtrate). The proximal renal tubule is near the glomerulus and is responsible for renal filtrate transport to blood. Hence, it can be affected by accumulation of a high quantity of circulating drugs or xenobiotics. The nephrons are connected to the ureter via a collecting duct.

Renal damage can be done at many locations such as the glomerulus, collecting duct or proximal tube. Renal injury such as acute kidney injury (AKI), end stage renal disease (ESRD) or also chronic kidney disease (CKD) demands kidney regeneration treatment such as dialysis, transplantation and bioprinting. The number of kidney donors is not sufficient to fulfill the requirement of the increased patients. The ability of distinction regarding stem cells through to renal cell kinds helps to prepare a bioengineered kidney using bioprinting technique.

As stated earlier, proximal tubules are more prone towards renal injury. Hence, many researchers have worked on them. King *et al* bioprinted a 3D proximal tubule containing interstitial cell types containing a renal proximal tubule epithelial cells (RPTEC) model to study epithelial–interstitial interaction and human nephrotoxicity (King *et al* 2017). Homan *et al* developed a 3D convoluted renal proximal tubule with a kidney-on-chip device by printing firstly the fugitive ink on a gelatin-fibrinogen ECM layer deposited at the gasket (figure 3.10). The 3D proximal tubule is formed by extruding fugitive ink from the nozzle of a 3D bioprinter. Then additional ECM is casted on the top of the printed fugitive ink. A model placed inside the perfusable chip and liquefied to 4 °C to remove the fugitive ink results in a

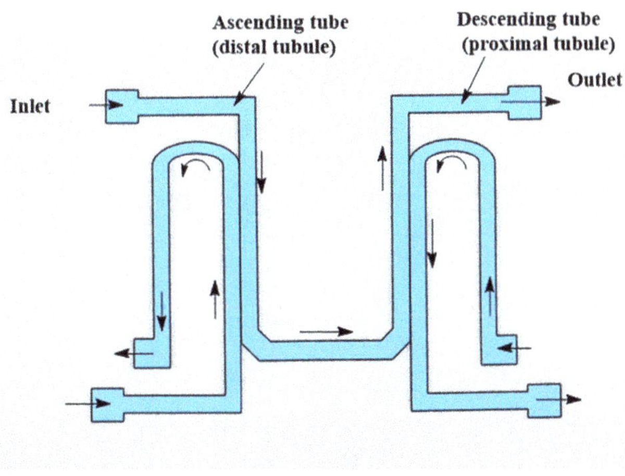

Figure 3.10. Kidney-on-a-chip *in vitro* model.

tubular channel. Within this tubule, PTEC cells are introduced which after a period of time become a polarized epithelium (Homan *et al* 2016).

3.5.3 Bladder

Total substitution of the bladder is required for patients undergoing cystectomy. Many problems like bladder cancer, trauma, infection, inflammation, iatrogenic injury and interstitial cystitis can lead to bladder injury (Gürpinar and Griffith 1996). Problems associated with the artificial bladder are deposition of calcareous material, urine leakage resulting in peritonitis, encrustation and infection. Hence, an artificial bladder should be made up of inert material, must be non-irritating and non-allergic. Bladder reconstruction can be done using a prosthetic bladder. The bioprinting technique has also been able to develop an artificial bladder to treat bladder damage (figure 3.11). There is still limited literature available on the printing of the bladder, but yet this 3D bioprinting technique is of great significance for bladder regeneration. Xu *et al* developed a cell-encapsulating droplet generation system to print building blocks within a sterile hood to control humidity and temperature. These building blocks assemble together to create a smooth muscle cell (SMC) patch. This patch was cultivated up to 51 days to form a 3D tissue assembly similar to native rat bladder (Xu *et al* 2010). Recently, Zhang *et al* used an integrated bioprinting technique for fabricating the urethra along with poly(ε-caprolactone) (PCL) also Poly(lactide-co-caprolactone)(PLCL) thermoplastic polymers along with bioink made of fibrin, gelatin, hyaluronic acid as hydrogel with urothelial cells (UCs) together with smooth muscle cells (SMCs) as the cellular component to imitate the natural urethrae of rabbits. With the printing of a tubular scaffold, simultaneously the bioink was delivered into the scaffold layers formed (Zhang *et al* 2017). Imamura *et al* reconstructed a urinary bladder using a 3D bioprinting robot system. The biofabricated structure contains bone marrow-originated cells, which disintegrate as SMCs to fabricate urinary bladders (Imamura *et al* 2018). More development is still required to meet the ideal requirements of fabrication of the

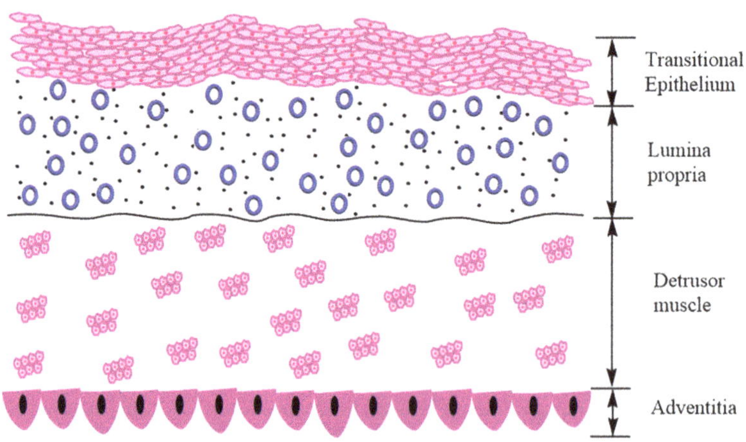

Figure 3.11. Diagrammatic representation of a 3-D model of the bladder.

artificial bladder. This bioprinting approach using inert biomaterials would resolve the problem of calcareous deposits.

3.5.4 Retina

The retina, responsible for the perception of sensory information, is a nervous tissue that contains rods and cone photoreceptors. The human retina has immense structural complexity, as shown in figure 3.12; it is highly vascularized tissue containing different cell types such as horizontal cells, amacrine cells, retinal pigment epithelial cells, retina ganglion cells, bipolar cells, glial cells as well as Müller cells. There are 55 different cell types and 10^5 per square millimeter photoreceptor cells in the retina (Peterman *et al* 2003). All these retinal cells must act in harmony with each other to transmit visual information to the brain. Retinal pigment epithelium (RPE) is the outermost layer of retina containing a single layer of hexagonal pigmented polarized epithelial cells. These RPE cells are responsible for the proper functioning of visual process and are difficult to duplicate. Any malfunction of RPE cells may cause retinal disease. The choroid is a layer providing vascular structure for the supply of nutrients and oxygen to other layers of the retina. In between RPE and choroid lies the Bruch's membrane (Komez *et al* 2016). The ganglion cells and the photosensors lies in the innermost and outermost portion of the retina, respectively. Light travels from the retina, strikes the photoreceptors and activates it. The photons are absorbed by the visual pigments of photoreceptor and translated into electrical signals. The electrical signals simulate the neurons of the retina and are transmitted to the brain from the spike discharge pattern of ganglion cells. For vascularization of retinal layers, blood vessels pass through the retina. In some retinal disease, only particular cells need to be replaced such as retina ganglion cells in glaucoma, but in others the retinal area needs to replaced. The retinal cells deteriorate with age or due to diabetes or occlusion and fail to regenerate, which

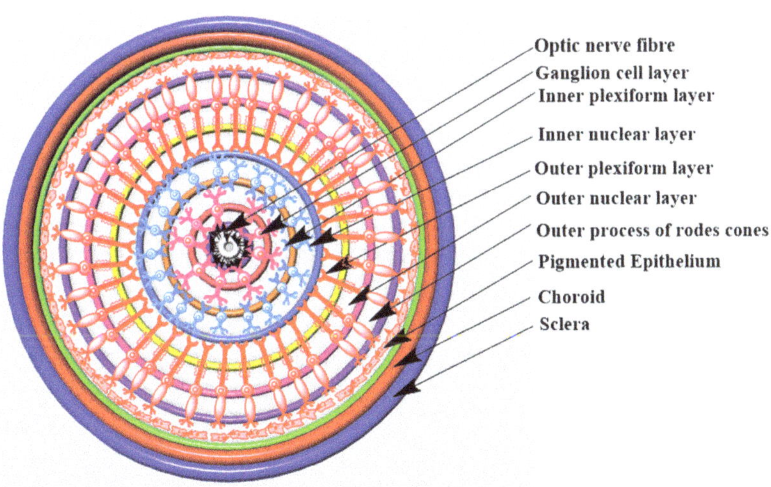

Figure 3.12. Schematic diagram of a 3-D model of the retina.

promotes the need for cell transplantation and 3D bioprinting techniques. Cell transplantation has a limitation of axion orientation in *in vivo* conditions. Bioprinting of a complex retina containing many cell types too is very challenging. The scaffold for retinal tissue engineering must be porous, biocompatible, not induce foreign body response, mechanically stable and must be thin.

With advancement in stem cell biology, retinal equivalent construct can be bioprinted by using stem cells responsible for retinal tissue engineering like embryonic stem (ES) cells along with induced pluripotent stem cells (iPSCs). Reconstruction of a destroyed retina can be done by using the differentiation capability relating to induced pluripotent stem cells towards different types of retinal cells (retinal pigment epithelial cells, retina ganglion cells, etc) (Yang *et al* 2017).

The retinal tissue engineering may include development of Bruch's membrane, printing of the retinal pigment epithelium, retinal glial cells and retinal ganglion cells. The bioprinted scaffold acts as a substitute to attach various kinds of retinal cells.

Among different types of retinal cells, retinal ganglion cells (RCGs) placed at ganglion cell layers transmit a visual message from photoreceptors to the brain, but they fail to regenerate. An important consideration for RGC development is to maintain the axon orientation as compared to existing axon orientation, RGC positioning and cell survival rate. These challenges can be met by providing physical support to the cells by means of a scaffold. The scaffold must have the potential to guide axon orientation, cell adhesion, cell differentiation, cell proliferation and cell migration. Kador and co-authors combined a thermal inkjet 3D bioprinting technique to place RGCs precisely on an electrospun scaffold surface to guide RGC neurite outgrowth (Kador *et al* 2016). Lorber *et al* studied the influence of inkjet printing methods upon retinal ganglion cells (RGC) and glia cells. They found that even though there is a reduction in cell population, there is no effect on the survival rate of cells during jetting (Lorber *et al* 2014)

The Bruch membrane is a specialized membrane which acts as a thin (2–4 μm) barrier between choroid and retina. It is an important component of retina tissue engineering as it acts as a scaffold for cell growth. Its key characteristics are that it is: ultrathin, porous (to control diffusion rate) and permeable (to allow nutrient exchange). To address the pore-size distribution limitation of Bruch's membrane, Tan *et al* developed an ultrathin PCL membrane with regular and interconnected pores instead of the larger pore and thick film reported earlier. The low cytotoxic membrane developed showed improved ARPE-19 cell arrangement forming a well-developed barrier to allow homeostatic functions (Tan *et al* 2017). To extend the application of this ultrathin membrane Shi and co-authors combined two cell lines which are the human retinal pigmented epithelial cell line (ARPE-19) and the human retinoblastoma cell line (Y79) containing alginate or pluronic bioink to prepare a retina model using a microvalve based bioprinting method. Here the bioprinted retina model prepared using an ARPE-19 cell predefined on ultrathin membrane epitomises the RPE monolayer along with Bruch's membrane into which Y79 cell-containing bioink is printed (Shi *et al* 2017).

3.6 Conclusion

The printing of functional human organs to overcome the issue of organ shortage is still required to revolutionize the healthcare industry. There remains a need to develop 3D bioprinted functional organs of relevant dimensions. The combination of imaging techniques, cell-laden inks and 3D bioprinting technologies aims to bring a novel concept within tissue engineering and regenerative medicine. The 3D printed tissue or organ based on imaging data are prone to imaging errors. Hence, imaging must be done accurately to create a patient specific construct. It is essential to check the reactions of cells post-printing as cells are subjected to shear stress either at the nozzle or at the orifice and also experience pressure, high frequency or heat to start the process of bioprinting. In an organ-derived scaffold, no cells should remain at the time of transplantation to avoid an immunological rejection.

Despite so many researches, so far no ideal material is available for 3D bioprinting of the bladder. The biomaterial for bladder reconstruction must be able to contract and expand as the walls of the bladder undergo mechanical deformation during the normal function of the bladder.

Also, it is important to perform a detailed research on printing of different types of retinal cells and even photoreceptors using a multi-head bioprinter to allow fabrication of a functional retina. The main challenge of retinal bioprinting is to maintain cell density to develop the layers of retina, which can be addressed by the use of multiple passes of stem cells. Different retinal cells must show synaptic stimulation among each other allowing proper transmission of the visual signal. Fabrication of stem cell based organoids has the ability to represent complex architecture, a multilayer and multicellular organ model to resemble different types of organs similar to *in vivo* counterparts. With these organoids, it is easy for researchers to develop new disease treatment.

References

Abbadessa A, Mouser V H M, Blokzijl M M, Gawlitta D, Dhert W J A, Hennink W E, Malda J and Vermonden T 2016 A synthetic thermosensitive hydrogel for cartilage bioprinting and its biofunctionalization with polysaccharides *Biomacromolecules* **17** 2137–47

Apelgren P, Amoroso M, Lindahl A, Brantsing C, Rotter N, Gatenholm P and Kölby L 2017 Chondrocytes and stem cells in 3D-bioprinted structures create human cartilage *in vivo PLoS One* **12** e0189428

Bendtsen S T, Quinnell S P and Wei M 2017 Development of a novel alginate-polyvinyl alcohol-hydroxyapatite hydrogel for 3D bioprinting bone tissue engineered scaffolds *J. Biomed. Mater. Res.* A **105** 1457–68

Bhise N S *et al* 2016 A liver-on-a-chip platform with bioprinted hepatic spheroids *Biofabrication* **8** 014101

Binder K W, Zhao W, Aboushwareb T, Dice D, Atala A and Yoo J J 2010 *In situ* bioprinting of the skin for burns *J. Am. Coll. Surgeons* **211** S76

Byambaa B *et al* 2017 Bioprintedosteogenic and vasculogenic patterns for engineering 3D bone tissue *Adv. Healthcare Mater.* **6** 1700015

Cui H, Zhu W, Nowicki M, Zhou X, Khademhosseini A and Zhang L G 2016 hierarchical fabrication of engineered vascularized bone biphasic constructs via dual 3D bioprinting: integrating regional bioactive factors into architectural design *Adv. Healthcare Mater.* **5** 2174–81

Demirtaş T T, Irmak G and Gümüşderelioğlu M 2017 A bioprintable form of chitosan hydrogel for bone tissue engineering *Biofabrication* **9** 035003

Faulkner-Jones A, Fyfe C, Cornelissen D-J, Gardner J, King J, Courtney A and Shu W 2015 Bioprinting of human pluripotent stem cells and their directed differentiation into hepatocyte-like cells for the generation of mini-livers in 3D *Biofabrication* **7** 044102

Gao G, Schilling A F, Hubbell K, Yonezawa T, Truong D, Hong Y, Dai G and Cui X 2015 Improved properties of bone and cartilage tissue from 3D inkjet-bioprinted human mesenchymal stem cells by simultaneous deposition and photocrosslinking in PEG-GelMA *Biotechnol. Lett.* **37** 2349–55

Gao G, Schilling A F, Yonezawa T, Wang J, Dai G and Cui X 2014 Bioactive nanoparticles stimulate bone tissue formation in bioprinted three-dimensional scaffold and human mesenchymal stem cells *Biotechnol. J.* **9** 1304–11

Gao G, Yonezawa T, Hubbell K, Dai G and Cui X 2015 Inkjet-bioprintedacrylated peptides and PEG hydrogel with human mesenchymal stem cells promote robust bone and cartilage formation with minimal printhead clogging *Biotechnol. J.* **10** 1568–77

Gupta S, Bissoyi A and Bit A 2018 A review on 3D printable techniques for tissue engineering *Bionanoscience* **8** 868–83

Gürpinar T and Griffith D P 1996 The prosthetic bladder *World J. Urol.* **14** 47–52

Homan K A, Kolesky D B, Skylar-Scott M A, Herrmann J, Obuobi H, Moisan A and Lewis J A 2016 Bioprinting of 3D convoluted renal proximal tubules on perfusable chips *Sci. Rep.* **6** 34845

Imamura T, Shimamura M, Ogawa T, Minagawa T, Nagai T, SilwalGautam S and Ishizuka O 2018 Biofabricated structures reconstruct functional urinary bladders in radiation-injured rat bladders *Tissue Eng. Part* A

Jia W *et al* 2016 Direct 3D bioprinting of perfusable vascular constructs using a blend bioink *Biomaterials* **106** 58–68

Kador K E, Grogan S P, Dorthé E W, Venugopalan P, Malek M F, Goldberg J L and D'lima D D 2016 Control of retinal ganglion cell positioning and neurite growth: combining 3D printing with radial electrospun scaffolds *Tissue Eng. Part* A **22** 286–94

Keriquel V *et al* 2017 *In situ* printing of mesenchymal stromal cells, by laser-assisted bioprinting, for *in vivo* bone regeneration applications *Sci. Rep.* **7** 1778

King S M *et al* 2017 3D proximal tubule tissues recapitulate key aspects of renal physiology to enable nephrotoxicity testing *Frontiers Physiol.* **8** 123

Kolesky D B, Homan K A, Skylar-Scott M A and Lewis J A 2016 Three-dimensional bioprinting of thick vascularized tissues *PNAS* **113** 3179–84

Komez A, Baran E T, Erdem U, Hasirci N and Hasirci V 2016 Construction of a patterned hydrogel-fibrous mat bilayer structure to mimic choroid and Bruch's membrane layers of retina *J. Biomed. Mater. Res.* A **104** 2166–77

Kundu J, Shim J-H, Jang J, Kim S-W and Cho D-W 2015 An additive manufacturing-based PCL-alginate-chondrocyte bioprinted scaffold for cartilage tissue engineering *J. Tissue Eng. Regen. Med.* **9** 1286–97

Kuss M A, Harms R, Wu S, Wang Y, Untrauer J B, Carlson M A and Duan B 2017 Short-term hypoxic preconditioning promotes prevascularization in 3D bioprinted bone constructs with stromal vascular fraction derived cells *RSC Adv.* **7** 29312–20

Lee K W, Kim D-H, Lee J H and Youn Y-N 2018 The effect of pulsatile flow on bMSC-derived endothelial-like cells in a small-sized artificial vessel made by 3-dimensional bioprinting *Stem Cells Int.* **2018** 7823830

Lorber B, Hsiao W-K, Hutchings I M and Martin K R 2014 Adult rat retinal ganglion cells and glia can be printed by piezoelectric inkjet printing *Biofabrication* **6** 015001

Ma X *et al* 2016 Deterministically patterned biomimetic human iPSC-derived hepatic model via rapid 3D bioprinting *PNAS* **113** 2206–11

Markstedt K, Mantas A, Tournier I, Martínez Ávila H, Hägg D and Gatenholm P 2015 3D bioprinting human chondrocytes with nanocellulose-alginate bioink for cartilage tissue engineering applications *Biomacromolecules* **16** 1489–96

Massa S *et al* 2017 Bioprinted 3D vascularized tissue model for drug toxicity analysis *Biomicrofluidics* **11** 044109

McBeth C, Lauer J, Ottersbach M, Campbell J, Sharon A and Sauer-Budge A F 2017 3D bioprinting of GelMA scaffolds triggers mineral deposition by primary human osteoblasts *Biofabrication* **9** 015009

Mironov V, Reis N and Derby B 2006 Review: bioprinting: a beginning *Tissue Eng.* **12** 631–4

Mouser V H M, Melchels F P W, Visser J, Dhert W J A, Gawlitta D and Malda J 2016 Yield stress determines bioprintability of hydrogels based on gelatin-methacryloyl and gellan gum for cartilage bioprinting *Biofabrication* **8** 035003

Müller M, Öztürk E, Arlov Ø, Gatenholm P and Zenobi-Wong M 2017 Alginate sulfate-nanocellulosebioinks for cartilage bioprinting applications *Ann. Biomed. Eng.* **45** 210–23

Murphy C, Kolan K, Li W, Semon J, Day D and Leu M 2017 3D bioprinting of stem cells and polymer/bioactive glass composite scaffolds for tissue engineering *Int. J. Bioprinting* **3** 54–64

Nguyen D G, Funk J, Robbins J B, Crogan-Grundy C, Presnell S C, Singer T and Roth A B 2016 Bioprinted 3D primary liver tissues allow assessment of organ-level response to clinical drug induced toxicity *in vitro PLoS One* **11** e0158674

Peterman M C, Mehenti N Z, Bilbao K V, Lee C J, Leng T, Noolandi J, Bent S F, Blumenkranz M S and Fishman H A 2003 The Artificial Synapse Chip: a flexible retinal interface based on directed retinal cell growth and neurotransmitter stimulation *Artif. Organs.* **27** 975–85

Poldervaart M T, Goversen B, de Ruijter M, Abbadessa A, Melchels F P W, Öner F C, Dhert W J A, Vermonden T and Alblas J 2017 3D bioprinting of methacrylated hyaluronic acid (MeHA) hydrogel with intrinsic osteogenicity *PLoS One* **12** e0177628

Shi P, Edgar T Y S, Yeong W Y and Laude A 2017 Three-dimensional (3D) bioprinting of retina equivalent for ocular research *Int. J. Bioprinting* **3** 138–46

Skardal A, Mack D, Kapetanovic E, Atala A, Jackson J D, Yoo J and Soker S 2012 Bioprinted amniotic fluid-derived stem cells accelerate healing of large skin wounds *Stem Cells Transl. Med.* **1** 792–802

Sun T, Yao S, Liu M, Yang Y, Ji Y, Cui W, Qu Y and Guo X 2018 Composite scaffolds of mineralized natural extracellular matrix on true bone ceramic induce bone regeneration through smad1/5/8 and ERK1/2 pathways *Tissue Eng. Part* A **24** 502–15

Tan E Y S, Agarwala S, Yap Y L, Tan C S H, Laude A and Yeong W Y 2017 Novel method for the fabrication of ultrathin, free-standing and porous polymer membranes for retinal tissue engineering *J. Mater. Chem.* B **5** 5616–22

Wang Q, Xia Q, Wu Y, Zhang X, Wen F, Chen X, Zhang S, Heng B C, He Y and Ouyang H-W 2015 3D-printed atsttrin-incorporated alginate/hydroxyapatite scaffold promotes bone defect regeneration with TNF/TNFR signaling involvement *Adv. Healthcare Mater.* **4** 1701–8

Wang X, Tolba E, Schröder H C, Neufurth M, Feng Q, Diehl-Seifert B and Müller W E G 2014 Effect of bioglass on growth and biomineralization of SaOS-2 cells in hydrogel after 3D cell bioprinting *PLoS One* **9** e112497

Xu F, moon S J, Emre A E, Turali E S, Song Y S, Hacking S A, Nagatomi J and Demirci U 2010 A droplet-based building block approach for bladder smooth muscle cell (SMC) proliferation *Biofabrication* **2** 014105

Yang T-C, Chuang J-H, Buddhakosai W, Wu W-J, Lee C-J, Chen W-S, Yang Y-P, Li M-C, Peng C-H and Chen S-J 2017 Elongation of axon extension for human iPSC-derived retinal ganglion cells by a nano-imprinted scaffold *Int. J. Mol. Sci.* **18** 2013

Yuan S Y and Rigor R R 2010 Structure and function of exchange microvessels—regulation of endothelial barrier function—NCBI bookshelf

Zhang K, Fu Q, Yoo J, Chen X, Chandra P, Mo X, Song L, Atala A and Zhao W 2017 3D bioprinting of urethra with PCL/PLCL blend and dual autologous cells in fibrin hydrogel: an *in vitro* evaluation of biomimetic mechanical property and cell growth environment *Acta Biomater.* **50** 154–64

Zhong C, Xie H-Y, Zhou L, Xu X and Zheng S-S 2016 Human hepatocytes loaded in 3D bioprinting generate mini-liver *Hepatob. Pancreat. Dis. Int.* **15** 512–8

Chapter 4

Numerical analysis of blood flow in vasculature structure

Arindam Bit and Jasjit S Suri

This chapter describes the numerical analysis of blood flow in vasculature structures with pathological conditions. It comprises three parts: firstly, a comprehensive literature survey of numerical analysis of a healthy segment of blood vessels with blood flow dynamics. It is followed by an illustration of the methodology adopted for evaluating the numerical model of blood flow in diseased phantom blood vessels; and thirdly, the outcome of the study is critically analysed for the purview of additional knowledge inputs.

4.1 Introduction

The contribution of Womersley number (α) for evaluating blood flow in a vessel was considered, while blood was assumed to be non-Newtonian in nature (Rohlf and Tenti 2001). The Womersley number is a cumulative effect of Re and St numbers. In physiological flows, St is insignificant. Thus, $\alpha < 1$, even for a very high Re value. Numerical integration of the full model thus validated perturbation theory.

Numerical investigation of blood flow in a porcine aorta bypass graft was conducted using image based segmentation (Favier and Taylor 2002). Segmented images were used to reconstruct the 3D model of the aorta. The level set method was employed in every stacked 2D slice for the above outcome. Velocity and pressure were solved using a stabilized finite element method. A traction-free boundary condition was used at the outlet. Two different pulsatile velocity profiles were imposed at the inlet of the calculation domain. Fast Fourier transform was used to extract the frequency content of the flow waveform.

Grigioni *et al* (2005) had conducted a numerical investigation of flowing blood in twisted arterial branches. Spiral flow structure was evaluated by an introductory method. It is based on the cluster dynamics employing particle trajectories. This technique estimates the spiral flow in quantitative manner. Flow was evaluated to

doi:10.1088/978-0-7503-2088-7ch4

ascertain the development of atherosclerosis in prone/critical zones within the twisted 3D geometry. This technique ensures the efficacy of Lagrangian analysis for evaluating vascular blood dynamics. The method is able to quantify spiral flow analysis of blood as well as tracking the progress in growth of stenosis in such a vessel.

Putter *et al* (2006) worked on mesh generation for generating meshes for deformable surfaces of vascular structure. Development of a cascade of algorithms leads to effective implementation of image segmentation of a deformable model followed by 3-D mesh formation. The outcome of the process is robust and results in accurate geometry reconstruction with high precise mesh density for computation of blood flow in diseased vessels. All these algorithms were evaluated on a wide range of deformed segment of blood vessels. Kim *et al* (2008) had evaluated the influence of multi-phase non-Newtonian properties of fluid on pulsating hemodynamics of a coronary artery. Variable hematocrit level of blood was used to evaluate the relationship between hematocrit level and viscosity of red blood cells (RBCs), shear stress, generation of secondary flow, and velocity of inlet flows. It was found that hematocrit level promotes secondary flow on the inner curvature of the vessel. A low value of wall shear stress (WSS) was found in the region of high value hematocrit. RBCs were found to have maximum residence time on curvature of the arterial wall.

Grinberg *et al* (2009) had conducted large-scale simulation of the human arterial tree. A three-level hierarchical approach based on vessel size (macro vascular network, MaN) was used for the simulation by solving the three-dimensional Navier–Stokes equations on arterial networks with tens of arteries and bifurcations. Morbiducci *et al* (2011) had evaluated the significance of hematocrit rheology for bulk flow in a hemodynamic model of bifurcated carotid arteries. This study employed image-based segmented 3-D reconstruction of the carotid artery. Rheology of blood was considered to be non-Newtonian (Carreau and Power Law). The Newtonian model was considered to be dependent on characteristics of shear rate. WSS and Lagrangian based metrics were employed as monitoring probe for quantified analysis of helical flow and vorticity dynamics.

Benim *et al* (2011) emphasised the influence of outlet boundary conditions with numerical analysis of blood flow in the aorta. Application of steady-state analysis was made corresponding to extra corporeal circulation (ECC). Newtonian property of blood was used. The proposed model was prescribed as a loss-coefficient model. In the case of ECC, retrograde and antegrade perfusion were evaluated. In this circulation, WSS predicted were of a much larger magnitude. This shows that the dissociation of atherosclerotic plaques is a very important phenomenon in ECC. In another study conducted by Han (2011), energy equations were solved for determining critical bulking pressure of blood vessels. The buckling equation was employed to assess the critical pressure (based on potential energy). It was observed that energy approach yielded the equivalent equilibrium approach, the differential equation and critical pressure in a cylindrical blood vessel. Buckling of blood vessel can be determined with the energy approach, and thus it could have potential to handle various irregularities of vessel geometry.

Helical flow of blood was also observed at the bifurcated region of the carotid artery. This phenomenon is also known as a surrogate marker which determines the distribution of disturbed shear. A new descriptor known as helicity-based bulk flow was also defined. A thorough analysis of the helicity based descriptor was done using the multivariate linear regression model for predicting the efficiency of the descriptor for determining disturbed shear. This statistical model consists of two bulk flow descriptors. The descriptors can be able to quantify the balance counter rotating helical flow pattern and helicity intensity. It defines the validity of a relationship with Pearson's correlation coefficient.

Currently, a significant number of numerical studies on a wide spectrum of blood vessel modelling and hemodynamics is being carried out. Bellini *et al* (2014) evaluated the arterial wall mechanics having a mechano-biological implication. Physical analyses were made by feeding microstructural details. Constituent specific deposition stretches were employed in the study. Simultaneously, a homeostatic phase at physiological blood pressure and axial stretch were formulated which directs as clinical configuration. This process was utilized for measurement of bi-axial pressure—diameter and axial force—length response of carotid arteries. It also predicts the associated intact and dissociate traction free configuration. The model revealed that different stresses associated with media and adventitia could be mostly homogeneous at physiological loads at distinct levels. Hence the load born by adventia increases at supra-physiologic pressure during safeguarding of media from abundant stress.

Fedosov *et al* (2014) computationally investigated the bio-rheology of human blood flow in healthy and diseased blood vessels. A numerical framework based on DPD was employed, and it has potential for intervention in cellular biophysics in the form of diagnostic, therapeutics and drug-delivery efficacy estimation for variety of diseases. The process started with experimental estimation of mono RBC to extract relevant parameters related to biophysical features. It uses the single-cell measurements method using optical tweezers and microfluidic devices (in the case of cell population study). Thereafter, these validated models were employed for prediction of biorheological behaviour of blood in healthy or diseased condition.

Saez *et al* (2014) developed a model using a numerical technique for prediction of hypertensive growth in the carotid artery. It presents a microstructural numerical model for assessment of chronic arterial hypertension using growth prediction of smooth muscle cells. The growth was modelled by multiplicative decomposition gradient of the growth part and the elastic part. The driving force for growth was assumed to be the stretch sensed by the smooth muscle cells. Results match with experimental data in qualitative as well as quantitative aspects. In another work, Sommer *et al* (2014) used contrast agent bolus dispersion in realistic coronary artery geometry, and obtained the influence of outlet boundary conditions on the numerical study of the same geometry. The influence of outlet boundary conditions on flow behaviour was evaluated for different outlets. Results revealed the presence of significant systemic difference in myocardial blood flow in the geometry of a realistic coronary artery. Outlet boundary conditions exhibited a significant impact on the flow structure of the fluid in upstream. These boundary conditions were shown to have an influence on the resulting MBF value.

Takizawa *et al* (2014) used a medical-image-based time-dependent anatomical model of the coronary artery, and performed numerical analysis based on zero-stress state estimates. The method had two aspects. The primary aspect comprised zero-stress state estimation, and it is an alternative for pre-stress evaluation. The secondary aspect is composed of a mixed zero-stress element for different units in the structural mechanics mesh with respect to cardiac instant imaging configuration. In another study, Qureshi *et al* (2014) investigated blood flow and pressure drop in pulmonary arterial and venous circulation. Multi-scale mathematical and computational model of the pulmonary circulation was presented in their study. MRI images were employed for performing patient-specific remodelling of three generation vessels within a pulmonary circulation circuit. Confirmation of the distinct vascular bed was made by employing an algorithm with temperature of the bifurcating structure of a stress and recursion model. Results obtained from the numerical study of the pathological boundary condition were found matching with the clinical outcome. This shows that the model has potential application for diagnosis of clinically relevant information of pulmonary circulatory disease.

In recent times, fluid structure interaction (FSI) analysis of flowing blood through thoracic cavity was also evaluated. The space-time framework variational multiscale method was employed for obtaining the solution of fluid mechanics equations. This method is a multi-scale version of the DSD/SST (deforming-spatial-domain/stabilized space-time) method. Smooth 3D models were reconstructed by projection of the original centreline to the average plane curvature. The effect of geometrical simplification was evaluated by checking the difference in flow field between original geometry and projected geometry. In another study conducted by Takizawa *et al* (2014), FSI estimation of the arterial flow system was made by zero-stress state analysis of the finite element based method. It has three segments. Firstly, it is an iterative method, initiating with an initial value of zero-state (ZS) state. It is employed for evaluating the element-based ZS state in the presence of applied load. It results in superimposing of image-based target matching. The second method is applied for linear geometries with single and multiple layers. It evaluates the element-based ZS state in order to matching the projected diameter and longitudinal stretch in the desired configuration. The final step involves mapping of the arterial and straight tube configuration using the element-based mapping technique. It also maps the estimated ZS state of the linear tube with respect to arterial configuration. It is employed as the primary estimation for the iterative method which superimpose the image-based target shape.

After an intensive survey of the above investigations, it is found that the choice of appropriate rheological model is still an open issue. A comparative assessment of different non-Newtonian models in conjunction with realistic flow profile are absent in the above field of literature.

4.2 Methodology

Most blood vessels experience a pulsating flow of blood, which is compensated by exerting micro-pumping by micro-compliances of vessels. The pulsating nature of this rheological fluid under the influence of diseased condition greatly predicts the vascular proliferation site, and it is also greatly influenced by the degree of deformation of the vessels by means of either an aneurysm or stenosis.

Typical pulsatile flow also maintains systolic and diastolic pressure throughout the circulatory system. The actual pulse profile at the inlet of the blood vessels of interest is difficult to obtain, and thus it has become one of the limitations for several researchers. Simultaneously, an effective study of viscosity of blood in diseased blood vessels is limited. Since the phenomenon of diseased condition of a blood vessel is a cyclic process as change in blood rheology induces the disease, so, it is still an open question: does actual pulsation enhance disease proliferation, or is it the change in blood rheology enhancing the process? In the previous chapter, the need for research effort in blood flow through a diseased artery was highlighted. In this section, the methodology of the numerical computation is discussed.

In the present work, mainly two types of geometry of blood vessels have been taken into consideration (i) two-dimensional axi-symmetric stenosed blood vessel, and (ii) two-dimensional axi-symmetric aneurismal blood vessel. Patient-specific pulsating flows are introduced at the inlet in the form of plug pulsating flow. The numerical model was initially tested at the developing flow region in a channel without pulsation at the inlet. Patient-specific pressure profiles have been considered for outlet boundary conditions in the case of three-dimensional patient-specific geometry. The length of the computational domain is varied for three different levels of Reynolds number. Two different non-Newtonian viscosity models are considered for realization of viscosity of flowing blood. Numerical simulations are performed for observing the effect of flow pulsation and viscosity model over enhancement of the disease in a diseased blood vessel, and to observe the spatial proliferation of the disease throughout the tunica intima lining of the endovascular region of vessel geometry. In this work, the presence of the vessel wall is incorporated by prescribing a no-slip boundary condition on the flow surfaces.

The analyses are carried out for fixed pulsating profiles-type 1, different Reynolds number (Re I, Re II) and rheological models (Carreau, and Casson) and the results are compared for wall shear stress distribution, vorticity, oscillatory shear index, relative residence time, pressure drop and overall performance factors.

4.2.1 Construction of geometry

Geometries used for numerical investigation are of two different categories. First, a two-dimensional axi-symmetric straight stenosed vessel is designed, details given later. Then, a two-dimensional axi-symmetric straight vessel containing an aneurysm is taken for the study.

The 2D computational domain of a stenosed blood vessel is shown in figure 4.1. The profile for defining stenosis was proposed by Neofytou and Drikakis (2003), as given by equations (4.1) and (4.2).

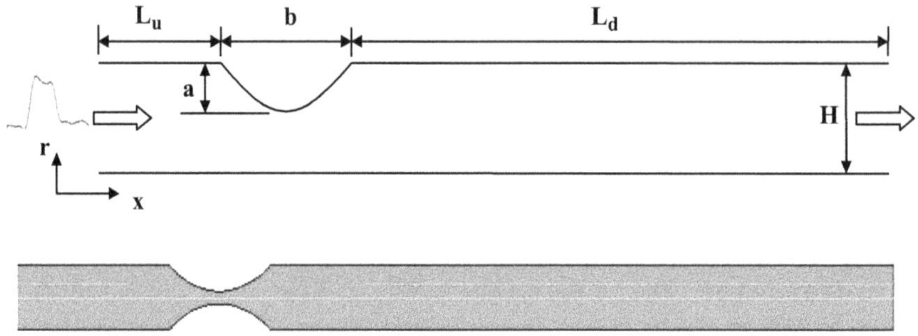

Figure 4.1. Plain/axi-symmetrical stenosed blood vessel.

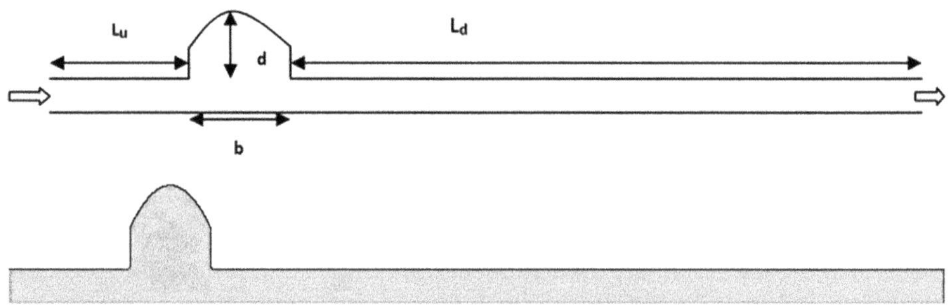

Figure 4.2. Plain/axi-symmetrical aneurismal blood vessel.

$$y = 0, \; y = 1 \quad \forall \; x \in [0,7] \cup [11.6,40] \tag{4.1}$$

$$y = 0, \; y = 1 - \left(a - R + \sqrt{R^2 - [b/2 - (x - L_u)]^2}\right) \tag{4.2}$$

where

$$R = \frac{[a^2 + (b/2)^2]}{2a} \quad \forall \; x \in (7,11.6) \text{ and } a = 0.56, \;\; b = 4.66.$$

Principle axes are defined by x and y, respectively. The stenosis is defined as the ratio of depth of local maxima of the stenosed part (a) to that of the mean height of the vessel (H). The depth of the stenosis is varied within 25%–80% in order to investigate the effect of degree of stenosis on the flow structure in downstream.

A two-dimensional axi-symmetric planar geometry of an aneurysm model of a blood vessel is developed following a design defined by Huang *et al* (2013), as shown in figure 4.2.

The height of the channel is considered to be 16 mm, while the total length of the vessel is considered as $L_u + b + L_d = 400$ mm, where L_u and L_d represent upstream and downstream length from the aneurysm, respectively. Width and depth of the aneurysm were considered as 40 mm. The shape of the aneurysm is considered to be of sigmoid function bulged over a height of 20 mm from the boundary wall. The height of the sigmoid curve is also limited to 20 mm. The direction of the arrow indicates the downstream of flowing blood through the vessel. An aneurysm of such dimensions may be considered a highly defined model.

Diseased blood flowing in a stenosed vessel enhances the presence of foreign ingredients in the blood stream, making it rheologically non-Newtonian. Thus, the Casson model was considered for projecting rheological behavior of blood (Casson 1959, Boyd *et al* 2007, Walawender *et al* 1975). A discrete form of the model was primarily introduced. Later, Papanastasiou modified the stress representation (as given by equations (4.3) and (4.4)):

$$\bar{\tau} = \left(\sqrt{\mu_\infty} + \sqrt{\frac{\tau_y}{|\dot{\gamma}|}}\left(1 - e^{-\sqrt{m|\dot{\gamma}|}}\right)\right)^2 \dot{\gamma} \tag{4.3}$$

and

$$\mu(|\dot{\gamma}|^*) = \mu_\infty\left(1 + \sqrt{\frac{Bi}{|\dot{\gamma}|^*}}\left(1 - e^{-\sqrt{m|\dot{\gamma}|}}\right)\right)^2 \tag{4.4}$$

where $m' = \frac{m}{l}/U_\infty$. Charm *et al* (1964), and Charm and Kurland (1965) had estimated τ_y =10.82 mPa, whereas, $\mu_\infty = 3.1 \times 10^{-3}$ Pa-s in the case of blood. Re_{CA} and Bi are the characteristic parameters of the Casson model.

Carreau had proposed a 4-parameter non-Newtonian rheological model (Bird *et al* 1987) which is distinct from other models, primarily in the curvature part near the transition points between Newtonian plateau and power region. It is given by equation (4.5).

$$\mu(|\dot{\gamma}|) = \mu_\infty + (\mu_0 - \mu_\infty)[1 + (\lambda|\dot{\gamma}|)^a]^{(n-1)/a} \tag{4.5}$$

where $\mu_0 = 0.056$ Pa$-s$. It is the viscosity of blood at null shear rate. $\mu_\infty = 0.0032$ Pa$-s$, $\lambda = 3.131$. It is a time constant, and it is a function of viscosity which alters with the shear rate, and $n = 0.3568$, $a = 2$.

The equation for mass conservation in a general form is given by equation (4.6) as,

$$\frac{\partial\rho}{\partial t} + \frac{\partial(\rho u_1)}{\partial x_1} + \frac{1}{r}\frac{\partial(\rho r u_2)}{\partial x_2} = 0 \tag{4.6}$$

where x_1, x_2, etc are defined as follows:

$x_1 = x$ for Cartesian coordinates

$= z$ for cylindrical axi-symmetric coordinates

$x_2 = y$ for Cartesian coordinates

$= r$ for cylindrical axi-symmetric coordinates

Moreover,

$r = 1$ for Cartesian coordinates

$= x_2$ for cylindrical axi-symmetric coordinates

The mass conservation equation for problems involving two-dimensional, Cartesian geometry may be written as in equation (4.7),

$$\frac{\partial \rho}{\partial t} + \frac{\partial(\rho u)}{\partial x} + \frac{\partial(\rho v)}{\partial y} = 0 \tag{4.7}$$

where x and y are the coordinate directions, u and v are the velocity components in those coordinate directions, respectively.

The xv momentum equation in Cartesian coordinates may be written by equation (4.8) as,

$$\frac{\partial(\rho u)}{\partial t} + \frac{\partial(\rho uu)}{\partial x} + \frac{\partial(\rho vu)}{\partial y} = -\frac{\partial p}{\partial x} + \frac{\partial(\tau_{xx})}{\partial x} + \frac{\partial(\tau_{yx})}{\partial y} \tag{4.8}$$

SIMPLE is the acronym for the semi-implicit pressure-linked equation, a procedure for solution of the governing equations of fluid transport. This scheme was used to solve continuity and momentum equations for evaluating shear stress distribution at the wall and the vorticity of blood flow.

In the case of two-dimensional axi-symmetric geometry, velocity inlet boundary condition is used at the entry section where fluid enters into the blood vessel. A pressure outlet boundary condition has been applied at the exit from where the blood is escaping to a bed of body fluid from inside of the blood vessel. The wall boundary condition has been applied to all other faces of the calculation domain.

In the case of a two-dimensional axi-symmetric stenosed blood vessel, a velocity profile of pulsating nature has been used. The profile is curve-fitted for transient responses. Magnitudes of velocity profiles are in m s^{-1}. An inlet velocity profile of type-1 was selected from patient-specific data at the entry point of an ascending aorta (Kleinstreuer 2006). The fitting of the curve was made using a Gaussian function, as described by equation (4.9).

$$\begin{aligned} u(t) = a_1 \times \exp\left[\left(\frac{-(t-b_1)}{c_1}\right)^2\right] + a_2 \times \exp\left[\left(\frac{-(t-b_2)}{c_2}\right)^2\right] \\ + a_3 \times \exp\left[\left(\frac{-(t-b_3)}{c_3}\right)^2\right] + a_4 \times \exp\left[\left(\frac{-(t-b_4)}{c_4}\right)^2\right] \\ + a_5 \times \exp\left[\left(\frac{-(t-b_5)}{c_5}\right)^2\right] + a_6 \times \exp\left[\left(\frac{-(t-b_6)}{c_6}\right)^2\right] \end{aligned} \tag{4.9}$$

Values of coefficients a's, b's, and c's are given in table 4.1.

The wall of a blood vessel has several special features, which makes it ideal for flow of blood through its lumen. There are three primary factors which protect the blood vessel from clotting phenomena: (1) activation of an intrinsic clotting system due to contact of blood with the endothelial layer of the vessel is prevented due to smoothness of the endothelium; (2) presence of glycocalyx layer on the surface of the endothelium prevents clotting activation by repelling clotting factors and platelets;

Table 4.1. Values of coefficients of equation (4.9).

Type-I waveform											
a1	−0.0214	**b1**	0.4558	**c1**	0.008 732	**a4**	0.0383	**b4**	0.6125	**c4**	0.243
a2	0.0842	**b2**	0.3734	**c2**	0.2184	**a5**	0.0932	**b5**	0.002 078	**c5**	0.1371
a3	0.0307	**b3**	0.7449	**c3**	0.2168	**a6**	0.0864	**b6**	1.012	**c6**	0.134

(3) binding of thrombomodulin protein on the surface of the endothelial layer. Binding of thrombin from clotting factors with thrombomodulin will reduce the concentration of thrombin from the pool of clotting factors. It reduces the rate of clotting. In addition to that, this complex of thrombin–thrombomodulin complex activates plasma protein-C which works as an anticoagulant due to its inactivating capabilities of activated factors V and VIII. Destruction of the endothelial layer leads to sacrifice of smoothness and a glycocalyx–thrombomodulin layer. It triggers the activation of factor XII and platelets, which in turn initiates the intrinsic pathway of clotting phenomena. Furthermore, the interaction of factor XII and platelets with sub-endothelial collageneous layer amplifies the intrinsic pathways.

A wall boundary condition is used to bind the fluid and the solid region. In the present investigation, a no-slip boundary condition is enforced at the walls (endothelial lining, also known as tunica interna) during fluid flow through a diseased blood vessel. It is given by equation (4.10).

$$u = v = w = 0 \tag{4.10}$$

A neuro-transmitter plays a significant role for vessel contraction. The vessel walls are elastic and thus the question of fluid-structure-interaction arises. Changes in blood flow can be caused by strong sympathetic stimulation, which constricts the blood vessels. Likewise, hormonal vasoconstrictors, such as nor-epinephrine, angiotensin II, vasopressin, or endothelin, can also reduce blood flow, at least transiently. Vasoconstriction process is being promoted by stimulating vascular smooth muscle. Vasoconstrictor effect can also be enriched in the presence of a high concentration of antidiuretic hormone (ADH). Regulating of adjacent tissues is achieved by paracrine regulators. The paracrine regulation phenomenon is present in a blood vessel. Contraction and relaxation of tunica media of the endothelial layer are due to production of paracrine regulators. Localized deflection of pressure produces compressibility in the fluid, and it is neglected.

However, it is observed that the difference of deflected diameter and resting diameter of the vessel is negligible in the presence of normal metabolic activities. The brain regulates the dilation of smooth endothelial cells during the hyperactive state in the presence of propagation of electro-potential using efferent neurons and synapses of the neuro-transmitter. Epithelial cells produce nitric oxide (NO) in the presence of neurotransmitter acetyl-choline. NO triggers vascular dilation and thereby promotes blood flow through the associated neurovascular vessel. A prolonged vascular dilation can be observed in this process in comparison to the rate-responsive cardiac cycle. Hence, the wall of the vessel is assumed to be stationary with respect to one cardiac cycle. Therefore, one can assume the carotid artery to be rigid momentarily. A rigid wall is considered neglecting the flexibility of the vessels in line with prior investigations.

Simulations of two-dimensional axi-symmetric stenosed geometry are executed with mesh resolution of 150 × 25. A grid-independence test was performed prior to optimizing the mesh size. Grid-independent extrapolated values of oscillatory-shear-index (OSI) and centerline velocity at multiple locations along the region of stenosis were evaluated. Evaluated values of interpolated parameters were found deviating among different mesh sizes by 1.5%–2.5%. Thus, the optimized mesh was finalized and employed for further study. Reproducibility of results is monitored over six consecutive time cycles. Axial velocity further downstream is monitored at different locations and perfect repeatability in time domain is observed.

Similarly, a quadrilateral geometrical mesh is used to discretize aneurysm geometry with the help of Pave meshing technique. The final number of elements used for achieving optimized mesh resolution for the geometry is 259 830. However, the region of aneurysm is enriched with a further finer mesh, a gradient geometrical increase in mesh size, while a linear constant mesh size is considered away from the aneurysm. A finer mesh size accumulates smaller time step size while solving the problem with the Lagrangian approach.

The governing equations of the viscous laminar transient flow were solved by employing SIMPLE formulation (Patankar 1980). A collocated grid arrangement was used to evaluate primitive variables of executed equations. Higher accuracy was achieved using the QUICK scheme. This also discretizes the higher-order equation. Fixed time step resolutions of 0.0005 s and 0.0004 s were employed for stenosis and aneurysm of 2D geometry, respectively. Computations are executed for multiple cardiac cycles to confirm the repeatability of flow (Stroud *et al* 2002). ANSYS FLUENT was employed to evaluate the solution of the system of equations.

Flow parameters like OSI, relative residual time (RRT), and WSS were computed numerically, and were found to be in good agreement with the existing solution in the literature. Wall shear stress, and axial velocity at different points along the centerline as well as near the wall at different flow rates (Re 700, 1000, and 1400) are compared with data obtained from experiment, and reasonably good agreement within 7%–9% is found in each case. During numerical study, results are executed iteratively for the grid independence test. Richardson extrapolation criteria are used for this grid independence test study.

Richardson extrapolation (Richardson 1927) was employed to evaluate the higher order estimate of flow field from a series of lower-order discrete values $(f1, f_2, \ldots, f_n)$. In grid independent study, the Richardson extrapolation technique estimates the zero-order values (grid size tends to zero) of the flow parameters. Extrapolated results were obtained from two consecutive grid resolutions. Convergence study used 3–4 grid solutions (Stern *et al* 2001).

A pth-order method was introduced to generalized Richardson extrapolation (Roache 1994):

$$f_{\text{exact}} \approx f_1 + \left[\frac{(f_1 - f_2)}{(r^p - 1)}\right] \tag{4.11}$$

In this study, the grid refinement ration r, is mean refinement ration r_{mean} of r_1 and r_2.

$$r = r_{\text{mean}} = \left(\frac{r_a + r_b}{2}\right) = \left(\frac{2.5 + 1.4}{2}\right) = 1.95 \tag{4.12}$$

where

$$r_a = \frac{\Delta_1}{\Delta_2} = \frac{0.05}{0.02} = 2.5 \text{ and } r_b = \frac{\Delta_2}{\Delta_3} = \frac{0.02}{0.014} = 1.4 \tag{4.13}$$

Here, subscripts 1, 2, and 3 represent cases of study with different grid sizes. From equation (4.11), the order of P varies the extrapolated values of a parameter. Thus, the order of accuracy can be evaluated by the Stern equation (4.14) (Stern *et al* 2001);

$$p = \frac{\ln(\varepsilon_{32}/\varepsilon_{21})}{\ln(r)} \tag{4.14}$$

where

$$\varepsilon_{i+1,\,i} = f_{i+1} - f_i \tag{4.15}$$

Estimation of extrapolated values requires determination of convergence conditions of the system. In this case, tentative convergence conditions are:

1. Monotonic convergence: $0 < R < 1$;
2. Oscillatory convergence: $R < 0$;
3. Divergence: $R > 1$.

where R is the convergence ratio and it is determined by equation (3.72):

$$R = \frac{\varepsilon_{21}}{\varepsilon_{32}} \tag{4.16}$$

Table 4.2 summarizes the order of accuracy for pre-stenosis axial mean velocity ($V_{\text{axial(mean)}}$), and absolute wall shear stress (WSS_{Abs}) from the simulation results on three grids. The convergence conditions for $V_{\text{axial(mean)}}$ and WSS_{Abs} are monotonic.

Table 4.2. Order of accuracy and grid convergence index for two variables.

	ε_{32}	ε_{21}	R	p	$\text{GCI}_{32}(\%)$	$\text{GCI}_{21}(\%)$
$V_{\text{axial(mean)}}$	0.0034	0.0017	0.513	0.9956	4.7125	2.486 87
WSS_{Abs}	0.0158	0.0076	0.483	1.0856	4.2408	2.090 35

The convergence of grid refinement study can be measured by grid convergence index (GCI). It is extracted from fractional error obtained from the generalized Richardson extrapolation. Resolution level is represented by GCI value. It also helps us to predict the degree of asymptotic solution of the convergence process. Evaluation of GCI can be made by equation (4.17):

$$GCI_{i+1,\,i} = F_s \frac{|\,\varepsilon_{i+1,\,i}\,|}{f_i\,(r^p - 1)} \tag{4.17}$$

The safety factor (F_s) selected for this study is 1.20 (Wilcox 2006).

A reduction in value of GCI for successive grid refinement ($GCI_{21} < GCI_{32}$) has been observed in table 4.3. It is observed that a finer grid has lower GCI than a coarser grid. It states that the numerical solution is now independent of grid-size. If the reduction of GCI value from coarser to finer is sufficiently high, one concludes that a grid-independent solution has been achieved. Thus, further refinement of the grid has a negligible effect on the computed solution.

4.3 Results and discussion

Prediction of transport phenomena in diseased vessels was done with physiological flow at the inlet of the vessels. The fluid considered in this study was rheologically considered as a non-Newtonian fluid i.e. blood. A series of numerical calculations has been performed by finite volume method for axi-symmetric and patient-specific non-axi-symmetric vasculature and the results are presented in order to show the effects of rheological property, inlet pulsation, and severity of vascular diseases on these vessels. The results are focused to localized WSS distribution and its gradients as well as harmonics operators, OSI and RRT further downstream of the diseased blood vessel. Furthermore, a comparative assessment is also made for severity of secondary disorder in blood vessels with stenosis and aneurysm, respectively. Computational results obtained with the pulsatile flow at different Reynolds numbers are compared simultaneously with results obtained from experimental study.

The numerical model was tested by computing an optimum rheological model to represent the diseased state of blood in a developing flow condition in a straight

Table 4.3. Recirculation length in a stenosed blood vessel.

		Degree of stenosis (m)	
Inlet flow	Viscosity models	56%	80%
I	Carreau model	0.046	0.12
	Casson model	0.044	0.29
II	Carreau model	0.14	0.265
	Casson model	0.176	0.286

blood vessel with stenosis ranges between 25%–75%, respectively. In the present work, investigations were performed within the range of $690 \leqslant Re \leqslant 1060$.

Blood flow in a stenosed vessel is mostly classified as vortex and post-stenotic recirculation length (Neofytou and Drikakis 2003). Figure 4.3 represents the contour of blood as a streamline representation at various time steps of cardiac cycle in a 56% stenosed vessel. Rheological properties of the blood were confined to a power-law model.

The efficacy of non-Newtonian models for evaluation of recirculation length has been investigated (Neofytou and Drikakis 2003). Development of vortices is quite rapid in a 56% stenosed vessel. A higher blockage ratio also shows an increase in recirculation length.

Table 4.3 represents mean circulation length in a stenosed vessel at various blockages at varying Re numbers of inlet flow. A 56% stenosed vessel shows that the development of recirculation length is independent of flow rate. Here, the Carreau model shows highest recirculation length. Recirculation length enhances further with an increase in stenosis from 56% to 80%.

The OSI distribution trend is governed by the velocity profile as well as degree of stenosis. The rheological model selected in the study also regulates OSI. The Carreau model (non-Newtonian) and Newtonian rheological model were found to predict OSI values different from the general trends. Significant variations of OSI pattern were found with varying the rheological model at the region of stenosis and post-stenosis. However, OSI values were found to be over-expressive at the region of

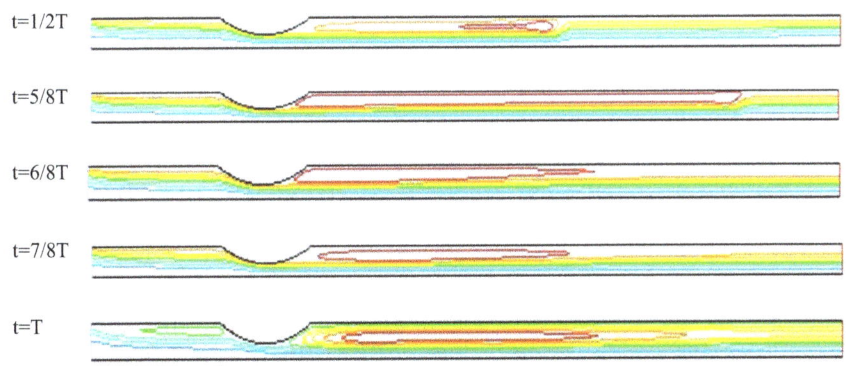

Figure 4.3. Streamline plot of velocity vector of blood in a stenosed vessel at different time steps. *T* indicates time period.

stenosis and the post-stenotic region while the Carreau model was implemented for rheology of blood in a 56% stenosed vessel.

On the other hand, OSI attenuates at the region of stenosis in an 80% stenosed vessel with the Carreau model as a rheologically predicting model. However, at the post-stenosis region, OSI magnitude remains higher. The Casson rheological model of blood also shows a similar trend at the post-stenotic region. OSI value becomes null at the stenosis region while implementing this model. A similar trend can be found at all degrees of stenosis by both the Carreau and Casson models.

With increase in stenosis, OSI value decreased at the region of stenosis while the Casson model was employed for blood rheology modeling. A similar trend of OSI with increase in degree of stenosis can also be observed at the onset of stenosis. At the lowest degree of stenosis (25%), a higher intensity of OSI was confined to a localized zone only. This zone of high intensity of OSI increases with decrease in intensity, as well as with increase in degree of stenosis. Flow-rate variations also influence the magnitude of OSI. With increase in flow rate of blood, intensity of OSI increases at the downstream region, while degree of stenosis decreases. A stenotic vessel with 80% blockage maintains a longer zone of high intensity OSI in the post-stenotic region. Thus, it was found that the location of high intensity of OSI drifted downstream with increase in degree of blockage. Therefore, it can be concluded that with increase in depth of stenosis: (1) length of region of high intensity of OSI increases, while (2) magnitude of high intensity of OSI decreases.

Figure 4.4 shows a contour plot of time averaged velocity stream function for type 1 inlet velocity profile at different degree of stenosis considering the Casson viscosity model. Comparison has been done at two different Reynolds number lines as in the above figure, described as Re I and Re II. At low Reynolds number, recirculation bubbles propagate away from the region of stenosis with time for a lower severity of stenosis (at 56%).

However, at higher Reynolds number the same severity of stenosis causes complex flow structure with multiple recirculation bubbles which move away from the region of stenosis with time. On increasing the degree of stenosis (of 56%) at low Reynolds number, recirculation bubbles oscillate in the axial direction and accumulate to form large eddies with time. The same phenomena occur while increasing the Reynolds number of inlet flow. At 80% stenosis recirculation bubbles drive inside towards the stenosis with time due to decrease in pressure shedding at low Reynolds number along the downstream with development of flow, and it dissipates into multiple recirculation bubbles forming a complex structure with increase in Reynolds number.

For type 1 inlet velocity profile both the viscosity models exhibit almost a similar nature of OSI curves with peak value of 0.5 at the post-stenosis region for 25% stenosis. In the case of higher Reynolds number the peak of OSI shifted slightly away from the region of stenosis as compared to the earlier case, because at higher Reynolds number, the dissipation of energy by the fluid at the wall immediately after stenosis reduces dynamic pressure further downstream during higher stress relaxation factor of the corresponding viscoelastic fluid. Shifting of peak value of OSI is also noticed for higher Reynolds number in the case of 56% stenosis, with a

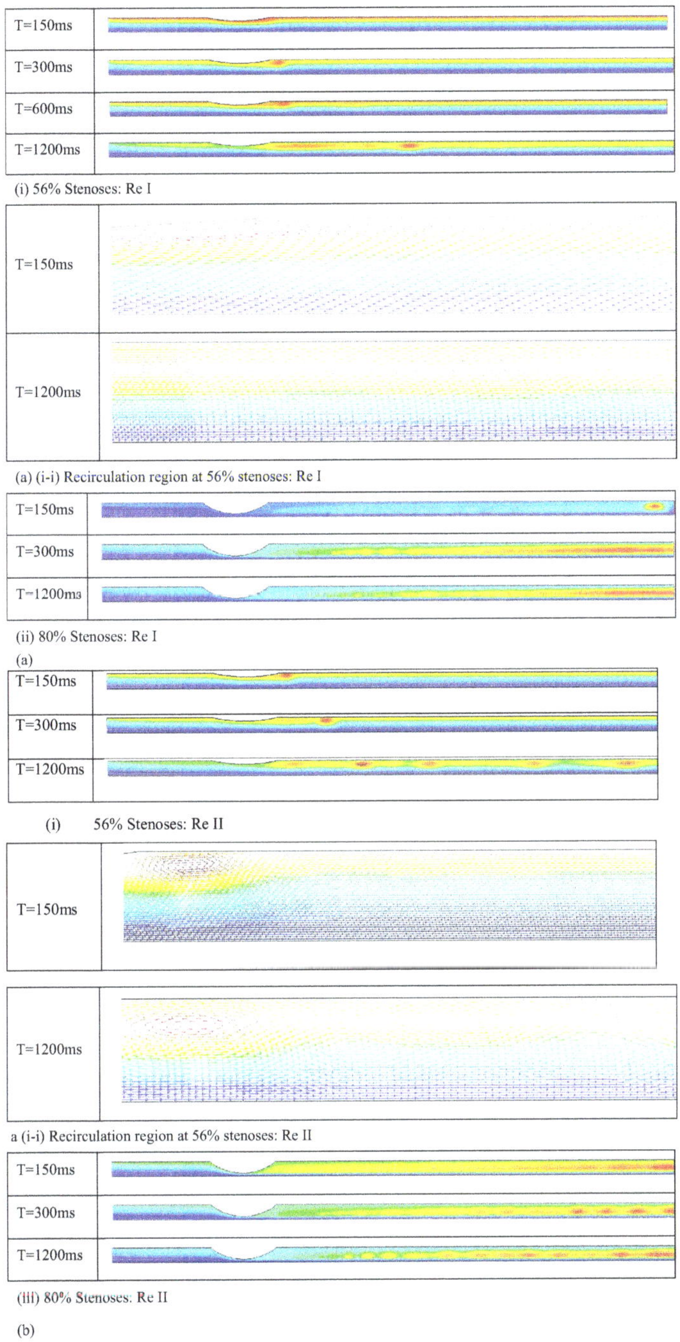

Figure 4.4. Time averaged stream function of velocity for Casson model with Type 1 inlet velocity at different degrees of stenosis (56% and 80%) at (a) Re I and (b) Re II, with inset of re-circulation zone in 56% stenosis case (i-i) for (a) and (b), respectively.

significant variation in OSI distribution along the length of the vessel at two different Reynolds numbers. The low shear region is linked to the existence of recirculation pattern at such locations. At a high degree of stenosis, the post-stenosis wall has a low value of wall shear stress (about 0.5), development of prone region for re-stenosis at downstream of flow. Development of wall shear stress at the post-stenosis wall has been found at low degree of stenosis (25%), indicating low susceptibility of re-stenosis further downstream. Re-stenosis can also occur in a vessel with a higher degree of blockage due to much higher gradient change wall shear stress across the region of stenosis, responsible for development of vortices near the immediate post-stenosis wall.

Temporal WSS at locations of 2 cm, 9 cm and 18 cm from the inlet (i.e. pre-, on- and post-stenotic zone) along the vessel length has been compared at different lengths of the vessel for two different viscosity models at 80% stenosis. It has been observed that WSS in the region of stenosis varies widely while applying the Quemada model at the initial stage of flow. While the Carreau model shows a negative WSS of −12 Pa initially at the region of stenosis, the corresponding WSS for the Casson model is −30 Pa. It is also noticed that variation of WSS in the pre-stenosis region for both the models is quite insignificant and very low, thus it can be neglected for further studies. A slight variation in WSS has been observed at 0.1 s of the flow cycle at the far-field post-stenosis region. It signifies that there may be some strong responses of variation of WSS at the post-stenosis region for lower severity of stenosis.

Temporal WSS variation for three different degrees of stenosis along the wall of the vessel for the power law model has been compared. For the type 1 velocity profile, while at the pre-stenosis region, WSS varies insignificantly at 80% stenosis, it varies from −1 to −0.25 for 25% stenosis with time, and the variation is dampened with the increase in degree of stenosis. It is concluded that even at the pre-stenosis region flow structure varies widely at the onset of stenosis. Interestingly at the region of stenosis, temporal WSS oscillates in the negative region (from −1.75 to −0.25) for lower order stenosis, and with the increase in degree of stenosis, WSS moves to the positive magnitude with time. It has been observed that WSS oscillates from negative to positive values at the end of the time cycle for 25% stenosis, and for 56% stenosis, it follows a symmetrical pattern. At 80% stenosis, WSS has a positive peak of 5 at the onset of flow cycle, and thereafter it dissipates with time.

A detailed study of axial velocity along the centerline has been done. Casson viscosity for blood is considered here with variation in degree of stenosis from 25%–80%. Velocity curve representation at each successive time step of T/8 depicts the resemblance of gradient variation of mean axial velocity about the centerline with that of inlet plug pulsatile velocity profile. Overshoot of temporal velocity at the region of stenosis (0.07–0.116 m) indicates the presence of gradient flow breaker in each type of stenosis.

From comparative study for different degrees of stenosis, it has been found that mean axial velocity increases (from 0.3 m s^{-1} to 4.5 m s^{-1} approximately) with an increase in degree of stenosis along the centerline. It indicates possible flow development from laminar to transition regime as kinetic energy of the flow structure increases. Thus it leads to vortex development further downstream of the stenosis due to formation of temporal-spatial low pressure region due to dissipation of energy by the fluid at those regions.

An overshoot of axial velocity from the region of stenosis has been found in both cases, followed by a stable velocity level throughout the downstream. Although the spatial and temporal distribution of magnitude of axial velocity increases two-fold with increase in Reynolds number, there are insignificant changes in the pattern of the same, except for an occasional overshoot of axial velocity further downstream at low Reynolds number, due to cumulative stress response of the fluid. As the fluid at wall has been considered to be of no-slip boundary condition, static wall pressure has been considered for comparative study of wall pressure at different degree of stenosis.

The physics behind this variation is not yet clear, and it has not been discussed in the literature yet. Interestingly, negative overshoot of wall pressure has been found further downstream of the stenosis, which increases with increase in degree of stenosis, confirming recirculation of flow. The phenomenon for recirculation associated with vortex formation had been already discussed. The effect of increase in flow rate on static wall pressure distribution is not of much significance. It has been found that wall pressure increases linearly with increase in flow rate. At the region of stenosis, the negative peak of pressure increases with increase in flow rate (about eight-fold). This is due to development of laminar to transition nature of flow at high flow rate, leading to higher vorticity in the flow field.

4.3.1 Hemodynamics of blood through axi-symmetric aneurismal blood vessel

The numerical study for aneurysm was performed by two rheological models, namely the Carreau and Casson models, to represent a diseased state of blood in a developing flow condition in a straight blood vessel with severe aneurysm of 250% of mean diameter of the vessel. In the present work, investigations were performed within the range of $690 \leqslant Re \leqslant 2760$. Responses of two different non-Newtonian viscosity models to different shear rate parameters near the boundaries are compared. Stream lines in the duct of the geometry have been shown in figure 4.5, which reflects the development of vortices considering the Carreau model as the viscosity model for blood. As the resolution of flow increases, the vortex size gets amplified, signifying serious impact of aneurysm on the flow of blood. The transient nature of flow development occurs in the duct of geometry. This phenomenon occurs due to development of the vortex initiating with offset damping, as depicted in figure 4.5.

Increased resolution of flow in time domain as well as elevated flow rates, double (denoted as Re II) and quadruple (denoted as Re IV), are simulated. Figure 4.6 shows the related stream line plots. A rapid development of vortices is observed with increased flow rate.

However, vortex shedding occurs at earlier time steps in the case of higher flow rate followed by dissociated monotonic vortex behavior. At higher Reynolds number, a radial up-liftment of the vortex towards the wall of the aneurysm is

(a) t = 0.2T (b) t = 0.8T

Figure 4.5. Stream line at different time steps at Re I (for Carreau model).

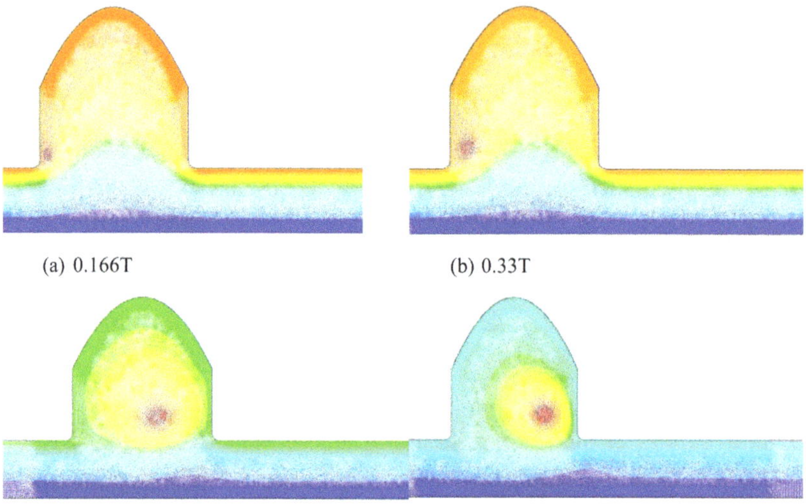

(a) 0.166T (b) 0.33T

(c) 0.833T (d) T

Figure 4.6. Stream line at Re II.

observed, indicating elevated shear stress experiences by the endothelial layer of the vessel at this region, while lower ones could be experienced further downstream. In order to validate the hypothesis based on the above observation, comparative studies on distribution of OSI and RRT for various flow rates have also been carried out. Monotonic vortex behavior after vortex shedding in figure 4.6 reflects with persisting aggregated hematocrit at higher Reynolds number, leading to endothelial lysis on the inner lining of the wall at the site of the aneurysm. However, vortex sustains at a much lower height, almost parallel to the axial flow, indicating induction of vortex shedding in the main course of blood flow along the length of the vessel. Thus, it also propagates aggregation or bolus formation of macro-proteins further downstream.

Behavior of the centerline velocity at the region of aneurysm is governed by influx flow rate. Insignificant velocity gradient is observed at low flow rates (Re I and Re II). It indicates gradual development of vortex within the region of aneurysm. Hence, it can be concluded that shear stress distribution at the wall of the vessel downstream of the aneurysm will be directly governed by the rate of inflow.

OSI are also computed at higher flow rates (Re II and Re IV) along the wall of the vessel. Higher flow rate exhibits higher value of OSI at the region of the aneurysm. Influence of flow rate on OSI distribution is also addressed. The power law viscosity model exhibits much greater response for OSI at higher flow rate (Re IV) in the region of the near downstream of the aneurysm. However, in the case of the Quemada viscosity model, OSI at Re II has wider response than Re IV, indicating multiple OSI peaks at near downstream of the aneurysm. This can be due to radial suspension of vortices along the direction of flow.

4.3.2 Comparative assessment of hemodynamics of blood in a stenosed or aneurismal vessel

A comparative assessment was made between hemodynamics of blood flowing through a severe stenosed vessel and an aneurismal vessel. The numerical models were tested by the most suitable optimal rheological models at diseased state in a developing flow condition in a straight blood vessel with severe aneurysm of 250% of mean diameter of vessel and another containing stenosis of 56%. Responses of the two different non-Newtonian viscosity models to different shear rate parameters near the boundaries are compared to address the above objective.

Development of flow is found to be instantaneous at 0.05 s in the case of vessels with blockages, whereas a progressive development in flow-vortices was found in the aneurismal vessel. A highly accelerating progress of velocity along the centerline was found in the aneurismal vessel when the flow rate is doubled. Diseases on the inner wall of the vessels have least effect on the centerline velocity at its vicinity downstream. A vortex started forming at the immediate downstream, and its intensity increased with increase in length of downstream. Thus, it has an increasing effect on the propagation of centerline velocity downstream (as reflected in figure 4.5). Random centerline velocity distribution in a stenosed vessel far downstream is due to induced disturbances by vortex shedding phenomena over time. It remains absent in a vessel with an aneurysm because shedding of vortices happens at the primary stage of downstream in the vessel. The centerline velocity of blood was found attenuated in an aneurismal vessel because of the occurrence of high intensity RRT phenomena which attenuate blood flow downstream.

A higher magnitude of centerline velocity is observed in the case of stenosis than in an aneurysm. However, the centerline velocity profile maintains a similarity in the pulsating profile with the input profile in the case of the aneurysm. The same have not been observed for stenosis, as random instantaneous peaks are found in the centerline velocity profiles. This highly random nature may eventually lead to a turbulent flow field which requires further investigation. The transient interval for flow development becomes longer in the case of lower flow rate (Re I) in a stenosed vessel.

An accelerating velocity field in the centerline is spotted at the post-aneurysm location at higher Re. The explanation for such an observation is not clear. At the post-stenotic region, the centerline velocity develops instantly, and it shows negligible effect of vortex formulation at the region. On the other hand, gradual development of the centerline velocity shows a significant effect of vortex formation and vortex shedding with time in the region.

Spatial distribution of higher values of OSI varies among vessels with aneurysms and stenosis. An intense level of OSI was found concentrated at the region of the aneurysm in an aneurismal vessel, whereas, it was found to be crowded in the post-stenotic region in vessels with stenosis. Persistence of higher OSI distribution is observed along the downstream of vessels with aneurysms. Abnormal geometry of a blood vessel with an aneurysm promotes oscillatory flow near the walls of the aneurysm. Endothelial proliferation can occur at higher intensity of OSI. It promotes elevation in RRT parameters of particulates flowing at the region of the aneurysm. Rapid increase in intensity of both these parameters at the site of the aneurysm makes it more vulnerable, and seeks intervention to avoid non-favorable circumstances. Magnitude of RRT was found to be higher by several times in a vessel with an aneurysm than one with stenosis. This kind of vessel shows elevation in RRT with increase in flow-rate. However, it decreases with increase in flow rate in stenosed vessels. Interestingly, the impact of viscosity model is found in the case of RRT distributions. Peak amplitude of RRT value in a vessel with an aneurysm was observed to be higher (about 300 ms) for higher flow rate (Re 2). A stenosed blood vessel reports higher RRT values (about 20 ms and 10 ms for Re 1 and Re 2, respectively). The peripheral end of an aneurismal vessel shows maximum intensity of distributed RRT parameters. However, a vessel with a stenosis has such a distribution of RRT amplitude at the immediate downstream of the stenosis section. A stenosed blood vessel with acuteness of the disease leads to a prolonged rise in OSI value near walls, which could result in the formation of a minute blockage (secondary) downstream. However, these vessels show least significant elevation of RRT value of flow particles further downstream. Thus, vulnerability of a stenosed vessel is less than that of an aneurismal vessel.

4.4 Conclusion

Two-dimensional axi-symmetric geometry of a blood vessel was made containing a stenosis profile which was pre-defined by Neofytou and Drikakis *et al* (2003). Severity of the stenosis was varied from 25%–80% and realistic pulsating flow profiles were used to realize influx of blood through the geometry. Different non-Newtonian rheological models were used along with a Newtonian model to represent rheological properties of blood in a diseased state. A comparative investigation was performed to identify the suitable viscosity models. Various shear stress parameters like WSS, OSI and RRT were used for qualitative analysis of higher stress distributions and their behavior near the wall of the vessel downstream of the stenosis. It was found that the Casson and Carreau models showed better response to shear stress markers near the wall in comparison to the responses shown by other non-Newtonian and Newtonian models at Re ranges between 690–1980.

Transport phenomena of flow structure in vessels containing severe aneurysms of saccular type were carried out. Two-dimension axi-symmetric geometry of a straight long vessel with a saccular aneurysm was considered. The aneurismal profile was adapted from a pre-defined aneurismal function proposed by Molla *et al* (2012). Pulsating flow profiles used for the study were varied to various Re and influences of flow rate on depth of aneurysm were measured in the form of distributions of defined shear stress markers at the wall, i.e., WSS, OSI and RRT. Zones were successfully identified downstream of the aneurysm which are vulnerable to occurrence of secondary stenosis.

An attempt was successfully made to identify the greater risk of secondary stenosis for a person suffering from dual symptoms of severe stenosis and severe aneurysm in two different blood vessels. A simplified form of two-dimensional axi-symmetric geometries of blood vessels containing an aneurysm and stenosis were used separately to measure flow behavior in terms of shear stress markers like WSS, OSI and RRT near the wall downstream of abnormalities in both vessels using best fitted non-Newtonian rheological models. It was concluded that vessels with a severe aneurysm are more rapidly prone to produce secondary stenosis downstream of abnormalities than in a vessel containing severe primary stenosis.

References

Bellini C, Ferruzzi J, Roccabianca S, Martino E S D and Humphrey J D 2014 A microstructurally motivated model of arterial wall mechanics with mechanobiological implications *Ann. Biomed. Eng.* **42** 488–502

Benim A C, Nahavandi A, Assmann A, Schubert D, Feindt P and Suh S H 2011 Simulation of blood flow in human aorta with emphasis on outlet boundary conditions *Appl. Math. Modell.* **35** 3175–88

Bird R B, Armstrong R C and Hassager O 1987 *Dynamics of Polymeric Liquids* vol 1 (New York: Wiley)

Boyd J, Buick J and Green S 2007 Analysis of the Casson and Carreau–Yasuda non-Newtonian blood models in steady and oscillatory flows using the lattice Boltzmann method *Phys. Fluids* **19** 1–14

Casson N 1959 *Rheology of Disperse System* (London: Pergamon)

Charm S E, Comis W and Kurland M G 1964 Rheology and structure of blood suspension *J. Appl. Physiol.* **19** 127–33

Charm S and Kurland G 1965 Viscometry of human blood for shear rates of 0–100,000 s^{-1} *Nature* **206** 617–8

Favier V and Taylor C A 2002 Image-based computational modelling of blood flow in a porcine aorta bypass graft *Center Turb. Res. Annu. Res. Briefs* 325–30

Fedosov D A, Dao M, Karniadakis G E and Suresh S 2014 Computational biorheology of human blood flow in health and disease *Ann. Biomed. Eng.* **42** 368–87

Grigioni M, Daniele C, Morbiducci U, Gaudio C D, D'Avenio G, Balducci A and Barbaro V 2005 A mathematical description of blood spiral flow in vessels: application to a numerical study of flow in arterial bending *J. Biomech.* **38** 1375–86

Grinberg L, Anor T, Madsen J R, Yakhot A and Karniadakis G E 2009 Large-scale simulation of the human arterial tree *Clin. Exp. Pharmacol. Physiol.* **36** 194–205

Han H C 2011 Determination of the critical buckling pressure of blood vessels using the energy approach *Ann. Biomed. Eng.* **39** 1032–40

Huang C, Chai Z and Shi B 2013 Non-Newtonian effects on hemodynamic characteristics of blood flow in stented cerebral aneurysm *Commun. Comput. Phys.* **13** 916–28

Kim Y H, VandeVord P J and Lee J S 2008 Multiphase non-Newtonian effects on pulsatile hemodynamics in a coronary artery *Int. J. Numer. Methods Fluids* **58** 803–25

Kleinstreuer C 2006 *Biofluid Dynamics, Principles and Selected Applications* (Boca Raton, FL: Taylor & Francis Group)

Molla M M and Paul M C 2012 LES of non-Newtonian physiological blood flow in a model of arterial stenosis *J. Med. Eng. Phys.* **34** 1079–87

Morbiducci U, Gallo D, Ponzini R, Massai D, Antiga L, Redaelli A, Deriu M A and Montevecchi F M 2011 On the importance of blood rheology for bulk flow in hemodynamic models of the carotid bifurcation *J. Biomech.* **44** 2427–38

Neofytou P and Drikakis D 2003 Effects of blood models on flow through a stenosis *Int. J. Numer. Meth. Fluid* **43** 597–635

Patankar S V 1980 *Numerical Heat Transfer and Fluid Flow* (New York: McGraw-Hill)

Putter S D, Laffargue F, Breeuwer M, Vosse F N and Gerritsen F A 2006 Computational mesh generation for vascular structures with deformable surfaces *Int. J. Comput. Assist. Radiol. Surg.* DOI 10.1007/s11548-006-0004-1

Qureshi A I and Caplan L R 2014 Intracranial atherosclerosis *The Lancet* **383** 984–98

Richardson L F and Gaunt J A 1927 The deferred approach to the limit. Part I. Single lattice. Part II. Interpenetrating lattices *Philos. Trans. R. Soc. A* **226** 299–361

Roache P J 1994 Perspective: a method for uniform reporting of grid refinement studies *J. Fluids Eng.* **116** 405–41

Rohlf K and Tenti G 2001 The role of Womersley number in pulsatile blood flow a theoretical study of the Casson model *J. Biomech.* **34** 141–8

Saez P, Pena E and Martinez M A 2014 A structural approach including the behavior of collagen cross-links to model patient-specific human carotid arteries *Ann. Biomed. Eng.* **42** 1158–69

Saez P, Pena E, Martinez M A and Kuhl E 2014 Computational modelling of hypertensive growth in the human carotid artery *Comput. Mech.* **53** 1183–96

Sommer K, Schmidt R, Graafen D, Breit H C and Schreiber L M 2014 Contrast agent bolus dispersion in a realistic coronary artery geometry: influence of outlet boundary conditions *Ann. Biomed. Eng.* **42** 787–96

Stern F, Wilson R V, Coleman H W and Paterson E G 2001 Comprehensive approach to verication and validation of cfd simulations. Part 1: methodology and procedures *J. Fluids Eng.* **123** 793–802

Stroud J S, Berger S A and Saloner D 2002 Numerical analysis of flow through a severely stenotic carotid artery bifurcation *J. Biomech. Eng.* **124** 9–20

Takizawa R, Maughan B and Arseneault L 2014 Adult health outcomes of childhood bullying victimization: evidence from a five-decade longitudinal British birth cohort *Am. J. Psychiat.* **171** 777–84

Walawender W P, Chen S and Cala D F 1975 An approximate Casson fluid model for tube flow of blood *Biorheology* **12** 111–24

Wilcox D C 2006 *Turbulence Modelling For CFD* 3rd ed (La Canada Flintridge, CA: DCW Industries Inc.)

Chapter 5

Numerical analysis of blood flow in a micro-capillary in *in vitro* conditions

Arindam Bit and Sonali Gore

Blood is a rich body fluid composed of red blood cells (RBCs). Movement and distortion of RBCs in a microchannel can be evaluated on the basis of their mechanical characteristics. Cell–cell interaction and deformability of cell have important effects on rheology of the blood, mainly in capillaries. In a semi-dilute regime, cells are considered to be showing pair-wise communications. The shear viscosity of the semi-dilute suspension increases proportionately to the volume fraction. Usually, plasma is considered as a Newtonian fluid and RBCs as a shear thinning fluid (non-Newtonian). Numerical investigations are often performed to study flow dynamics of plasma (or blood as a whole) in blood vessels. The observations reveal that the velocity profile of plasma alone is parabolic in blood vessels, whereas it becomes flattened across the central region in the case of whole blood.

5.1 Introduction

Blood behaves differently in micro-capillaries. It interacts with elastic valves of a capillary, and this phenomenon is highly governed by the apparent viscosity of blood. The viscosity of the blood while flowing through narrow glass tubes apparently decreases with respect to the decrease in the diameter of the tube within the range from about 300 μm to about 10 μm. This observed event is known as the Fahraeus–Lindquist effect, which takes place as blood is a rich suspension of red blood cells that are deformable. Comparative study of velocity and stress fields of the visco-elastic fluid flowing through the channel, taking into consideration the predictions for a Newtonian fluid, shows that the shear thinning is the key source rather than elasticity. Increasing elasticity parameter at the fluid–solid interface for a given state of stress because of the flow of viscoelastic fluid through the channel leads to larger strain within the solid with decrease in the shear rate modulus of the solid.

doi:10.1088/978-0-7503-2088-7ch5

However, in the case of vascular diseases, either blood or the capillary wall change in their inherent properties.

Most of the non-communicable diseases like diabetes, cancer, dementia and cardiovascular diseases are major causes of death, disability and reduced quality of life worldwide. Due to a fundamental flaw in our understanding of oxygen transport in tissue, we have so far overlooked capillary flow disturbances, so-called *capillary dysfunction*, as a source of hypoxia in human disease. Thorough literature reviews, animal studies, and a series of neuro-imaging studies in humans now support that this phenomenon plays a vital role for the oxygenation of normal tissue and also in the progression of ischemic heart diseases, acute stroke, tumour hypoxia, diabetes and diabetes complications, and in vascular as well as Alzheimer's dementia. Indeed, recent findings using a magnetic resonance imaging based method to assess local hemodynamics in the brain confirm that capillary flow patterns indeed predict tissue oxygenation in cerebrovascular disease, correlate with disease severity in Alzheimer's, and also predict disease progression in cerebral cancer. There is hence a need to develop a means of assessing the circulation of RBCs through micro and meso vessels in order to detect the diseases at a much earlier stage with a minimally invasive and cost effective technique.

One novel way of early detection of disease is by understanding the physico-chemical dynamic environment at capillary confinement responsible for change in viscosity of blood at the region and simultaneously measuring the viscosity of the same blood in that confined region in *in vivo* and *in vitro* testing situations. This objective can be addressed by developing a novel mathematical tool for visualizing the consequent events occurring at micro-confinement. Boundary conditions thus implemented while defining the mathematical tool/model will be helpful for interpreting the possible consequences.

In order to address the above challenges, the complete flow phenomenon has been modeled using numerical techniques, and the results thus obtained has been cross-validated by experiments. A micro-viscometer has been used for the experiment. It works on the falling ball viscometer principle. The falling ball viscometer assumes importance in the study of non-Newtonian fluids. The method uses Newton's law of motion on a falling sphere ball under force balance when it reaches a final velocity. The falling ball viscometer is most accurate for measurement of a fluid viscosity. The sample viscosity relates with the time required by the ball to drop a specific distance, and the final result is considered the dynamic viscosity. The instrument applies the first and second coils at different known positions over the falling tube and records the falling time to avoid any possible human error.

The circulation of RBCs in blood is a crucial phenomenon to the physiology and pathology of the cardiovascular system. Omori *et al* (2015) investigated diffusion of RBCs that is flow-induced in a semi-dilute system and also calculated the pair-wise interactions in between the RBCs in a simple shear flow. They modeled an RBC by using a capsule having a hyperelastic membrane. They resolved the deformation using the finite element model while the boundary element model was used for the fluid motion inside and outside the RBC. Diffusion of shear-induced RBC was found to be significantly anisotropic. Misalignment was found between velocity gradient direction component and vorticity direction. The ratio of viscosity of internal to external fluid governs the motion of RBCs. It was found that on increasing viscosity ratio, diffusivity decreases monotonically. When the capillary number is invariant and the suspension is in semi-dilute capillary confinement then the shear rate and hematocrit govern the diffusivity of the suspension. This results in suspension of shear induced diffusion in RBCs by elasticity of membrane. When coming to the membrane deformability, the trajectories have a displacement perpendicular to the flow direction, which induces shear-induced self-diffusion of RBCs. Anisotropic diffusion of RBCs should be calculated by solving individual behaviour of RBCs in fluid flow. All these findings are helpful in understanding the transport phenomenon in blood flow.

Guido and Tomaiuolo (2009) postulated that the highest deformability of RBCs is requisite *in vivo* for the optimal exchange of gas in between gas and the tissues in microcirculation. Results of the studies conducted on velocity and shape of RBCs were analyzed together with both model predictions and numerical simulations. Even though so much progress has been made so far, pathophysiological implications of deformability of altered RBCs still require further study to be completed. Apart from this, information and date from experiments that are relevant clinically and interpretation of modeling-based methods are also required.

RBCs play a vital role in maintaining the mechanical properties of blood. Stimulation of the mechanical behaviour of RBCs in micro-capillaries was developed and improved using several methods. Particularly on the microscopic scale, discrete methods are used over continuum methods. The moving particle semi-implicit (MPS) method was used in simulation of micro-fluidic flows which is majorly based on Navier–Stokes equations (Ahmadian *et al* 2012). The MPS method has emerged as a considerably effective tool for blood flow modelling in micrometer dimensions. A semi-implicit algorithm is developed by various modifications and used on motion of RBCs through microvessels. This particular algorithm helped in reducing time of calculation by more than a factor of twenty keeping the position and velocity error constant.

The functioning of capsule and biological cell have good importance in the field of bioengineering and chemical engineering. Omori *et al* (2012) studied regarding the movement of RBCs or a spheroidal capsule in a creeping shear flow. The orientations of a nonspherical capsule were found variant upon time reversal, whereas that of a spheroid that was rigid were found invariant. The arrangement of a nonspherical capsule for long time duration was found to exhibit a transition that depends on the shear rate. Jeffery's theory explains clearly about the rigid spheroid dynamics while the dynamics of a spheroidal capsule is still unclear. The non-Brownian rigid ellipsoid trajectory in Stokes flow was invariant under reversal of time and also the reorientation for over long time duration did not take place under the conditions of simple shear flow. A capsule was used as a model deformable object for the above investigation. Two fundamental laws for the membrane, namely: the neo-Hookean (NH) law and the Skalak (SK) law were used. Here, the NH law described isotropic volume of incompressible material properties like rubber, and the Skalak law expressed the incompressibility area of a biological membrane. A capsule having SK membrane tended to have a higher critical Ca_0 when compared with the NH membrane. This was because the SK law showed the property of strain-hardening, while NH law showed strain-softening.

Ellinor *et al* (2012) determined the membrane tension which contributes in releasing of the molecules by RBCs, and severely high tension might lead to haemolysis. Investigation on the membrane tension of RBCs with pair-wise communications in simple shear flow was made. RBCs were modeled as capsules having a hyperelastic membrane which is two-dimensional, and major deformations were solved by using the finite element method. Because of the RBCs' small size, Stokes flow was used for estimation of the surrounding fluid motion and it is further solved by using the boundary element method. The results revealed that the isotropic tension appeared maximum around the dimple of the biconcave surface but not around the rim. Comparison of the results showed that the isotropic tension and maximum principal tension were increased significantly by the effect of cell–cell interactions.

Lima *et al* (2008) introduced an approach using confocal micro-PTV for measuring the movement of labelled RBCs which flows in suspensions of normal RBCs that are concentrated. The capability of confocal systems in obtaining thin in-focus planes enabled measuring the radial position of individual RBCs accurately and consequently helped in measuring the interaction in between the multiple labelled RBCs. Whole measurements were performed in the center plane of both 50 and 100 mm glass capillaries at Reynolds numbers (Re) ranging from 0.003 to 0.005 by use of hematocrit (Hct) from 2% to 35%. Microscale phenomena of the flow behaviour of red blood cells, such as the RBC–RBC interactions, and tendency of the RBCs for radial migration were analyzed by the use of RBC radial dispersion coefficient (D_{yy}) approach. Results demonstrated that D_{yy} depends majorly on the Hct. The RBCs showed higher D_{yy} at radial positions in between 0.4–0.8 R and lower D_{yy} at places that are adjacent to the wall (0.8–1 R) and over the centre of the capillary (0–0.2 R). It also demonstrated that D_{yy} favors lowering with decrease in the diameter, while it increases with the Hct and levels off at Hct of about 25%. This suggests that the plasma layer development and effective decrease in the cell density that surrounds the RBCs may help in the enhancement of radial dispersion of RBCs. It also indicates the reduction in the radial dispersion of RBCs and the repeated multibody collisions are directly related to the decrease in apparent viscosity with decreasing diameter, known as the Fahraeus–Lindqvist effect.

The dimensional aspects of blood viscosity were studied by Kaliviotis *et al* (2011) by linking a modified equation with shear deformation fields that are calculated from measurements of velocity obtained by using a technique based on μPIV. Blood present at physiological hematocrit levels in the existence of aggregation was sheared in narrow gap one-plate geometry, while the characteristics of velocity and aggregation were confirmed from images that were captured using a high resolution camera. The flow characteristics were affected by the aggreggation caused that changed the microstructure of the blood. In the conditions of low shear stress, formation of network and high aggregation caused the movement of RBC in becoming essentially two-dimensional. It was also found that characteristic of blood that is non-uniform microstructure plays a role in influencing the viscosity distribution. Formation of a network was observed at low shear rates. It also affected the field of velocity and apparently the rate of shear deformation. The characteristics of the effective viscosity were found to differ from the core viscosity that is calculated with the help of zero-gradient core profile assumption. The variation in viscosity and also the differences between the core and effective viscosities are prominent for the haemodynamics of the microcirculation. Aggregation and network restructuring is likely to take place in various parts and can affect local viscosity and other fluid mechanics properties.

Omori *et al* (2014) conducted a numerical study on the rheological properties of an RBC suspension from the dilute to semi-dilute order. Here RBCs were structured as a capsule having a 2-D hyperelastic membrane. Major deformation of the thin membrane was calculated using the finite element method. Because of the small size of the RBC, fluid motion that is present surrounding the RBC was assumed to be following Stokes flow and was solved by use of the boundary element method. In the dilute limit, cell–cell interactions were deleted and the suspension having bulk stress was evaluated by the stresslet that is generated on a single RBC. The shear viscosity of the dilute suspension was found to be decreased with increase in the viscosity of the internal liquid. In the semidilute order, cells are considered to be showing interactions pairwise. The effective shear viscosity of the semi-dilute suspension observed a quadratic increase proportional to the volume fraction. On the other hand, in the dilute regime the effective shear viscosity was observed to be decreasing with increase in the λ, the reason behind it is the angle of the RBC reduced with increase of λ. They concluded λ as one of the vital parameters while discussing the rheological properties of the RBC suspensions in the order of the dilute to semi-dilute suspensions.

Zeng and Zhao (2011) examined the rheological behavior of blood with the help of a microfluidic device using various flow conditions, they also tested for the electrical behaviour with electrical impedance spectroscopy. It resulted that flow conditions of blood that is flowing in microfluid channels similar to vessels can be observed by measuring its electrical impedance and by that the blood viscosity can also be derived. when hematocrit value is maintained constant, RBC aggregation will become the predominant factor. Microfluidic channels were fabricated with rectangular cross sections and microelectrodes were placed on the lower area of the microchannels. The electrical impedance between the two electrodes was recorded while the blood was flowing through the channels. It is considered as rheological behaviour of blood. They performed the measurement with shear rate of 100 s^{-1} and stimulating frequency of 20 kHz. With increase in the hematocrit value, blood showed high electrical resistivity with the drop of electrical capacitance. These observations showed that electrical properties of various components of blood vary with flow conditions and can also be considered as a measure of the rheological behaviour.

Secomb and Pries (2013) observed that the apparent viscosity of blood that is flowing through narrow glass tubes decreased highly with decrease in the tube diameter in the range of about 300 to 10 μm. This phenomenon that is known as the

Fahraeus–Lindqvist effect, takes place because blood is a highly rich and concentrated suspension made up of deformable RBCs having a dimension of about 8 μm. The resistance to blood flow through the circulatory system is observed in microvessels having diameters in this range. Apparent viscosity of blood in microvessels *in vivo* was found to be significantly greater than in that of glass tubes with the same diameters. In this work, they reviewed the observations of apparent viscosity of blood both *in vitro* and *in vivo* and also observed its growth in respect of understanding the involved mechanisms.

Zeng and Zhao (2011) using a microfluidic chip examined the rheological properties of blood flow in various flow conditions. Electrical impedance spectroscopy was used to examine the electrical properties of blood. As per results the electrical impedance measurement and viscosity could derive the flow conditions of blood in a vessel like a microfluidic chip. RBC aggregation was a predominant factor under constant hematocrit value. Rectangular cross section microfluidic channels were fabricated. Placement of microelectrodes was done at the bottom surface of the microfluidic chip. The rheological behavior of blood was measured by recording the electrical impedance between the microelectrodes when the blood is allowed to pass through the channels at the shear stress of 100 s^{-1} and stimulating a frequency of 20 kHz. Increase in hematocrit value of blood leads to high electrical resistivity, whereas electrical capacitance declines. Varying the different shear stress and stimulating frequencies electrical measurements were recorded, which showed the electrical properties of different blood components in different flow conditions and rather could be used for measurement of rheological behaviour of blood.

Secomb and Pries (2013) observed the relation between the apparent viscosity of blood and diameter of glass tubes. They noticed that with decrease in the diameter of narrow glass tubes there is a fall in the apparent viscosity of blood flowing through glass tubes over a diameter range from 300 to 10 μm. This Fahraeus–Lindqvist phenomenon occurs because of concentrated suspension of RBCs of dimension 8 μm in blood. When blood flows through the circulatory system via micro-vessels this range of dimension leads to most of the blood flow resistance. In *in vivo* the apparent viscosity of blood is found to be higher than in glass tubes of corresponding dimensions. Progress towards the quantitative theoretical understanding of the mechanism involves reviewing of experimental observations of apparent viscosity measurement under both *in vitro* and *in vivo* conditions.

Afzal *et al* (2014) carried out numerical stimulation using a shear dependent viscosity model to investigate the flow dynamics and mixing behavior in serpantine and T-shaped microchannels using non-Newtonian fluid, the microfluidic transport of blood was taken into consideration. Non-Newtonian characterization was done using Carreau–Yasuda and Casson non-Newtonian blood viscosity models. Steady Navier–Stokes equations with diffusion convection model for species concentration were solved in mixing and flow analysis. Flow and mixing dynamics under similar conditions were compared using working fluids: water (a Newtonian fluid) and blood using the Carreau–Yasuda non-newtonian model. The mixing performance of both the considered fluids was observed to be equal, for mass flow rate of $\dot{m} < 10^{-2}$ kg h^{-1} with decreased flow rate. With increased flow rate, the mixing of water was improved significantly and negligible change in mixing performance with blood Carreau–Yasuda. The pumping power needed is found to be less for water (~0.40 bar) than blood (~1 bar) at the same flow rate. In order to study the flow dynamics and mixing performance a computational fluid dynamics (CFD) was performed.

Kim and Yoo (2009) proposed a method for three-dimensional analysis of RBCs in circular microcapillary. The pressure driven superimposed electrophoresis of RBCs was used for generation of an RBCs migration model. To locate and identify the red fluorescent labeled RBCs, a particle tracking scheme was implemented. The RBCs were transported through microchannels of diameter 85 μm. The tightness of cell focusing was investigated through designed probability density function (PDF). the obtained results showed that the nonspherical particles could be subjected to 3D focusing using electrical charges.

Ting *et al* (2010) designed a modified SIMPER algorithm for analyzing microfluidic effect on the motion and deformation of red blood cells in a capillary. With assumption of much reduced value of Re number in microfluidcs, the algorithm facilitated the convergence of the momentum equations by summing up the advantages of SIMPLEC and SIMPLER algorithm together. it also satisfied the continuity equation with high accuracy by modifying with fine adjustment techniques. Parachute-shaped structure of RBC was confirmed at a velocity of 0.1 mm s^{-1}. Validation of relation between deformation index and mean velocity was also performed. The effect of various essential parameters was investigated on motion and deformation of RBCs along with the radius of RBCs, bending stiffness and elasticity modulus of RBC membrane. The numerical study demonstrated the effect of various parameters on the behaviour of RBCs.

Based on the knowledge of the mechanical characteristics of blood, the motion and deformation of RBCs can be analysed. Accurate prediction of apparent viscosity can be done using an axis-symmetric model for single file motion of RBCs in the capillaries of diameter 8 μm (Secomb 2011). Two-dimensional numerical studies with representation of RBCs as at a set of interconnected viscoelastic elements predicted that centred RBCs take asymmetric shapes and drift towards the centre line. The trajectories predicted agreed with the observation obtained in microvessels of the rat mesentery.

Alizadehrad *et al* (2012) investigated RBC deformation in a microvessel with varying vessel diameters, shear rates and hematocrit values. A particle based model was used to analyse the deformation of RBCs in the circular channels having diameter ranging from 9 μm to 50 μm at shear rates ranging from 20–150 s^{-1} and with the hematocrits from 20%–45%. The deformation of RBCs is assumed to be a function of radial position. The radial variation of RBCs was classified into three regions: near centre, middle and near walls. Vessel diameters, shear rates and hematocrit values influence the local shear stress and wall effect. The response to the shear rate was found nonlinear. The deformation of RBCs exhibits the non-linear response to varied shear stress in the shear flow condition. With the increase in Hct the plug flow becomes more prominent at the centre, which results in higher shear stress near walls for a constant pseudo-shear rate.

Mauroy (2013) studied the pulmonary capillary which is an oxygen capturing site for RBCs. A camera method was proposed for the numerical study behaviour of RBCs along the whole capillary. Change in geometry of erythrocytes along the capillary affects its oxygen capturing capacity. The hyperelastic approach was used to model the complex membrane of RBCs. The length of vessel was in the order of mm and the radius within the range 5–15 µm. The plasma is a viscous Newtonian fluid like water but with twice its viscosity.

Xing *et al* (2014) investigated heme–dynamic interaction of RBCs in micro-capillary flow; based on the suspension flow the interaction was considered to be non-contact mutual interaction of cells.

To solve the suspension of RBCs The distributed Lagrange multiplier/fictitious domain method in the lattice Boltzmann framework was used. The modification of flow due to cells, dependence of deformation of cell on flow and cell–cell interaction via fluid were evaluated. It was analysed that the rheological properties of suspension such as mean velocity had been affected by long-range hydrodynamic interaction with long interacting distance. Short-range interaction affects the shape of the cell in micro-capillary flow. The flow velocity also plays an important role in the cell deformation.

5.2 Methodology

Numerical and theoretical methods were conducted to investigate the rheology of blood in a capillary.

5.2.1 Micro-viscometer study

Dynamic viscosity, kinematic viscosity and shear stress of blood flow in capillary were calculated using a Lovis-2000M micro viscometer (make: Austria). The length of the capillary vessel was 140 mm and diameter was 1590 µm. The Lovis micro-viscometer works on the principle of Newton's law of motion which implies the balancing force on a falling spherical ball when it reaches its terminal velocity. This viscometer is also called the 'falling ball viscometer'. the viscosity of the sample was correlated with the traversal time required by the ball to a specific distance. Test results were calculated and dynamic viscosity was evaluated, which is governed by equation (5.1).

$$\mu = K.\,(\rho_p - \rho).\,t \tag{5.1}$$

A schematic diagram of the Lovis micro-viscometer is shown in figure 5.1. The figure shows the various forces being applicable on the metallic ball inside a microvessel inclined at an angle θ. it also shows the position of inductive sensors 1, 2 and 3 with inter distances between sensors of 25 mm and 75 mm between 1 and 2, and 2 and 3, respectively. The travelling time of the metallic ball across sensors 2 and 3 is recorded with reference to the line crossing sensor 1. The logarithmic mean was evaluated for traverse time in order to calculate t.

Twenty blood samples of males and females (10 each) were prepared. 3 ml of blood from veins were collected using prefilled vasopressin injection (with a sterile

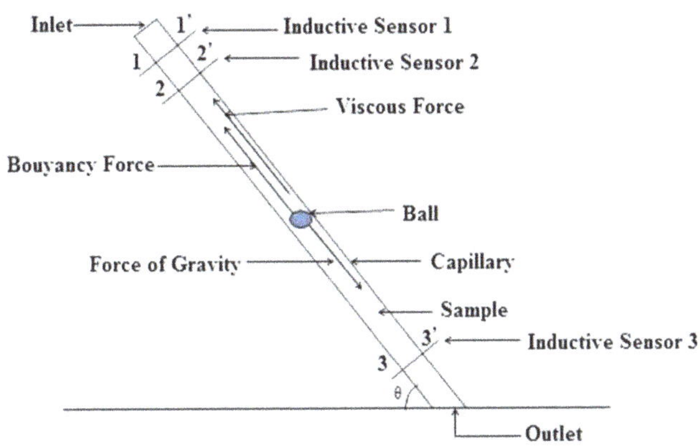

Figure 5.1. Schematic representation of a falling-ball viscometer with three inductive sensors.

needle) in antecubital fossa of subjects at 11 h of fasting condition at an institute-in-house dispensary facility. The collected blood samples were refilled in tubes containing anticoagulant (K_2EDTA). 15 min prior to the blood drawing from subjects they were instructed to be in a sitting position for 15 min. Male subjects were classified on the basis of cholesterol level. The HDL-cholesterol is analyzed using precipitation method and sample precipitation was obtained using centrifugation of the sample for 20 min at 1500 g at a temperature of 15 °C. Needles were disposed properly in a needle crashing system after collection of blood samples.

Prior to loading blood samples in a microviscometer, density of blood sample is measured using a density meter, DMA35 (make: Austria), using 2 ml of blood samples. Precaution should be taken to avoid bubbles in the sample while performing this procedure. An ionized water sample was used to calibrate the density meter. Density of all samples is measured which falls under the range between 1049.15 kg m^{-3} to 1049.73 kg m^{-3} at 20 °C.

Precautions have been taken while handling blood samples. Protective clothing, mainly disposable rubber gloves, has been used. After completion of the experiment hands were well washed with soap and warm water. The experimental area is cleaned with sodium hypochlorite solution containing 0.5% chlorine content.

100 μl of the blood sample was used for refilling the micro-capillary. Viscosity was measured at constant temperature of 20 °C with varying angle of inclination within range of 20°–70° with increments of 5°. A pulse is generated by the inductive sensors present in the inner walls of the capillary slot when the metallic ball (SS 1.4125, density = 7.69 g cm^{-3}) passes through each ring. Pulse transit time was recorded and the difference between each pulse was formulated. Shear rate, dynamic and kinematic viscosity obtained by a micro-viscometer are confirmed by evaluating backward and forward deviation to a value of 0.02%; variation coefficient is also obtained to a value of 0.01% and mean distance travelled (about 60°, 99.99 mm) by the ball for twelve iterations or lower.

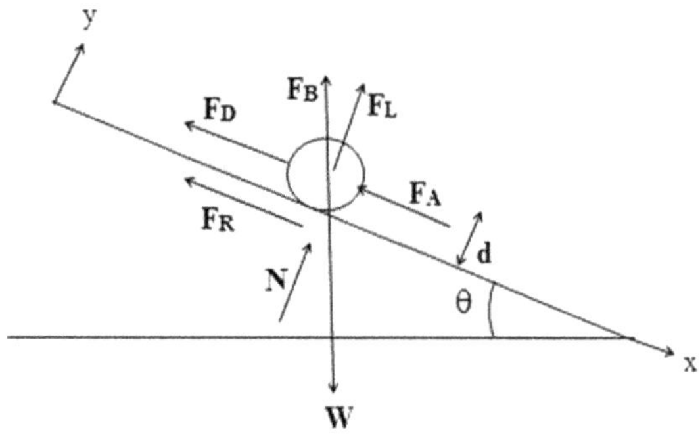

Figure 5.2. Free body diagram of falling ball over an inclined plane.

5.3 Numerical modeling

Numerical modeling has been carried out in an iterative manner. The ball is fixed at different lengths of the tube and relative motion of the blood sample is implied to evaluate the strain rate of fluid crossing to circumvent the ball. The force balance equation of rolling ball over an inclined plane is solved analytically.

Figure 5.2 represents the free body diagram of a falling ball rolling over an incline plane through non-Newtonian fluid. The equation of motion was solved analytically (table 5.1). Drag force, lift force weight, buoyant force and normal reaction force from plane, resistance and inertial force of added mass are acting on the rolling ball (Martino *et al*).

Force balanced equation of rolling ball over an inclined plane is given by equation (5.2) below:

$$-C_D\frac{1}{2}\rho U^2\pi r^2 + \Delta p\frac{4}{3}\pi r^3 g\sin(\Theta) - C_A\rho\frac{4}{3}\pi r^3\frac{dU}{dt} - F_R = \rho_p\frac{4}{3}\pi r^3\frac{dU}{dt} \tag{5.2}$$

Table 5.1. Inlet velocity of fluid with respect to position of the ball.

Angle of inclination (θ)	Relative velocity (m s^{-1})
20	0.000 45
25	0.000 65
30	0.0009
35	0.001
40	0.001
45	0.0012
50	0.0014
55	0.001 57
60	0.002
65	0.0021
70	0.0023

First term: drag force
Second term: submerged weight (W-Fa)
Third term: added mass
Fourth term: rolling resistance
$C_A = 2$ added mass coefficient considered in the reference of result of Jan and Chen.

Force due to rolling resistance is given by the equation (5.3):

$$F_R = \frac{2}{5}\rho_p \frac{4}{3}\pi r^3 \frac{dU}{dt} \tag{5.3}$$

Substituting equation (5.3) in equation (5.2), we get,

$$\frac{\Delta_\rho}{\rho}\frac{4}{3}dg\,\sin(\theta) - C_D U^2 = \frac{4}{3}d\left(\frac{7}{5}\frac{\rho_p}{\rho} + C_A\right)\frac{dU}{dt} \tag{5.4}$$

Drag coefficient C_D can be formulated by equation (5.5).

$$C_D = A + \frac{B}{R_e} \tag{5.5}$$

We can use ICEM-CFD (Ansys 14.5) for the finite element modeling of the geometry. Richardson extrapolation criteria is used for the study of grid convergence. The highlights of the grid-convergence test are described in the following section. A capital quality mesh has quadrilateral elements (13 110), and these have 15 500 elements and 14 343 nodes. For the calculation of elements and node we use the paving mesh technique. Figure 5.3 represents the finite element modeling of quadrilateral mesh elements containing a fine vessel fluid. Lagrangian method is widely used to resolve the issue. In our problem or issue a small content mesh accumulates lower content time-step.

Due to fluid—structure communication it produces local compressibility of fluid, and these cause step-up of localized pressure deflection. These events are not considered in our study because of the rigid structure of the metallic ball in the boundary. Our problem becomes easier if we consider the nature of blood is incompressible. Therefore, the reduced form of the mass conservation continuity equation is represent in equation (5.6) as:

$$\frac{1}{r}\frac{\partial}{\partial r}(rv) + \frac{d}{dx}(u) = 0 \tag{5.6}$$

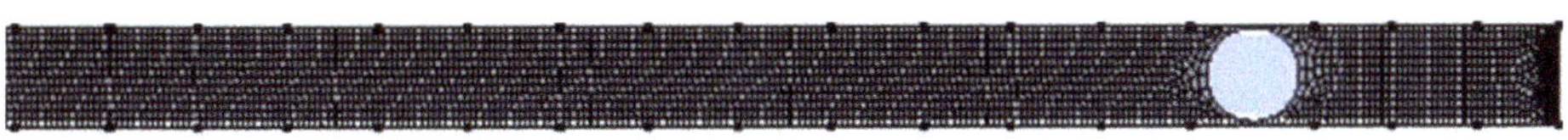

Figure 5.3. Finite element modeling of capillary fluid.

Similarly, the Navier–Stokes equation (equation for momentum) along with the axis of symmetry includes a transient term, a negative pressure term, convective term, and diffusion term, and all these are represented by equation (5.7).

Axial momentum equation:

$$\frac{du}{dt} + \frac{1}{r}\frac{\partial}{\partial r}(ruv) + \frac{d}{dx}(u^2) = \frac{1}{\rho}\frac{dP}{dx} + \frac{1}{r}\frac{\partial}{\partial r}\left(\frac{\mu_{\text{app}}}{\rho}\left(\frac{\partial u}{\partial r} + \frac{dv}{dx}\right)\right) + \frac{\partial}{\partial x}\left[\frac{2\mu_{\text{app}}}{\rho}\frac{du}{dx}\right] \tag{5.7}$$

All the walls of the vessels are to be considered as rigid, and the slip boundary condition will be assumed to be zero at the walls. In our study we consider a segment of finite length capillary or vessel so these assumptions create zero gauge-pressure at the outlet of vessels.

The dynamic viscosity of a Newtonian fluid are collected from experimental results. Here we consider the blood viscosity is available in as a non-Newtonian form. Casson (1959), Walawender *et al* (1975), Misra and Ghosh (2000) and Boyd *et al* (2007) all discussed the Casson model. All these used this model to show the rheological behaviour of blood using equations (5.8) and (5.9).

$$\bar{\tau} = \left(\sqrt{\mu_\infty} + \sqrt{\frac{\tau_y}{|\bar{\gamma}|^*}}(1 - e^{-\sqrt{m|\bar{\gamma}|}})\right)^2 \bar{\gamma} \tag{5.8}$$

and

$$\mu(|\bar{\gamma}|^*) = \mu_\infty\left(1 + \sqrt{\frac{Bi}{|\bar{\gamma}|^*}}(1 - e^{-\sqrt{m|\bar{\gamma}|}})\right)^2 \tag{5.9}$$

From previous study papers, Charm *et al* (1964) and Charm and Kurland (1965), we take the value of $\tau_y = 10.82$ mPa and $\mu_\infty = 3.1 \times 10{-}3$ Pa-s for blood.

Quemada (1977) proposed a model based on shear rate and hematocrit. The equation of shear stress and effective viscosity in tensorial and dimensionless form are described by equations (5.10) and (5.11)

$$\bar{\tau} = \mu_F\left(1 - \frac{1}{2}\frac{k_0 + k\sqrt{|\bar{\gamma}|/\gamma_c}}{1 + \sqrt{|\bar{\gamma}|/\gamma_c}}\varphi\right)^{-2} \bar{\gamma} \tag{5.10}$$

and

$$\mu(|\bar{\gamma}|^*) = \mu_F\left(1 - \frac{1}{2}\frac{k_0 + k\sqrt{|\bar{\gamma}|^*/\gamma_c^*}}{1 + \sqrt{|\bar{\gamma}|^*/\gamma_c^*}}\varphi\right)^{-2} \tag{5.11}$$

where $\gamma_c{}^* = \frac{\gamma_c}{U_\infty/1}$ viscosity of plasma is $\mu_F = 1.2 \times 10^{-3}$ Pa-s, and for hematocrit $\varphi = 0.45$, $\gamma_c = 1.88\ s^{-1}$, $k_\infty = 2.07$ and $k_0 = 4.33$.

5.4 Grid independence study

Higher order flow-fields are calculated from a series of lower-order discrete values ($f_1, f_2, \ldots, f_n$) using the Richardson extrapolation method Richardson and Gaunt (1927). According to Grid independence study, the value calculated from Richardson extrapolation criteria is obtained at the smallest size of grid ($h \rightarrow 0$). Grid convergence index (GCI) is evaluated using pth-order methods (Roach 1994). First we calculated the convergence conditions of the system and then we calculated the extrapolated values from solutions. R is used for the convergence ratio. Then, solutions have monotonic convergence when $0 < R < 1$, while it has oscillatory convergence for $R < 0$, and divergence for $R > 1$. Convergence ratio R is obtained from the below equation as:

$$R = \frac{\epsilon_{21}}{\epsilon_{32}} \tag{5.12}$$

Table 5.2 compiles the level of accuracy of mean axial-velocity ($V_{\text{axial(mean)}}$) at 2 mm pre-ball position, and we use a capillary vessel at three kinds of grid size to calculate an absolute strain rate (σ_{Abs}). We use monotonic convergence conditions for mean axial-velocity and mean radial-velocity. Convergence is obtained from the GCI for the grid.

GCI was observed for a fine grid solution with the help of equation (5.13):

$$GCI_{i+1,i} = F_s \left| \frac{\epsilon_{i+1,i}}{f_i (r^p - 1)} \right| \tag{5.13}$$

F_s is safety factor, and its value is settled at 1.25.

GCI reduces in successive refined grids ($GCI_{21} < GCI_{32}$) for each parameter. GCI_{21} is used as notation term for the GCI representation of finer grid, and its value is lower than the GCI of a coarser grid (GCI_{32}). These all show that the numerical simulation is self-reliant of cell size.

5.5 Results and discussions

Experimental results of dynamic viscosity, kinematic viscosity and shear rate with respect to angle of inclination have been presented. This study has been carried out

Table 5.2. Degree of accuracy and GGCI for two parameters.

	ε_{32}	ε_{21}	R	P	GCI_{32} (%)	GCI_{21} (%)
V_{axial}(mean)	0.0035	0.0018	0.514	0.9957	4.7126	2.486 88
σ_{Abs}	0.015 90	0.0077	0.484	1.0857	4.2409	2.090 36

over a different class of population whose details have been given in subsequent section. Experimental findings have been interpolated to obtain relationship between flow parameters and angle of inclination. Numerical results on strain rate and velocity with respect to time as well as angle of inclination for various non-Newtonian models (Quemeda, Carreau, Casson and power law) have also been discussed in detail. The contour plots at different angles of inclination and at different position in the tube have also been presented.

5.5.1 Evaluation of fluid flow parameter in a microviscometer

Parameters like dynamic viscosity, kinematic viscosity, shear-stress, strain rate, pressure magnitude, wakes of vorticity, and spatial distribution of velocity magnitude are calculated from an experimental and numerical approach. We must do experimental calculation before numerical calculation. We also measure the kinematic viscosity with respect to angle of inclination. We observed an inverse relationship between dynamic viscosity of blood and angle of inclination for the female population. Variations in dynamic viscosity of the female population have been validated by re-considering six samples with post-parasitic plasmodium phalsiferum affected disease (5–8 days post disease condition) and find a similar property. But for the male sample we find the existence of an inverse relationship between dynamic viscosity of blood and angle of inclination. Samples 5 and 6 represent the maximum slope of dynamic viscosity at lower angle of inclination (20° to 40°). Samples 5 and 6 are related to high level of cholesterol, i.e. dynamic viscosity of loaded cholesterol is validated by considering another set of 6 samples from the same group.

We found Shear rate and it varies in range from between 40 s^{-1} and 260 s^{-1} and it increases to 210 s^{-1} in angle of inclination of the plane for female samples at pre- and post-disease condition, respectively. For the post-diseased conduction female sample the shear rate varies between 80 s^{-1} and 120 s^{-1} with angle of inclination of capillary changes from 30° to 70°. The diseased female samples' shear rate is justifies by the same characteristics of the female population. Observations have been taken out from providing variation in shear rate of the diseased female sample.

Male shear stress variation of RBC inside a capillary and female shear stress variation of RBC inside a capillary are compared. The magnitude of shear stress is found higher in female samples than male samples, and angle of inclination is directly proportional to the difference. We also find the linear relationship between shear stress and angle of inflation of the capillary. We are getting identical shear rate of males and females at a lower angle value of inclination (25°).

5.6 Numerical assessment of the rheological model for blood flowing in an inclined plane

Numerical examinations are shown by the temporal and spatial plot. The temporal plot provides the relationship between change of velocity magnitude and strain rate magnitude with respect to various viscosity models, angle of inclination, and position of the ball inside the inclined capillary.

We had plotted the centre line velocity for the metallic ball for Newtonian and non-Newtonian fluid flowing through the inclined capillary at 70°. We obtained constant velocity for Newtonian fluid irrespective of time, whereas we recorded a small variation of velocity for non-Newtonian fluid. Similar kinds of observations are recorded for the cases of maximum, average, and minimum velocities.

Earlier we observed that the response of Newtonian fluid has been found to be constant. Therefore, it is necessary to conduct a study considering the blood to be non-Newtonian. We have plotted temporal variation of the strain rate for a non-Newtonian viscosity model for blood flowing through a capillary which is situated at an angle of 35°. We monitored the strain rate at 15 mm and 20 mm downstream. We plotted the strain rate in three modes called high, average, and low value. And we found that Casson, Carreau and power law models do not depend upon the strain rate. However, the Quemada model highly affects the strain rate because of consideration of hematocrit of blood, which plays a very critical role for determining blood viscosity. Strain rate varies from range 12 s^{-1} to 300 s^{-1} for blood operated in the Quemada model with 35° of capillary inclination.

Figure 5.4 (a) shows averaged strain rate plot of fluid around the boundary of the ball (15 mm downstream from the inlet of the capillary at various angles of inclination. We obtained the highest strain rate (510 s^{-1}) at 45° of inclined capillary and lowest strain rate (125 s^{-1}) at 25°. If the position of the ball was 25 mm downstream from the inlet of the capillary we found highest strain rate (600 s^{-1}). From figures 5.4(a) and (b) it is clearly shown that the strain rate stepped up abruptly and drastically by providing variation to the ball from 15 mm to 25 mm downstream from the inlet of the capillary. From the above conclusion we found that the strain rate with increased angle of inclination is achieved with the help of shifting the ball to different positions downstream from the inlet of the capillary. We also achieved higher strain rate, 1250 s^{-1}, at 70° inclination when the position of the ball was 10 mm downstream of the inlet.

Spatial representation of velocity distribution and strain rate distribution was plotted by a contour plot. Figure 5.5 shows the contour plot of strain rate of blood

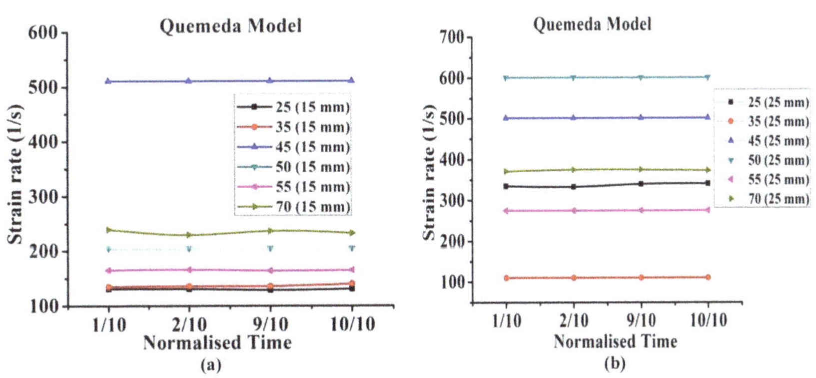

Figure 5.4. Variation of strain rate of non-Newtonian fluid (Quemada model) with varying angles of inclination at (a) 15 mm, and (b) 25 mm position of the ball downstream from the inlet of the capillary.

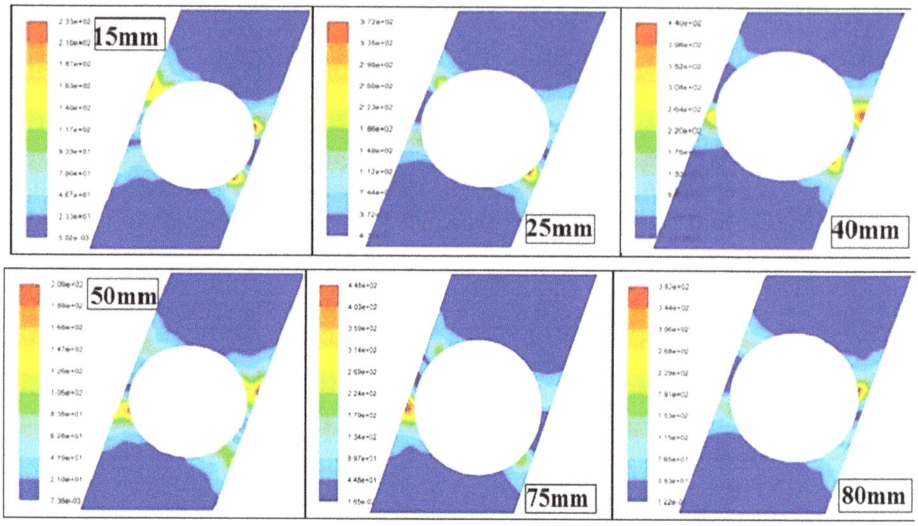

Figure 5.5. Contour plot of strain rate distribution at the boundary of the ball for different positioning of the ball.

live through a capillary situated at 70° of inclination verses the position of the ball varied from 15 to 80 mm away from the inlet of the vessel.

The regional or spatial representation of speed distribution as well as strain rate distribution have been created through a contour scheme of both the standards. Figure 5.5 depicts the contour schematic of strain rate of blood streaming from the capillary at a slant of 70°, by situation of the ball changing from 15 mm to 80 mm farther from the creek of the vessel. The advancement of vortexes changes with the situation of the ball, and its magnitude tends to increase with the downstream movement of the ball. The vortex tends to fade or vanish at 80 mm downstream after primary increase of about 15 mm to 40 mm. The advancement of the subsidiary vortex encapsulating the circumference of the ball tends to be independent about the situation of the ball.

The speed contour being formed through the enhancing depth of the ball within the capillary displays advancement in vortex with backward stream. But durability of this may not be ensured from the surveillance. Having the ball's situation at 40 nm and 75 nm downstream of the creek the elementary vortex fades away; at the same time there is a significant reduction in the size of the subsidiary vortex. Also, the delaminating of the subsidiary vortex is seen when the ball is at 15, 25 and 80 mm downstream. Nevertheless, this delaminating fluctuates with the enhancement in the position of the ball. We notice that at 50 mm there is no vortex delaminating, but at this situation four elementary reverse flow vortexes develop, also the subsidiary vortexes are at peak. This formation and de-lamination of vortexes results in creation of a dead space area on the reverse side of the ball, which progresses to a nearby layer of the capillary. Creation of this boundary layer levelled through the raise in the depth of the ball downstream to the creek conduit.

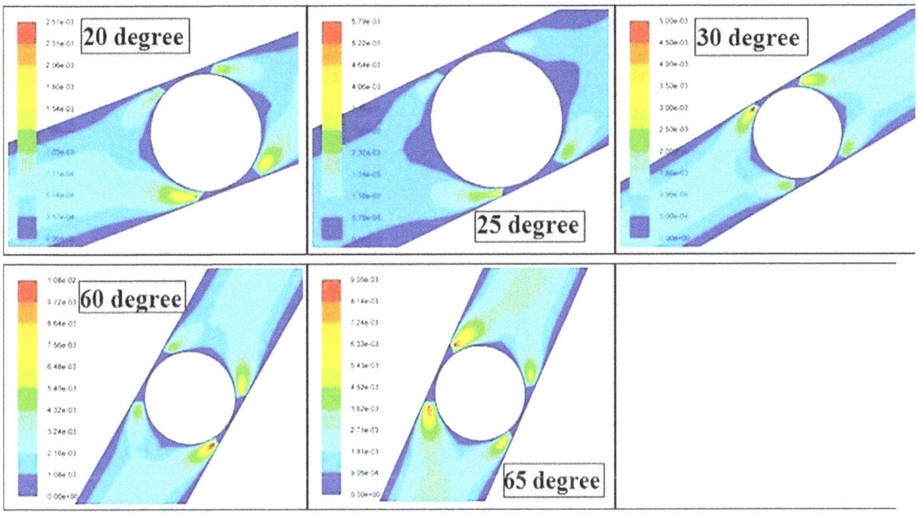

Figure 5.6. Contour plot of velocity distribution at the boundary of the ball for different degrees of inclination of the capillary.

Figure 5.6 depicts a contour schematic of the fluid encapsulating the ball at various degrees of slant of the capillary in relation to the horizontal axis. Durability of vortex establishment of the velocity contour can be seen at a further degree of slant from 60° to 70°, whilst at lower degrees of slant only temporal vortex creation has been seen. The subsidiary vortex was also created and broke down with the raise in slant angle. Also, regional vortices are created at lower slant angles that later extend to the subsidiary vortex by decreasing the presence of boundary layers. However, constancy of boundary layers between the wall and the ball exist throughout all the angles of inclination.

Considerations of the attributes such as viscosity (dynamic as well as kinematic), velocity strain rate and shear rates as well as conformation schematic of velocity along with strain rate resulting from empirical as well as numerical study provide vital results related with viscosity of blood in a micro-capillary.

5.7 Discussion

The momentum or propulsion of the ball rises along with increasing slant angle, this also hinders the accumulation of the hematocrit portion of blood. This causes reduction in the volume and dynamic viscosity of blood. Nonetheless, unhealthy blood contains a high level of hematocrit, which causes a miscellaneous spread of hematocrit in the micro-capillary. This makes a non-linear reaction about dynamic viscosity constant for angles of inclination. It has been observed that because of the enhanced molecular extent of extra-cellular protein's portion in blood, the existence of negative elevation on a 40° slant angle and positive elevation on a 55° slant angle is present in the event of male (affected) subjects. However, identical trends of dynamic viscosity at a lower slant angle are not evident. Inside the most sophisticated and balanced inclined planes the boundary layer incident about the capillary wall as well as the metallic ball connection may hinder the reactions. In the event related to female (sick) blood samples, oscillatory reactions about the dynamic viscosity with increasing slant angle may have a correlation as the centre of oscillation is seen within 35° to 65° of inclinations of the capillary. Further, corroborative examinations on a different group of specimens have been done to rule out any possibility of experimental mishaps. The reaction about kinematic viscosity at elevated slant angle about the capillary can be determined on the basis of former arguments; also, the reaction of sick blood samples (both male and female) upon attributes such as kinematic viscosity as well as its variations with elevated slant angle of the capillary can be properly described as being conducted for dynamic viscosity.

Further, we have seen viscosity values of male blood samples are higher than female blood samples at the full range of angles of inclination. As of now there is no absolute explanation for such an observation but research is being conducted in order to find the answer.

Increased slant angle causes the momentum in the stream of the ball via the blood sample within the capillary, which causes advancement of high shear rate. Hence there is a linear association between slant angle and shear rate. However, this association can become non-linear if the constituents of blood specimen alter its concentration with having either a large quantity of haematocrit or macromolecular protein in the extra-vascular fluid of blood. With the advancement in slant angle, the increased consolidation of hematocrit may resist the movement of the ball, consequently originating an oscillatory reaction of the shear rate with increased slant angle. However, samples of males with a high amount of cholesterol impinge momentum upon a rolling ball, which has a momentum less than the regress of cholesterol at lower slant angle, which in turn becomes symmetrical with gaining momentum of the rolling ball with increased angle of inclination.

Within the micro-capillary blood has very low Re value (0.001 to 0.24) at which more blood flow happens because of creeping event instead of pressure driven flow. In a micro-capillary blood acts like a non-Newtonian fluid which becomes a reason for autonomous association of velocity and shear rate of blood at a high slant angle. In this the latter is taken to be Newtonian in nature. As blood is non-Newtonian, the Quemeda model was found to be more suitable than the other non-Newtonian models. It is also a reason for its being affected by increased slant angle at centerline velocity along with strain rate of blood within the micro-channel.

At times of examination of strain rate modification with increased slant angle while blood inside the capillary is said to be non-Newtonian (Quemeda model), it is learned that for the ball's situation from 5 to 25 mm downstream of the creek, strain curves can be found to initiate with a plateau ensuing by a sigmoidal-shaped curve. The largest reaction of the strain rate at 70° slant angle is ensured by the sigmoidal shape of the curve. The magnitude of the sigmoid curve reduces due to downstream remoteness of the ball. This happens because of the situation of the ball inside the flow development length, the environment could not attain enough momentum to beat the circumference of the ball, which causes high strain rate in the environment surrounding the ball. Again, the opposite of this happens if the ball is far distant from the creek, creating backward flow of the fluid (when the ball is 25 mm downstream) because of advancement in momentum of the fluid in the middle. This can also explain the variation in velocity with raised slant angle of the capillary for various ball positions (Bit and Chattopadhyay 2014).

At fixed 70° slant angle of the capillary, the advancement of the vortex is seen for varying the situation of the ball from 15 mm to 80 mm. There are two causes for which the oscillatory phenomenon of forming a vortex happens. The first is that the fluid gets the momentum with the furtherance of the ball downstream, which causes substitution of boundary layers between the ball and the wall of the capillaries. The second is that division of secondary vortex happens due to fluid pressure head, which initiates abrupt immediate zones of low pressure, letting in the high speed vortex organisation at the bypass of the ball near the wall of the capillary.

The slant angle of the capillary shifts from 20° to 70° through the horizontal axis as the ball is situated 15 mm downstream of the creek of the capillary, here the origination of vortex can be seen, and conversely the primary vortex reduces with increased slant angle. This event takes place as the radial-component of velocity is reinstated with enhancement of slant angle. This event enhances the flow control toward the ranges of the capillary, heading to the further the axis of the capillary and hindering the boundary layer constitution. But when the slant angle is reduced, fluid is going to be alongside the radial axis due to the impact of gravity. This causes formation of un-symmetric types of boundary layers within the circumference of the ball along with the wall of the capillary.

5.8 Conclusion

In this thesis, empirical as well numerical studies have been conducted to understand the influence of a ball's situation and slant angle of the capillary upon blood viscosity. It has been evident that blood acts more pronouncedly as a non-Newtonian fluid, especially when hematocrit constituent of the blood is taken into account. It has been also noticed that blood viscosity is hugely dependent on the above stated differential geometric constants. It leads to development of a new non-Newtonian viscosity model of blood within the micro-capillary. Our study confirms the variability of viscosity reading of blood (non-Newtonian fluid) in a micro-viscometer. Also, interesting features within the viscosity of blood have been obtained in populations with pre-disease, and post-disease conditions. This brings more interesting directions to the investigation.

References

Afzal S, Brøndum-Jacobsen P, Bojesen S E and Nordestgaard B G 2014 Genetically low vitamin D concentrations and increased mortality: Mendelian randomisation analysis in three large cohorts *BMJ* **349** g6330

Ahmadian M T, Firoozbakhsh K and Hasanian M 2012 Simulation of red blood cell motion in microvessels using modified moving particle semi-implicit method *Sci. Iran.* B **19** 113–8

Alizadehrad D, Imai Y, Nakaaki K, Ishikawa T and Yamaguchi T 2012 Quantification of red blood cell deformation at high-hematocrit blood flow in microvessels *J. Biomech.* **45** 2684–9
Bit A and Chattopadhyay H 2014 Numerical investigations of pulsatile flow in stenosed artery *Acta Bioeng. Biomech.* **16** 33–44
Boyd J, Buick J and Green S 2007 Analysis of the Casson and Carreau–Yasuda non-Newtonian blood models in steady and oscillatory flows using the lattice Boltzmann method *Phys. Fluids* **19** 1–14
Casson N 1959 *Rheology of Disperse System* (London: Pergamon)
Charm S and Kurland G 1965 Viscometry of human blood for shear rates of 0–100,000 s^{-1} *Nature* **206** 617–8
Charm S E, Comis W and Kurland M G 1964 Rheology and structure of blood suspension *J. Appl. Physiol.* **19** 127–33
Ellinor P T *et al* 2012 Meta-analysis identifies six new susceptibility loci for atrial fibrillation *Nat. Genetics* **44** 670–7
Guido S and Tomaiuolo G 2009 Microconfined flow behavior of red blood cells *in vitro* *C. R. Phys.* **10** 751–63
Kaliviotis E, Dusting J and Balabani S 2011 Spatial variation of blood viscosity: modeling using shear fields measured by a μPIV based technique *Med. Eng. Phys.* **33** 824–31
Kim Y W and Yoo J Y 2009 Three-dimensional focusing of red blood cells in microchannel flows for bio-sensing applications *Biosens. Bioelectron.* **24** 3677–82
Lima R, Ishikawa T, Imai Y, Takeda M, Wada S and Yamaguchi T 2008 Radial dispersion of red blood cells in blood flowing through glass capillaries: the role of hematocrit and geometry *J. Biomech.* **41** 2188–96
Mauroy B 2013 Following red blood cells in a pulmonary capillary *ESAIM Proc.* **23** 48–65
Misra J C and Ghosh S K 2000 Flow of a Casson fluid in a narrow tube with a side branch *Int. J. Eng. Sci.* **38** 2045–77
Omori T *et al* 2015 Pulmonary resection for lung cancer in patients with idiopathic interstitial pneumonia *Ann. Thoracic Surg.* **100** 954–60
Omori T, Ishikawa T, Imai Y and Yamaguchi T 2014 Hydrodynamic interaction between two red blood cells in simple shear flow: it's impact on the rheology of a semi-dilute suspension *Comput. Mech.* **54** 933–41
Omori T, Imai Y, Yamaguchi T and Ishikawa T 2012 Reorientation of a nonspherical capsule in creeping shear flow *Phys. Rev. Lett.* **108** 138102
Quemada D 1977 Rheology of concentrated disperse systems III. General features of the proposed non-Newtonian model. Comparison with experimental data *Rheol. Acta* **17** 643–53
Richardson L F and Gaunt J A 1927 The deferred approach to the limit. Part I. Single lattice. Part II. Interpenetrating lattices *Philos. Trans. R. Soc.* A **226** 299–361
Roache P J 1994 Perspective: a method for uniform reporting of grid refinement studies *J. Fluids Eng.* **116** 405–10
Secomb W T and Pries A R 2013 Blood viscosity in microvessels: experiment and theory *C. R. Phys.* **14** 470–8
Secomb T M 2011 Mechanics and computational simulation of blood flow in microvessels *Med. Eng. Phys.* **33** 800–4
Ting H J and Khasawneh F T 2010 Glybenclamide: an antidiabetic with *in vivo* antithrombotic activity *Eur. J. Pharmacol.* **649** 249–54

Walawender W P, Chen S and Cala D F 1975 An approximate Casson fluid model for tube flow of blood *Biorheology* **12** 111–24

Xing L *et al* 2014 Alopecia areata is driven by cytotoxic T lymphocytes and is reversed by JAK inhibition *Nat. Med.* **20** 1043–9

Zeng H and Zhao Y 2011 Rheological analysis of non-Newtonian blood flow using a microfluidic device *Sens. Actuat.* **166** 207–13

Chapter 6

Experimental analysis of blood flow in vessels with pathological conditions

Arindam Bit and Jasjit S Suri

This chapter describes the experimental analysis of blood flow in vessels with pathological conditions. It is divided into three segments: firstly, a comprehensive literature survey of experimental analysis of diseased vessels with blood flow dynamics; secondly, an illustration of the conducted experimental techniques; thirdly, representation of the results of these activities.

6.1 Introduction

One of the first experimental studies was performed by Berry *et al* (1931) for investigating the relationship between the pulmonary and bronchial vascular systems. Animal experiments were conducted to show that two thirds of bronchial arterial flow goes into pulmonary circulation. Later, in another experiment carried out by Heimburg (1964), bronchial collateral circulation was computed for a stenosed pulmonary artery. He carried out experiments on eleven mongrel dogs by anaesthetizing them with pentothal (30 mg kg^{-1}). Collateral flow of blood to the lung was mildly enhanced in dogs with stenosed pulmonary arteries. Collateral flow increases with increased degree of stenosis in the right pulmonary artery. Stenosis at such a critical anatomical standpoint leads to development of pathological phenomena in collateral vessels with both stenosed lobes and other healthy lobes. This experiment had the limitation of observing the flow structures inside the vessels, thereby limiting observations to gross structures only.

Talukder (1975) carried out an experiment creating a model of bifurcated geometry of arterial branching using flexible PVC tubing. A Womersley inlet velocity profile was considered, and a flush mounted hot-film probe and a DISA anemometer system are used to measure the flow rate of blood—equivalent flowing through the bifurcating region. High values of oscillatory wall shear stress (WSS) were observed at bifurcation zones, as well as changes in direction of flow observed

doi:10.1088/978-0-7503-2088-7ch6

at the same region, which were considered to be factors regulating the zone of stenosis near bifurcation. However, the study was restricted to the laminar flow regime.

In another study, flow visualisation at the region of bifurcation of the same geometry was made using a high speed camera. While illumination of the aperture zone was made using laser light, laser Doppler anemometer technique was used here for flow visualisation. The order of particle residence time was relatively long, and a correlation between loci of such particles and pathological patterns of atherosclerotic involvement of the sinus area may shed further light on atherogenic factors. Thus, an extensive work on particle path mapping and residence time determination for pulsatile flow was of most importance.

Later on, Davies *et al* (1986) had investigated turbulent fluid shear stress inducing endothelial cell turnover. Contact-inhibited cell monolayers are considered for *in vitro* study. These cells are exposed to regulated flow condition in a modified cone-plate viscometer. A cone angle of 5° generates turbulent flow. Characteristics of shear stress were difficult to quantify. Thereafter, it was found that the flow disturbances influence endothelial turnover and hence destroy the integrity of the endothelial monolayer. Focal atherosclerosis is linked with degraded endothelial monolayers.

Minsky *et al* (1988) had validated experimentally the Doppler echo-cardiographic assessment of volumetric flow through a stenosed aortic valve. In a canine model, the chest was opened and leaflets were sutured, creating variable aortic stenosis. A micromonometer was used to measure the aortic and connected ventricular pressure. A roller pump was calibrated and employed to resend the circulated blood into the atrium while regulating volumetric flow of venous blood. The outflow tract cross-sectional areas calculated from long-axis diameters were nearly identical to areas measured by planimetry of short-axis images. Alignment of short-axis images perpendicular to the outflow tract is more difficult, and misalignment resulted in an oblique tomographic plane; the absence of long-axis reference points results in greater difficulty in acquiring the tomographic plane just below the aortic leaflets.

Subsequently, Zheng *et al* (1993) estimated the errors present in calculation of arterial WSS while using curve-fitting criteria for determining velocity profiles. They considered the fluid to be Newtonian, with a non-elastic wall, and linear accumulation of harmonic components. Quantitative outcomes of the analysis were found independent of shear stress at the flow separation region. WSS assessment was made by using linear and quadratic interpolation equations over the fluid flow function near the wall of the vessel. Shear stress was damped while employing a pulsatile flow function at the region of bifurcation and bends.

Deplano *et al* (1998) used a pulsed Doppler ultrasonic velocimeter for evaluating stenosis influence length in a severely blocked blood vessel. The vessel was made of plexiglass, and experimental flow rate varied between $368 < Re < 984$. Types of inlet flow were found governing velocity profiles and frequency parameters of the experimental fluid. Hence, different flow-types, namely pulsatile flow, physiological flow and oscillatory flow were incorporated in the study. Observation made by the study concluded that the arterial wall submitted to opposite forces due to oscillatory behaviour of WSS, which might make an elastic wall inelastic in nature. This will then lead to the development of an aneurysm at the post-stenotic region.

Development of blood vessels in order to carry out experimental investigation was made with tissue engineering by Musey *et al* (2002). They had used biodegradable membranes of polylactide-co-glycolide (PLAGA) for fabricating a base for cell growth. Bovine aortic endothelial and smooth muscle cells were then cultured, and cell seeding followed. Metabolic breakdown products from PLAGA were found to be lowering the pH of the environment. Finally, it was found that sustained cell colonization of PLAGA copolymers without effective control of localized pH impaired their use as scaffolds.

Talukder (2003) conducted an experiment with hot-film technique. It reveals the development of flow separation, secondary flow, and oscillatory wall shear. Stenosed vessels induced disturbances in flows which were captured using Doppler ultrasound. Increase in depth of stenosis lead to enhancement of magnitude of disturbance index (relative disturbance: RD). Geometry of the flow domain was symmetrically bifurcated.

Arterial pulse wave velocity in coronary arteries was proposed to be correlated to arterial stiffness. Angiographic study was conducted for the assessment of pressure and velocity of blood in the vessel. Two collocated wires with pressure and velocity transducers fitted at their tips were employed for the measurement of the flow parameters in the vessel. Data were acquired along with an ECG signal in a sequence of left arterial descending, circumflex left main stem, and right coronary artery. Forward and backward propagated waves were evaluated at different time-steps by wave-intensity analysis technique. Wave separation in coronary arteries allowed distinguishing the original root of each wave, which determines the characteristic waveform of a blood vessel. Multiple waves were seen in coronary arteries in a complete cardiac cycle. They vary with respect to position of vessel and propagation of cardiac muscle contraction.

Nam *et al* (2009) had conducted experimental studies on carotid arteries of a mouse. They characterised partial carotid ligation for developing a model of disturbed flow with features of low and oscillatory WSS. It was observed that partial ligation causes up-regulation of pro-atherogenic genes, down-regulation of anti-atherogenic genes, rapid atherosclerosis and endothelial dysfunction in a fortnight, and advancement in the lesion in four weeks. Such an observation can be made in the presence of a physiologically relevant model of disturbed flow. Rapid intimal RNA isolation is also possible in the above condition. This approach is effective for genome study. It estimates molecular mechanisms underlying flow development control of vascular biology.

Particle image velocimetry (PIV) method was introduced by Brunette *et al* (2008) for investigating three-dimensional flow in a mildly stenotic coronary artery phantom. Experimental velocity information was formulated to evaluate the 3D shear-stress distribution and secondary flows within the test volume. At an early stage of stenosed pathology, the shear stress accompanied above normal and athero-protective range at an early segment of the stenosed condition. At this stage, secondary flows were also observed. The outcomes hold true to the argument of a vena contracta which is associated with a jet in a coronary blocked stenosis. It results in elevated shear stress in the post- stenotic region.

Talukder *et al* (2009) studied arterial flow disturbance in the presence of a stenosed artery by employing a pulsed Doppler ultrasonic system. A silicone tube equivalent to artery dimensions and fed with various heights of blockages was considered as a test bench in the study. Pattern of blood flow was visualised by video camcoder. RD increases with the increase in depth of stenosis, linearity at lower order stenosis, and logarithmically at higher order stenosis. They claimed it might lead to a new non-invasive method of assessing degree of stenosis in peripheral arteries.

PIV method was employed for analysis of blood flow in a non-axisymmetric stenosed model (Chua *et al* 2009). A long cylindrical vessel with 50% blockage was subjected to steady flow. Time-step analysis of a 3D velocity field and shear stress distribution were made in this study. Emergence of a Womersley flow profile indicated slight pulsatility from the working fluid pump's operation. However, full implementation of 3D PIV technique was not made here.

ST-segment elevated myocardial infarction had also been found to be useful tool for determining lesions in coronary arteries. Quantitative coronary angiography was used to measure the distances from the vessel ostium, major bifurcations, and major curvatures. It was found that patients with ST-segment elevation myocardial infarction (STEMI) are susceptible to frequent lesions immediately distal to bifurcations, also it was found in proximity to major curvatures where disturbed flow was known to occur. Hence the role of WSS in the pathogenesis of STEMI was confirmed. In another study, Chatterjee *et al* (2011) had used coil migration through skin after posterior tibial artery pseudo aneurysm trapping.

Conductance catheter technique was employed for measuring the flow parameters in the lumen of a stenotic artery. Ionic solvent, namely saline solution was injected in the flow field for determining the performance of conductance catheter. Saline fraction and blood temperature were found to govern electrical conductivity in the spatial domain. A measure of the electric field in a stenosed artery was employed for conductance measurement, and for flow recirculation bubble.

In recent years, experimental techniques moved to a new height. Chong *et al* (2014) found that pro-atherogenic shear stress and HIV proteins synergistically up-regulate cathepsin K in endothelial cells. Experiments involved several steps such as preparation of HIV-transgenic mouse model, localization of cathepsin protein expression in the aortic wall, cathepsin and MMP zymography, endothelial cell culture, and shear stress condition reproduction at the site of cell culture using a shear stress bioreactor. The athero-protective and pro-atherogenic waveforms were generated using a cone-and-plate shear system consisting of poly-tetra-fluoro-ethylene cones at an angle of 0.5°.

In another study conducted by Pavlides *et al* (2014), endothelial caveolin-1 was found to play a vital role in the development of atherosclerosis. Experimental investigation was carried out for study of transportation of low density lipoprotein

(LDL) through the arterial wall. Experiments of LDL transport were performed on the isolated rabbit common carotid arteries acquired from fifteen rabbits after 12 weeks of high-fat diet. Energies that give the best binding were computed. Experimental results showed a large uptake for shear stress below 1 dyn cm^{-2}.

In another study, Afara *et al* (2014) had developed a sensor to measure WSS in minor channels with stenosis. WSS and its statistical moments along with its PSD were evaluated at different lengths of stenosis segment. Flow rate was maintained at $100 < Re < 900$. It was observed that there is generation of a significant number of flow disturbance phenomena such as flow separation, and re-attachment due to shifting of flow from the laminar regime to laminar-turbulent regime in the observed vessel. The latest experimental study carried out by Bouillot *et al* (2014) investigated flow analysis in the presence of steady and pulsatile flow in an aneurysmal vessel. Studies were conducted using multi-time lag PIV setup. Uncertainty analysis was performed on the functioning components of PIV in order to ensure high precision of measured velocity field of blood flowing through an intra-cranial (IC) aneurysmal vessel. On increasing the frequency of physiological flow, it was observed that several vortices were generated in the flow field of blood in IC.

After the above survey of experimental investigations, a few lacuna are identified from the survey, such as determination of WSS, oscillatory shear index, relative residual times are absent in most of the studies. Effects of influence length are also absent in any of the experiments. Influence of gravity is also neglected in these studies, which needs special attention, especially for pulmonary blood vessels.

6.2 Methodology

This section describes the overview of the laser Doppler principle, followed by a brief discussion of the laser Doppler velocimetry (LDV) specification and illustrations of its various components with respective settings according to the current experiment. It is followed by a detailed illustration of the construction of the test bench where a pulsatile flow generator was used to observe flow in constricted tubes resembling blocked arteries. Then, complete analysis of the physiological-pulse flow generator is made, and a measure of similarity of pulse produced by the generator is made with actual pulse generated by blood flowing through blood vessels in *in vivo* conditions.

6.2.1 Overview of experimental table

The experimental setup consists of the following parts: laser source, reservoir, pulsatile pumping system, test bench, data acquisition and processing unit.

The LDV set up is supplied by TSI Inc., USA. Here, the laser source used is a gaseous laser of argon—neon components. It is producing a dual beam of green light (540 nm) and blue light (490 nm). A dual beam helps in acquiring 2D flow physics of any event occurring at the field of view of the focusing point of laser beams. Energy of the radiated laser is 18 W, powerful enough to burn tissues. However, operating range has been kept to 500 mW only. Laser beams on production from the coherent lead chamber are fed inside an optical bench of a beam splitter. The laser beam gets split into two phases shifted by 90°. One phase of each beam directs to a transmitter (trans-receiver) to participate in creating the field of view. The other phase of each beam performs Bragg cell shifting, and gets connected to a data acquisition system in order to act as reference signal for the Doppler shift reflected beam of light. Production of a laser beam of high energy creates intense heat inside the vacuum lead chamber. Overshoot of heat may produce hindrance in production of the high energy coherent laser beam. Thus, a cooling pipe circuit is used to maintain the temperature inside the vacuum tube for efficient production of laser light. Cooled water from a temperature controlled chiller absorbes heat produced in the chamber, thus maintaining ambient temperature of 24°C inside the laser production chamber.

The reservoir is made up of polymer. It contains the fluid to be used for the experiment. It has a capacity of 10 l, with a valve at the bottom. However, feedback delivery to the reservoir is done by returning the delivery pipe inside the tube from the top. The fluid used in the experiment is a homogeneous mixture of 60% water, and 40% glycerine. The seeding material made of borosilicate powder (5 micron grain size) is mixed well by continuous stirrer in order to maintain homogeneity of the seeding particle inside the fluid; at the same time it will restrict higher density glycerine from settling down. Seeding material is used to track the trajectory of fluid flowing through the field of view of laser light. The seeding material used is optically sensitive, and consists of fine particles moving with the fluid without disturbing the fluid motion; the laser beams gets Doppler shifted by those particles, and actual information about the fluid flow can be transmitted.

6.2.2 IDV principle

The experiment uses the Doppler effects to calculate the velocity of particles in fluids. Light scattered on moving particles experiences a shift in frequencies according to equation (6.1).

$$f_r = f_b \frac{1 - \frac{\vec{e_b} \cdot \vec{v_P}}{c}}{1 - \frac{\vec{e_{Pr}} \cdot \vec{v_P}}{c}} \approx f_b + \frac{\vec{v_P}}{\lambda_b} \cdot (\vec{e_{Pr}} - \vec{e_b}) \tag{6.1}$$

where f_b and f_r are the frequencies of the beam and the receiver, respectively. $\vec{v_P}$ represents particle velocity, and $\vec{e_b}$ is the unit vector in the direction of the beam. $\vec{e_{Pr}}$ represents the unit vector directed from particle to receiver. λ_b denotes wavelength of the beam. The second formula is derived based on the approximation $|\vec{v_P}| \ll c$. Light is shifted once at the emission point and next at the absorption point. Optical beating phenomena were employed to measure these shifts. According to these phenomena, the angle produced by two incident beams at their point of intersection due to scattering in the presence of particles is measured for the above evaluation. Then, these beams were heterodyned on a detector to fetch an optical beating frequency in measurable length. Equation (6.2) is derived from equation (6.1) and a few geometrical arguments. Equation (6.2) is obtained as

$$f_d = f_2 - f_1 = \frac{2 \cdot \sin\left(\frac{\theta}{2}\right)}{\lambda_b} \cdot |v_{pt}| \tag{6.2}$$

It can be observed that f_d is not influenced by the positioning of the receiver. It provides an advantage while setting up the experiment. The velocity component perpendicular to the optical axis of two beams can be evaluated in this process. However, measurement of such parameters is insignificant as long as the flow of blood remains laminar in a straight vessel.

The intersection point of two fringes produces an interference fringes pattern by coherence of laser light. A fringe pattern is often known as an alternative pattern to Doppler ultrasound frequency. It also illuminates flowing particles in the blood stream. Fringe spacing (d_f) within an LDV system relies on the wavelength of light (λ), and half angle (k). Amplitude variation of fringes results in Gaussian distribution of the intensity profile. The collector (recipient) of the dual beam system can be placed at an arbitrary angle. Identical frequency can be obtained. In addition to that, the signal received from the terminal end varies in its quality and intensity with respect to angle of incident beam.

$$d_f = \frac{\lambda}{2 \sin \kappa} \tag{6.3}$$

$$u_y = f_d d_f = f_d \frac{\lambda}{2 \sin \kappa} \tag{6.4}$$

Thus, the velocity u is evaluated by product of frequency and fringe spacing d_f, as given by equation (6.5).

$$u = d_f f_D \tag{6.5}$$

Behaviour of the seeding particles flowing through working fluid is critically analysed in order to evaluate the frequency response of it with respect to later behaviour. Aerodynamics diameter is a characteristic of a particle which follows the flow of the medium in which it is transported. This diameter is defined as the diameter of a unit density spherical volume which had similar setting velocity to that of seeded particles. It depends on shape, density, and size of seeded particles. Once a step-change in fluid velocity occurs, the corresponding particle velocity will be given by equation (6.6) ($\rho_f \ll \rho_P$):

$$\frac{V_f - V_P}{V_{fi} - V_{Pi}} = e^{-t/\tau} \tag{6.6}$$

where:

V_f = fluid velocity after a step change
V_P = particle velocity at time t.

V_{fi} and V_{Pi} = fluid and particle velocity before the step change, and $\tau = \frac{\rho_P d_P^2}{18\mu}$ = relaxation time of the particle. The frequency response can be calculated from τ by equation (6.7):

$$f = \frac{1}{2\pi\tau}\left[\frac{1}{A_r^2} - 1\right]^{1/2} \tag{6.7}$$

where A_r = particle oscillation amplitude divided by fluid oscillation amplitude.

Optimum density of seeding particles also plays a significant role in two ways. Firstly, density of seeding particles may alter the dynamic behaviour of them from that of working fluid. Secondly, counts of seeding particles in the working fluid at measuring volume alter flow dynamic information of working fluid statistically. Thus, the particles producing high intensity measurable signals are of interest. Small particles produce noise in the signal, and thus they are discarded. At the same time, significantly large particles deviate their path from mainstream flow, and thus, should be avoided. Uniformity in particle size can be measured by mono dispersity. It is expressed by σ_g, as described by equation (6.8):

$$\sigma_g = \frac{d_{84\%}}{d_{50\%}} \tag{6.8}$$

The diameter of 84% and 50% particles are smaller than $d_{84\%}$ and $d_{50\%}$, respectively. In ideal condition, $\sigma_g = 1$, whereas, in mono-disperse condition, $\sigma_g \leqslant 1.2$. The particle concentration in the flow is calculated by equation (6.9):

$$N_P = \frac{Q_S * N_S}{Q_T} \tag{6.9}$$

where:

Q_S = aerosol output of the seeders
N_S = particle concentration of the seeders
Q_T = total flow rate.

Multi-coloured pairs of laser light (green 514.5 nm, and blue 488 nm) are used for 2D measurements of flow dynamics characteristics of working fluid. An unshifted beam, and a frequency shifted beam (of 40 MHz) made a pair. Shifted beams are increasing its frequency by addition of 40 MHz from Bragg cells. Hereby, 40 MHz is added to the Doppler burst. A stationary particle present in measuring volume thus generates a 40 MHz signal. It facilitates the determination of reversal flow and also activates a wide range of frequencies in compressed form. As the working fluid traverses through the cylindrical tube (tangential to the curved surface), measuring volume might not be concurrent to the defined focal point of the laser beam. A shift in focal point of the laser beam is possible depending on various parameters like numerical aperture of the curved surface normal/tilted to laser beam, refractive index of two mediums across the two sides of the curved surface.

The ratio of measuring volume of working fluid (blood flowing through synthetic glass tube) to the surface area of the curved window normal to the laser beams is of order 0.001, and ratio of depth of penetration of measuring volume to the surface area of the curved window normal to the laser beams is of order 0.2. It signifies that the curved surface area normal/tilted to the incident laser beam can be assumed to be an infinitesimally planar surface with respect to the spatio-temporal aspects of measuring volume. Thus, the fundamental equation of Snell's Law in this case is simplified as equation (6.10):

$$N_1 \sin K_1 = N_2 \sin K_2 \tag{6.10}$$

where N_i represents the refractive index of the corresponding medium, and K is the half angle between a laser beam pair.

From equation (6.11), it follows that

$$N_A \sin K_A = N_W \sin K_W = N_F \sin K_F \tag{6.11}$$

where N_A, N_W, N_F represent refractive index of air, window (plane surface), and fluid, respectively. K_A is incident beam angle with respect to optical axis. K_W is beam angle within the window of optical axis, and K_F is beam angle within the fluid to the optical axis.

With the increase in refractive indices, angle of beam with the normal at the interface attenuates. K_F is independent of the window thickness, and it is expressed by equation (6.12):

$$K_F = \sin^{-1}\left[\frac{N_A}{N_F}(\sin K_A)\right] \tag{6.12}$$

For air, $N_A = 1$. Therefore,

$$\sin K_F = \frac{\sin K_A}{N_F} \tag{6.13}$$

The distance between fringes is given by

$$d_f = \frac{\lambda}{2 \sin K} \tag{6.14}$$

λ is the laser-light wavelength, d_f = inter-fringes distance, K is the half angle of intersection beams. The function of medium wavelength is given by equation (6.15) as

$$\lambda_f = \frac{\lambda_A N_A}{N_F} = \frac{\lambda_A}{N_F} \tag{6.15}$$

Therefore, for the new medium (working fluid), distance between the fringes is modified, which is given by equation (6.16) as:

$$d_f = \frac{\lambda_f}{2 \sin K_F} = \frac{\lambda_A / N_F}{2(\sin K_A)/N_F} = \frac{\lambda_A}{2 \sin K_A} \tag{6.16}$$

Now, using the trigonometric principle, the new focal distance F is calculated from two similar triangles by equation (6.17) as:

$$F = F_D \frac{\tan K_A}{\tan K_F} + t\left[1 - \frac{\tan K_W}{\tan K_F}\right] + d_L\left[1 - \frac{\tan K_A}{\tan K_F}\right] \tag{6.17}$$

where F and F_D are actual focal distance and focal distance of lens, respectively. Thickness (t) of the curved surface is made uniformly homogenous. And d_L is the distance from the front of the lens to the window.

In the present experiment, $F_D = 512$ mm, $K_A = 2.795°$, $d_L = 507$ mm, $t = 1$ mm, $N_A = 1.00$, $N_W = 1.49$, and $N_F = 1.384\ 13$. Therefore,

$$K_W = \sin^{-1}\left[\frac{N_A}{N_W}\sin K_A\right] = \sin^{-1}\left[\frac{1.00}{1.49}\sin 2.795\right] = 1.875$$

$$K_F = \sin^{-1}\left[\frac{N_A}{N_F}\sin K_A\right] = \sin^{-1}\left[\frac{1.00}{1.38413}\sin 2.795\right] = 2.019$$

Therefore, the new distance from the lens is

$$\begin{aligned} F &= F_D \frac{\tan K_A}{\tan K_F} + t\left[1 - \frac{\tan K_W}{\tan K_F}\right] + d_L\left[1 - \frac{\tan K_A}{\tan K_F}\right] \\ &= 512\frac{\tan 2.795}{\tan 2.019} + 1\left[1 - \frac{\tan 1.875}{\tan 2.019}\right] + 507\left[1 - \frac{\tan 2.795}{\tan 2.019}\right] \\ &= \mathbf{513.91\ mm} \end{aligned}$$

6.3 Laser head

The laser head houses the plasma tube, the axial field magnet, the resonator structure, various electrical components, and the optical controls and indicators used in the operation of the laser. The Innova 70C series ion laser incorporates a solid, 'L'-shaped anodized aluminium resonator. The cooling water hoses and power cables which extend from the laser head to the power supply are housed inside an umbilical cord.

6.4 Plasma tube

The plasma tube generates the amplification for the laser light. To function effectively the tube must withstand current densities of 700 A cm^{-2} and dissipate heat at a rate of over 200 W per linear centimetre. To dissipate the heat and achieve low temperature operation, the Innova 70C series ion laser plasma tube features a construction, in which a unique tungsten-ceramic combination and an internal gas return enable the tube to handle higher discharge currents while producing greater power output and reliability.

6.5 Power supply

Electrical power to operate the tube is drawn from an incoming three-phase power line, and regulated power supply. A three-phase rectifier and LC ripple filter were employed to generate DC supply to the tube and magnet. A control board of power-supply commissioned fireware controls the laser system. This operating system is accessed by either the remote control module or the optional RS-232C interface. A phone jack for the remote control module is located on the rear panel of the power supply. The head control port is also located on the power supply as is a dip switch for configuration selection. There are three additional ports for computer interfaces.

6.6 Multi-colour beam splitter

The fiber-light multi-colour beam splitter is a unique part which produces a laser beam which is required for 1D, 2D or 3D flow analysis. As the argon ion laser beam is transmitted through an optical fiber splitter, it results in generation of three pairs of beams: green (514 nm), blue (488 nm) and violet (476 nm). Each pair of beams is composed of the native beam and a 40 MHz frequency shifted beam. Frequency down-mixing of the processor provides the facility of a wide range of frequency selection, enhancing the accuracy of the system, for low Re and low density particle size assessment.

6.7 Fiber optic transmitter probe

A fiber optic transmitting probe is one of the segments of the LDV apparatus. It transits the laser beam to the experiment table for formulation of zone of control volume. It also performs assessment of the scattered light from the flowing particles across the control volume. In an LDV system, the primary connection of the fiber optic probe is to the fiber-light multi-colour beam splitter. If the probe is used to connect the light scattered by the particles moving through the probe volume, there is a secondary connection of the probe to the photo detector module (PDM). Light from an argon-ion laser is separated by a fiber-light beam separator into blue, green, and violet beams. The green and blue beams are connected with a probe of fiber optic using fibers which is connected to a fiber-coupler which is mounted on the fiber light separator. At the PDM, the scattered light from the particles is distinguished by colours and transformed into electrical signals. A flow size analyzer (FSA) signal processor performs the Fourier discretization of these received signals, extracting frequency components from the incoming signal for frequency analysis.

6.8 Photo detector module

The PDM contains one to five photomultiplier tubes depending on the configuration for each velocity or phase signal. Velocity channels are labelled 1, 2, and 3. The three-phase-Doppler signals are labelled 1A, 1B, and 1C. In a phase system, channel 1 serves as one of the three-phase channels and as a velocity channel. The PDM also contains high-voltage supplies for the photo multiplier tube (PMT) that range from 0 to 1000 V. Channel 1, 1B, and 1C share one high voltage supply. The high voltages to the PMTs must match each other so that the PMT phase delays match as the high voltage is changed. This is because the phase delay through the PMT is a function of its high voltage setting. The scattered light collected by the probes comes into the PDM, through the back-panel connectors to the PMTs. The signal from each PMT is amplified and then split-one is sent through a high-pass filter and the other is sent to the connectors bayonet Neill–Concelman (BNCs) on the front panel. The signal that is sent through the high-pass filter has the pedestal removed. This signal then comes out through the BNCs, on the back panel of the PDM to go to the FSA.

The high-pass filter can be 5 or 20 MHz and can be selected through the FLOWSIZER software. The signal that does not go through the high-pass filter goes to a high-voltage clamping circuit with the pedestal still present. When light levels to a PMT are very high, this circuit holds the high voltage to the PMT to a level lower than that selected in the software and the light indicator, labelled saturated on the PDM front panel turns on. When the PMT is saturated, the FLOWSIZER displays the actual high voltage. For the phase Doppler system, PMTs 1A, 1B, and 1C share a common high-voltage supply and clamping circuit. Therefore, too much light into any one of these PMTs, clamps the high voltage for the other two as well. The PDM for phase Doppler systems contain an additional module with a calibration laser diode and an intensity measurement circuit. The calibration laser diode is used to calibrate out-phase delay differences between channels 1, 1B, and 1C. The laser creates an artificial Doppler burst that is introduced into the fiber bundle in the phase Doppler receiver probe. The intensity measurement circuit measures the peak height of the pedestal of each burst for channel 1 only. This intensity-based size measurement is used in the software as additional validation for the phase-based size for each droplet. The intensity voltage levels (0–1 V) given in the software, correspond to the voltage levels seen on the channel 1 front panel detector out BNC when connected to a 50-Ω input.

6.9 FSA signal processor

The FSA signal processor receives and processes analog burst signals from the PDM and sends the results to a peripheral computer. The FSA can process from one to five Doppler signals, depending on the signal processor configuration (number of channels). Velocity channels are labeled 1, 2, and 3. The three-phase Doppler signals are labeled 1A, 1B, and 1C.

6.10 Down-mixer

The signal coming from the PDM, with the pedestal removed, goes through a down-mixer. The down-mixer changes the frequency shift that was introduced by the Bragg cell in the fiber-light multicolour beam generator. The down-mix frequency can be any value between 0 and 40 MHz, which means some or all of the 40 MHz frequency shift can be left on the signal. The process of down-mixing is equivalent to multiplying the incoming signal with present down-mixed frequency. Thus, the output signal possesses two frequency components: sum and difference of downmix frequency and incoming signal's frequency. The 'difference' frequency component is the desired signal while the 'sum-term' frequency is not needed and is thus removed by the bandpass filters. After going through the down-mixer, the signal passes through one of the 20 bandpass filters to remove noise and also, if down-mixing was performed, the high-frequency sum-term component. The resulting signal can be observed at the signal out BNC connector on the front panel of the FSA.

6.11 Burst acquisition system

Signal passes through band pass filter (BPF) and splits into two fragments: the first one goes to a burst detector, and the second to a burst sampler. The burst detector identifies and tags the beginning and end of a Doppler burst signal from spontaneous background noise. Discrimination of Doppler signal from background noise was performed by the burst detector with the help of real-time monitoring of the signal-to-noise (SNR) value of the incoming signal. It detects a Doppler burst if the SNR value is greater than a pre-defined threshold value. For the above task, burst signal is first digitized followed by real-time DFT, and then, employing a look-up table. If a signal exceeds the SNR level, the burst gate will start and the timer of the burst detector evaluates the burst-signal duration. The burst gate is stopped at the end of the burst signal, and the burst duration is employed for determining central burst point.

Since the frequency of the signal after going through the bandpass filters is always within a maximum range of 1–10, the burst detector samples the signal with three sampling frequencies so that the DFT lookup tables can be optimized over a small range. The burst detector also has an option for an amplitude-based thresholding for identification of larger amplitude signals with superior SNR. Opening and closing of the burst gate of the burst detector are being regulated by control logic of all these parameters. Real-time value of time-stamp counter and requested input data from the burst gate are stored in the FSA. The data are paired up with other processed burst-source data at a later stage. On-time of the burst gate was also being monitored by the burst detector.

A high-speed, multi-bit AD converter was used for the sampling of the incoming signals. Sampling of the signal is done over multiple sampling frequencies. Sampling frequency information is provided by the burst detector for obtaining the optimum sampling frequency. The signal processor automatically picks the sampling rates based on the bandpass filter range such that there are 10–25 samples per cycle. As with the burst detector, the 1–10 frequency range is divided into three sub-ranges so there are three possible sample rates for each burst. This allows the optimum number of samples per cycle to be used for each burst. The burst detector determines which of the three sub-ranges each burst's frequency is in. When the burst gate comes on, 8-bit samples are stored in the sample memory until the burst gate goes off. The sample memory can store samples for 15 separates bursts. This ensures that a series of bursts arriving close together in time are not missed even though the burst-processing section may be busy. The burst gate selects the optimum region of burst for sample collection. Output of burst gate is convoluted with optimum sampling rate to select the best samples to pass on to the burst processing unit.

The burst processor starts reading samples from memory, and processes the data subsequently as soon as the burst gates goes off. Processing is carried out by four digital signal processing chips for distinct channels. It involves evaluating frequency of a signal using an autocorrelation technique. In a phase Doppler system with phase channels 1A, 1B and 1C, the phase difference between signals A, B and A, C is determined using a cross-correlation technique. Validity of burst is also ensured during burst processing based on the selected SNR threshold. A peripheral computer receives the burst data which passes a validation step. Measure of number of valid bursts per second is recorded and displayed by the flow-sizer. The burst efficiency displayed in the FLOWSIZER is the ratio of valid bursts to its gates. It also gives a measure of the noise of the signal. After processing a burst, it is transmitted to the computer with its features such as frequency, transit time, time stamp, and channel number.

6.12 Photo multiplier tube voltage

The PMT provides voltages to each channel. Voltage gain has a linearly direct relationship with gain of PMT. The gain of PMT amplifies both signal and associated noises. With increase in PMT voltage, smaller particles will emit sufficient signals to be detected by the processor. Hence it will promote data rate. On further increasing voltage, data rate starts reducing due to reduction in SNR (due to amplification of noise).

6.13 Burst threshold

Burst threshold is a significant parameter in an LDV experiment. LDV data can be optimized by adjusting between PMT voltage and burst threshold. Burst threshold value ranges between 30 to 300 mV. The value reduces to 30 mV for smaller particles (<10 μ), and enhances to 100–200 mV for large particles. In the presence of background noise (in the case of near-wall measurement) or in a dense medium, its value enhances above 500 mV, which in turn provides best burst data rate.

The ratio of particle velocity and fringe shift gives rise to signal frequency. An effective shifted frequency of 40 MHz is added to the input signal for estimating the frequency of incoming signal. Proper selection of band pass filter range is necessary to get a good frequency histogram.

6.14 SNR and downmixing frequency

SNR fixes the threshold for validation which is used by FSA's burst sub-system for computing burst which passes through the burst detector. High value SNR allows best quality burst to pass the test, which results in more precise frequency and phase data, but at a very low data rate. On the other hand, a lower value of SNR results in allowance of noisy burst at a high data rate. Setting of SNR value does not influence the burst detector functionality of FSA, but only the burst processing validation. Therefore, an optimum value of SNR needs to be selected for best outcome of the burst detector, and it requires the first few cycles of the burst detection mechanism to understand the characteristics of the received scattered signals. Then, an SNR setting of 'Very Low' is used.

The fiber-light's Bragg cells add 40 MHz to the shifted beam and make the fringes shift at 40 MHz. This results in addition of 40 MHz to the Doppler frequency of each burst. A stationary particle within control volume will therefore generate a 40 MHz signal. It helps in measuring the flow reversal phenomenon, which is associated with all particles which generate <40 MHz signals. Downmixing phenomena can be employed in the FSA to limit or eliminate the signals of stationary particles. Down-mixing subtracts the down-mix frequency from the input signal. Setting 40 MHz will attenuate all 40 MHz Bragg cell shifts, leaving only Doppler frequency which is directly proportional to the velocity of dynamic particles in the control volume. Fringe spacing corresponds to flow-time parameter. Down-mixing usage is optimum for Doppler frequency of 20 MHz or lower. It permits the usage of the narrowest BPF range on the signal at post-down-mixing to remove most of the noise.

6.15 Noise in LDV

Light approaching the photo-detector originates from incident laser light scattered from the signal produced by particles in the control volume. Various other light sources induce reflection and scattering of laser light from the wall, room light, optical components, and radiation from the fluid. Together they form short noise of random signals at the photo-detector. In addition to these, the photo-detector also produces short noise in the presence of dark current in the absence of light. An approach to attenuate light of extraneous nature will be include:

1. Incorporation of high quality and clear optics and a visual platform for test flare;
2. Minimizing direct reflection using masking by black background or by using minimally reflective orientation of optics;
3. Incorporation of a protective aperture on the received optical signal which will act as a spatial filter;
4. Use of optical filter of cut-off frequency analogous to laser frequency. It will limit the entry of scattered light to the photo-detector;
5. Highly reflecting mirror usage for delimiting the scattering of light from the wall to random direction.
6. Experiment to be conducted in a dark room.

6.16 Test bench of blood vessel

This section discuss the development of geometry and physiological pulse generator, Doppler working principle for fluid in motion, and focal length correction of laser light due to curve wall for incidence. The test bench comprises a cylindrical tube of internal diameter = 1 cm, and length = 40 cm, with a stenosis of 4.5 cm length. Optical visibility is ensured by keeping the tube transparent by fabricating it out of an acrylic sheet (of refractive index 1.52). Rigidity of the tube makes it restricted for a compliant wall by following equations (6.18) and (6.19), as used by Bit and Chattopadhyay (2014).

$$y = 0,\ y = 1\ \forall\, x \in [0,7] \cup [11.5,40] \tag{6.18}$$

$$y = 0,\ y = 1 - \left(a - R + \sqrt{R^2 - [b/2 - (x - L_u)]^2}\right) \tag{6.19}$$

where $R = \frac{[a^2 + (b/2)^2]}{2a}\ \forall\, x \in (7,\ 11.5)$ and $a = 0.50,\ b = 4.50$.

The stenosis is defined as the ratio of blockage height (a) to the total height of the vessel (H). Different volumetric flow rate was introduced to the study section by employing three peristaltic pumpings system in parallel connection. Mixing is done for generating physiological pulses. The test bench is shown by the below schematic diagram (figure 6.1).

40% glycerine solution was considered as analogous to blood in the experiment. Density of working fluid was maintained at 1050 kg m^{-3}, and its viscosity was maintained at $2.02 \times 10^{-6}\ m^2\ s^{-1}$ at 32 °C. Borosilicate glass sphere of 10 μ diameter was used as seeding material to make the solution optically sensitive. Doppler shift of the incident laser beam on these particles will provide the information related to the localized trajectory of the flowing fluid.

6.17 Result

In this work, as already mentioned, transport phenomena of blood in a diseased vessel were analysed in the presence of physiologically relevant pulsatile pressure of translumen flowing blood. Blood was also considered to be non-Newtonian in nature, and the geometric constraints were maintained similar to a patient-specific model. A series of experimental calculations were performed using LDV and a customized work-bench. Results are presented in order to show the effects of rheological property, inlet pulsation, and severity of vascular diseases on these vessels. Results obtained are mainly focused on WSS parameters and its harmonic and gradient form in the post-stenotic downstream section of the vessel, the

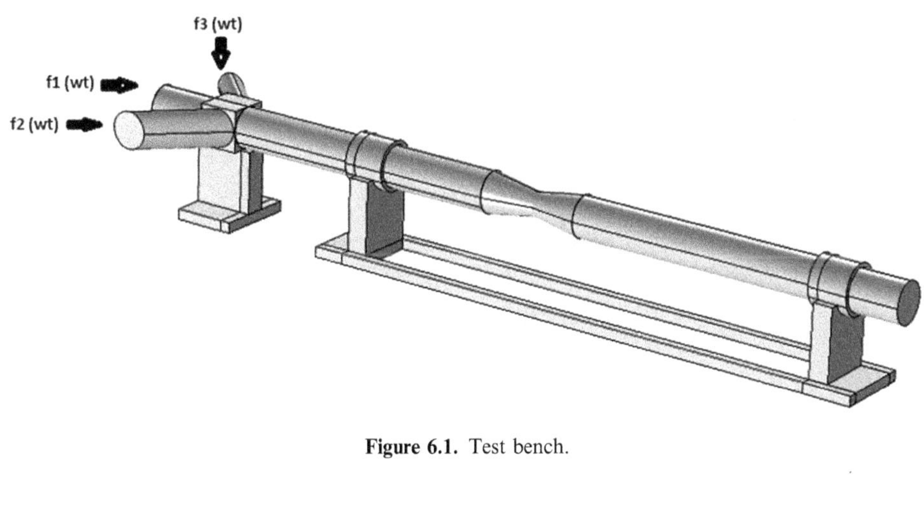

Figure 6.1. Test bench.

oscillatory-shear-index and relative residual time of oscillatory flow and further general discussion. Furthermore, a comparative assessment is also made for severity of secondary disorder in blood vessels with stenosis and aneurysm, respectively. Computational results obtained with the pulsatile flow at different Reynolds numbers are compared simultaneously with results obtained from experimental study. Uncertainty analyses of experimental parameters are made to check accuracy of results obtained through experiments.

The primary data of the temporal domain obtained from the receiver burst signal were used to evaluate shear stress, velocity distribution, pressure profiles and vorticity distribution in the workbench section. Two higher order flow parameters (time-averaged wall shear stress (TAWSS) and oscillatory-shear-index (OSI)) are of significant importance and of clinical relevance to the diseased vessel (Tan *et al* 2008). Derivation of these higher order parameters from primary parameters is given by equation (6.20)

$$OSI = \frac{1}{2}\left(1 - \frac{\tau_{\text{mean}}}{\tau_{\text{mag}}}\right) \tag{6.20}$$

where τ_{mean} is transient mean WSS, and τ_{mag} is the time-mean magnitude of WSS, and they are evaluated by equations (6.21) and (6.22), respectively.

$$\tau_{\text{mean}} = \left| \frac{1}{T}\int_0^T \tau_w dt \right| \tag{6.21}$$

$$\tau_{\text{mag}} = \frac{1}{T}\int_0^T |\tau_w| dt \tag{6.22}$$

A low value of TAWSS (lower than 0.4 Pa) stimulates pro-atherogenic endothelial phenotype (Huang *et al* 2013), and perturbed endothelial alignment on the wall of the vessel, and it is localized in regions associated with frequent WSS derivatives within a complete cardiac cycle. These locations are prone to high OSI (Himburg *et al* 2004). OSI varies between 0 to 0.5, where 0.5 represents highest oscillatory behaviour of the stress. Regions exposed to high OSI are prone to endothelial dysfunction and atherogenesis which makes them clinically the most critical zones. relative residual time (RRT) is derived by combining OSI and TAWSS in a function as given by equation (6.23).

$$RRT = \frac{1}{(1 - 2 \times \text{OSI}) \times \text{TAWSS}} \tag{6.23}$$

Several studies (He *et al* 1996, Lee *et al* 2009, Morbiducci *et al* 2010) recommended RRT as a single metric of low and oscillating shear. 128 channel data acquisition system was used with LDV to obtain transient information of all flow dynamics parameters (like real-time velocity and its mean and average, velocity histogram, skewness and flatness of velocity, turbulent intensity of flow) of the fluid. Blockages of the tube were varied between 12.5%–50% and the influence of severity of blockages on shear stress near the wall, stenosis influence length, pressure-drop coefficient and pressure recovery factors were evaluated along the downstream of the blockage.

6.17.1 Analysis of stenosis influence length

Influence length of stenosis is defined as the maximum downstream length of the post-stenotic segment until the magnitude of velocity of working fluid does not deviate more than 2% of Womersley's velocity profile at the inlet of the tube. This length can evaluate by the following steps.

Let at a given instant t, the volume fraction of fluid evolved through the stenosed section of stenosis reach the axial location $\bar{l} = l/R_0$ in such a way that

$$\int_{t_0}^{t} Q(t)dt \leqslant V_{sl} \tag{6.24}$$

V_{sl} is the confined volume between height of stenosed section and axial location l. In this study $Q(t) = u_{\exp t}(t)$ as defined by equation (6.10). Simultaneously, on considering the equal sign in equation (6.24), and by fixing a lower bound on l, V_{sl} can be expressed by equation (6.25) as:

$$\int_0^l \pi R(z)^2 dz = \frac{\varepsilon}{\omega_{ei}}(\omega_{ei}t - \omega_{ei}t_0) + \frac{F_{ijP}}{\omega_{eij}}\{(-\cos(\omega_{ei}t) + \cos(\omega_{ei}t_0)) + (\sin(\omega_{ei}t) - \sin(\omega_{ei}t_0))\} \tag{6.25}$$

A comparative study was conducted between transient velocity profiles obtained from experiment at different radial lengths of the tube with the theoretical profiles obtained at respective positions of the tube. Theoretical velocity profiles imply velocity information obtained from 4D Phase Contrast MRI data and being used as inlet-boundary conditions for conducted numerical studies. Figure 6.2 shows

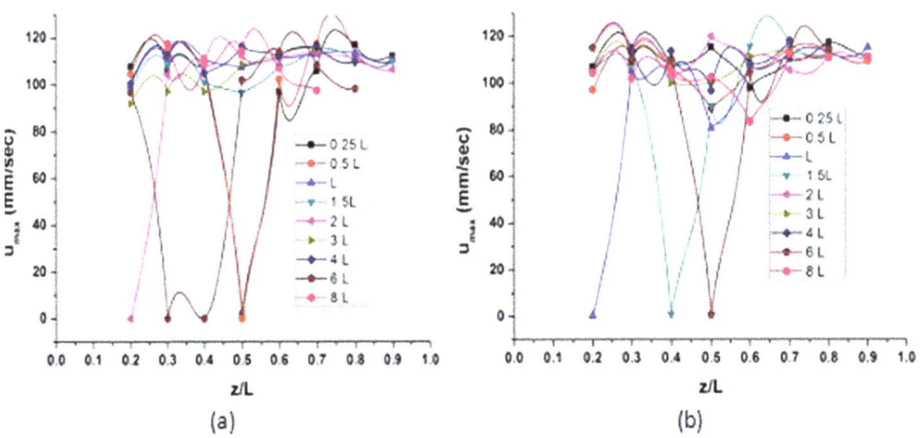

Figure 6.2. Maximum axial velocity(mean) plot at different length scale of the tube with (a) 25% blockage, and (b) 50% blockage.

significant overlapping of experimental and theoretical waveforms at different depths of cross section (center, 23% away from center and 46% away from center) of the vessel. It confirms closeness of results obtained from experimental to numerical results (Bit *et al* 2017).

Another comparative study shows the mean axial velocity between vessels with different blockages of 12.5% and 25%, respectively. Results are obtained at different length scales of the tube. Mean velocity varies between 7.5 mm s^{-1} – 9.0 mm s^{-1} and 6 mm s^{-1}–8.5 mm s^{-1} at a distance of 2 cm pre-stenosis in vessels with 12.5% blockage and 25% blockage, respectively. However, greater variations in magnitude of mean value of velocity along the radius of the tube are observed in the case of 25% blockage. Therefore, severity of blockage confirms the induction of disturbances to flow in upstream at a distance of 2 cm. On approaching closer to the region of blockage (1 cm), dynamic variations of mean velocity decrease with severity of the blockage. Surprisingly, pulsating flow developed in vessel with 25% blockage at 0.5 cm pre-stenoses region, and it remain absent in vessel with 12.5% blockage. Dynamic variations of mean velocity of fluid (7.5 mm s^{-1}–9.0 mm s^{-1}) are observed in a vessel with 12.5% blockage, in contrast, a vessel with 25% blockage shows less variation in mean velocity (8.5 mm s^{-1}–9 mm s^{-1}) at a close proximity of 0.5 cm downstream from the blockage. Flow becomes laminar further downstream. However, disturbance persists further downstream (up to 1.5 cm) of the blockage in a vessel with 25% blockage.

Variations of mean axial velocity at different lengths of the vessel at different time steps in the case of 25% stenosis were evaluated. A wide fluctuation in velocity along the z-direction is observed 1 cm upstream of the blockage. Two non-symmetric peaks are observed from the velocity profile across the cross-section of the tube. It indicates formation of vortices near one wall. A reverse velocity profile is observed in the opposite wall of the tube at the same point. Therefore, flow disturbances are enhanced 1 cm upstream from the blockage. Immediately pre-stenosis, 0.5 cm upstream from the blockage, the flow-field becomes highly oscillating, whereas at a point immediately downstream (0.5 cm post-stenosis) sizes of vortices are greater than those 1 cm upstream from the stenosis.

Dynamic variations in flow also get arrested in the immediate downstream in contrast to upstream flow. The structure of vortices persists further downstream (1 cm and 3 cm post-stenosis), and dynamic variations of velocity of fluid are also regained further downstream. Flow profiles also become laminar further downstream. However, the laminar regime of flow is limited in space across the cross-section of the channel far downstream. It is also noted that laminar behavior of the flow profile is achieved with respect to progress in time steps. It can be concluded that disturbance in flow is solely due to severity of the blockage.

Figure 6.2 shows the variations of maximum axial velocity at different lengths of the vessel at 25% stenosis and 50% stenosis. Maximum velocity (mean) is found around 100 mm s^{-1} in both the vessels. It was also sustained at such high value for the entire length downstream from the blockage in both cases. A few under-shoot plots of maximum velocity are also observed, which can be treated as independent of flow behavior.

However, at a length of 0.25 L downstream from the stenosis, flatness of velocity is minimum. It indicates the re-development of flow after blockages. Higher peaks of flatness of velocity at 3 L and 6 L across the cross-section of the vessel with 25% blockage indicate the probability of development of localized vortices. A similar trend is also observed in a vessel with 50% blockage at a length of 0.5 L downstream from the stenosis.

6.18 Uncertainty analysis

6.18.1 Shifted frequency of reflected light at the receiver (f_r)

Shifted frequency of reflected light at the receiver is re-written by equation (6.26)

$$f_r = f_b + \frac{v_p}{\lambda_b}(e_{pr} - e_b) \tag{6.26}$$

f_b is the frequency of beam at the receiver; v_p is velocity of the particle; λ_b is wavelength of the beam; e_b is unit vector in the direction of the beam; e_{pr} is unit vector towards the receiver from the plane of the particle to the receiver. Therefore, the uncertainty in a shifted-frequency light wave at the receiver can be calculated from equation (6.27).

$$\begin{aligned}\frac{\Delta f_r}{f_r} &= \frac{1}{f_r}\left[\left\{\frac{\partial f_r}{\partial(f_b)}\cdot\Delta f_b\right\}^2 + \left\{\frac{\partial f_r}{\partial(v_p)}\cdot\Delta v_p\right\}^2\right.\\ &\left.\quad + \left\{\frac{\partial f_r}{\partial(e_{pr})}\cdot\Delta e_{pr}\right\}^2 + \left\{\frac{\partial f_r}{\partial(e_b)}\cdot\Delta e_b\right\}^2\right]^{0.5}\\ \equiv \frac{\Delta f_r}{f_r} &= \frac{1}{f_r}\left[\left\{\frac{\Delta f_b}{(f_b)}\right\}^2 + \left\{\frac{\Delta v_p}{(v_p)}\right\}^2 + \left\{\frac{\Delta e_{pr}}{(e_{pr})}\right\}^2 + \left\{\frac{\Delta e_b}{(e_b)}\right\}^2\right]^{0.5}\end{aligned} \tag{6.27}$$

6.19 Conclusion

Experimental analyses were made in three-dimensional optical sensitive cylindrical closed channels containing blockages of various severities. An axi-symmetric blood vessel containing stenoseis was made of plexi-glass, and was kept transparent. LDV technique was used to measure flow behavior of analogous blood flowing through the vessel. Depth of stenosis was varied between 12.5% and 50%. Various parameters like mean velocity, rms velocity, turbulent intensity, velocity flatness, velocity skewness, maximum velocity and minimum velocity over every cardiac cycle were measured across the cross-section of the vessel along the axial length of the vessel. Several derivative parameters like WSS, stenoses influence length, pressure drop coefficient, and pressure recovery factors were calculated, and the results were successfully validated with the results obtained from numerical investigation of a stenosed blood vessel in two-dimensional form. A pulsatile flow generator was found to be a better approximation for evaluating flow behavior of blood in an experimental test phantom.

References

Afara S, Medvescek J, Mydlarski L, Baliga B R and MacDonald M 2014 Development of a wall-shear-stress sensor and measurements in mini-channels with partial blockage *Exp. Fluids* **55** 1–10

Berry J L and Daly I D B 1931 The relation between the pulmonary and bronchial vascular systems *Proc. R. Soc. Lond. B* **109** 319–36

Bit A, Ghagare D, Rizvanov A A and Chattopadhyay H 2017 Assessment of influences of stenoses in right carotid artery on left carotid artery using wall stress marker *BioMed Res. Int.* **2017** 2935195

Bit A and Chattopadhyay H 2014 Numerical investigations of pulsatile flow in stenosed artery *Acta Bioeng. Biomechan.* **16** 33–44

Bit A and Chattopadhyay H 2014 Assessment of rheological models for prediction of transport phenomena in stenosed artery *Prog. Comput. Fluid Dynamics Int. J.* **14** 363–74

Bouillot P, Brina O, Ouared R, Lovblad K O, Farhat M and Pereira V M 2014 Particle imaging velocimetry evaluation of intracranial stents in sidewall aneurysm: hemodynamic transition related to the stent design *PLoS One* **9** e113762

Brunette J, Mongrain, Laurier J, Galaz R and Tardif J C 2008 3D flow study in a mildly stenotic coronary artery phantom using a whole volume PIV method *Med. Eng. Phys.* **30** 1193–200

Chatterjee M S 2011 Systems biology of blood coagulation and platelet activation *Publicly Accessible Penn Dissertations* 348 http://repository.upenn.edu/edissertations/348

Chong W, Zhang Y, Qian Y, Lai L, Parker G and Mitchell K 2014 Computational hemodynamics analysis of intracranial aneurysms treated with flow diverters: correlation with clinical outcomes *Am. J. Neuroradiol.* **35** 136–42

Chua C S, Sheard G J, Higgins S, Jamison R A, Fouras A and Hourigan K 2009 Particle image velocimetry of non-axisymmetric stenosis models *8th Int. Symp. on Particle Image Velocimetry (Australia)*

Davies P F, Remuzzi A, Gordon E J, Dewey C F and Gimbrone M A 1986 Turbulent fluid shear stress induces vascular endothelial cell turnover *in vitro Proc. Natl Acad. Sci. USA* **83** 2114–7

Deplano V and Siouffi M 1998 Wall shear stress analysis of pulsatile flows through stenosis *J. Biomech.* **1001** 19

Heimburg P 1964 Bronchial collateral circulation in experimental stenosis of the pulmonary artery *Thorax* **19** 306

Himburg H A, Grzybowski D M, Hazel A, LaMack J A, Li X M and Friedman M H 2004 Spatial comparison between wall shear stress measures and porcine arterial endothelial permeability *Am. J. Physiol. Heart Circul.* **286** 1916–22

Huang C, Chai Z and Shi B 2013 Non-Newtonian effects on hemodynamic characteristics of blood flow in stented cerebral aneurysm *Commun. Comput. Phys.* **13** 916–28

Lee S W, Antiga L and Steinman D A 2009 Correlations among indicators of disturbed flow at the normal carotid bifurcation *ASME J. Biomech. Eng.* **131** 7

Minsky B D, Mies C, Recht A, Rich T A and Chaffey J T 1988 Resectable adenocarcinoma of the rectosigmoid and rectum. II. The influence of blood vessel invasion *Cancer* **61** 1417–24

Morbiducci U, Gallo D, Ponzini R, Massai D, Antiga L, Redaelli A and Montevecchi F M 2010 Quantitative analysis of bulk flow in image-based haemodynamic models of the carotid bifurcation: the influence of outflow conditions as test case *Ann. Biomed. Eng.* **38** 3688–705

Musey P I, Ibim S M and Talukder N K 2002 Development of artificial blood vessels: seeding and proliferation characteristics of endothelial and smooth muscle cells on biodegradable membranes *Ann. NY Acad. Sci.* **961** 279–83

Nam K D, Ishihara A, Matsuzawa K, Mitsushima S and Ota K I 2009 Partially oxidized niobium carbonitride as non-platinum cathode for PEFC Electrochem *Solid State Lett.* **12** B158

Pavlides S, Gutierrez-Pajares J L, Iturrieta J, Lisanti M P and Frank P G 2014 Endothelial caveolin-1 plays a major role in the development of atherosclerosis *Cell Tissue Res.* **356** 147–57

Talukder G 1975 The effects of some drugs of the phenothiazide group on human chromosomes *in vitro Indian J. Med. Res.* **63** 457

Talukder N K, Ing D, Shujaee K A and Mohammadi S 2009 Model studies on arterial flow disturbance related to severity of arterial stenosis *Int. J. Cardiovasc. Med. Sci.* **6** 33–7

Talukder N K 2003 Flow characteristics related to arterial diseases *Proc. Int. Conf. on Mechanical Engineering (Dhaka)*

Tan F P P, Soloperto G, Bashford S, Wood N B, Thom S, Hughes A and Xu X Y 2008 Analysis of flow disturbance in a stenosed carotid artery bifurcation using two-equation transitional and turbulence models *J. Biomech. Eng.* **130** 061008-1–11

Zheng W and Worek W M 1993 Numerical simulation of combined heat and mass transfer processes in a rotary dehumidifier *Num. Heat Transfer* A **23** 211–32

Chapter 7

Biomaterials for a synthetic and tissue engineered blood vessel

Arindam Bit, Jasjit S Suri and Sharda Gupta

Organs of cardiovascular networks are critical in their physical characteristics due to extensive shear-stress bearing capacity of these organs. Hence it is crucial to consider the proper materials for tissue reconstruction of these organs from a tissue engineering perspective. In this chapter, a systemic representation of the selection of specific biomaterials for the synthesis of blood vessels has been executed. It is a trivial task to correlate the microscopic histological features of a blood vessel or its segment while re-engineering the same in *in vitro* conditions. One can approach the most significantly similar properties provided the microscopic physical and mechanical features of the native materials have been correlated to make hybrid biomaterials. In this chapter, such development is showcased in subsequent sections.

7.1 Introduction

Cardiovascular diseases like malfunction of the coronary artery condition or myocardium seem to be the most common cause for death worldwide. A therapeutic approach to address those diseases is the use of a healthy autologous vascular graft. However, the only limitation is its availability. The appropriate solution for overcoming this challenge is to develop synthetic and tissue-engineered blood vessels. The demand for tissue-engineered blood vessels has driven scientists to discover new biomaterial as a vascular graft to meet the demand of such a large number of patients. The new biomaterial was evaluated just as a vascular graft material. The challenges in preparation of an artificial blood vessel are to fabricate a scaffold of appropriate biomaterial that can allow vascular cell growth, proliferation, and attachment. It must be able to integrate with host tissue and must have good strength.

The artificial vascular graft has become an attractive alternative to damaged blood vessel replacement; however, proper research must be done for determining

doi:10.1088/978-0-7503-2088-7ch7

which biomaterial would be optimal. Inappropriate biomaterial selection has been associated with graft failure. For the determination of best-suited biomaterial for vascular replacement, its hemodynamics and physico-chemical properties must be taken into consideration. The biomaterials employed for artificial vascular graft must have excellent mechanical properties. Vascular cells also adapt as per mechanical stimuli and remodel themselves by changing their properties, structure, and geometry.

The process behind the degradation of biomaterials *in vivo* can be attributed to immune cells penetrating scaffold, and especially to phagocytes (neutrophils along with macrophages). The macrophages are considered important in biomaterial degradation as well as tissue repair in the early stage of vascular regeneration. The process of formation of neo-tissue and degradation of vascular graft are dependent on macrophages. On scaffold, macrophages release hydrolytic enzymes as well as reactive oxygen species (ROS) resulting in enzymatic and oxidative biomaterial degradation, respectively (Wissing *et al* 2019). Due to changing hemodynamic load, macrophages' response is influenced in an undetermined manner. However, researchers have tried to regulate macrophage polarization by continuous generation of crucial signaling molecule nitric oxide (NO) in the vascular graft and also incorporating Tyr-Ile-Gly-Ser-Arg (YIGSR) with polyethylene glycol (PEG) polymer in order to improve its organization as well as speed of endothelialization (Tang *et al* 2018).

The preparation of functional blood vessels involves the mechanisms of vasculogenesis, arteriogenesis and angiogenesis. For the preparation researchers tried to mimic these mechanisms *in vitro* to recapitulate the native blood vessel, for which the selection of biomaterials that promote vasculature is necessary. The vascular biomaterials should mimic the necessary physical and chemical signals for development or recovery of the damaged vascular constructs. The characteristics like viscoelasticity, stiffness, degradation, density and integrin-binding motifs must be considered to create a fatigue-resistant long-term vascular construct (Crosby and Zoldan 2019). As per the authors' conclusion, the lower the ECM protein concentration, the more the multiplication will be, as well as relocation of endothelial cells to formulate a vascular network. They also explained that with greater matrix density, pore size will decrease resulting in an increase of diffusion time hence limiting the angiogenic growth factor penetration depth.

The native human blood vessel is made of different layers: tunica intima, tunica adventitia, along with tunica media. The exterior most layer consists of fibroblasts, the middle layer has smooth muscle cells along with endothelial cells in the lumen. Depending upon the site of the vessel, vessels have different architecture and compositions as per the mechanical stress applied. The walls of arteries are thick and elastic as they transport blood under high pulsatile pressure. Hence, the fabricated vascular graft essentially contains mechanical attributes similar to patients' indigenous blood vessels for having improved patency rates. The relation between strain to intraluminal pressure in an open-ended vessel is defined by an elastic modulus E_{p} as given in equation (7.1) (Tai *et al* 1999).

$$E_p = \frac{P_s - P_d}{\text{Strain}} \tag{7.1}$$

The strain is defined as the change in pulsatile diameter when an artery is exposed to changes in intraluminal pressure and is given by equation (7.2):

$$\text{Strain} = \frac{D_s - D_d}{D_d} \tag{7.2}$$

where D_s—vessel diameter during systole;
D_d—vessel diameter during diastole;
P_s—vessel pressure during systole;
P_d—vessel pressure during diastole.

The arterial wall thickness index (β) describing the elasticity of an artery is given as in equation (7.3):

$$\beta = \frac{\log_e\left(\frac{P_s}{P_d}\right) X\ D_d}{D_s - D_d} \tag{7.3}$$

The above-mentioned equations are used to calculate the vessel wall properties.

7.2 Blood–biomaterial interaction

When biomaterials appear to be in relationship with blood, plasma proteins get adsorbed rapidly into the surface, which is followed by platelet adhesion and activation. With activation of platelets, coagulation process starts resulting in clot formation. The various factors that affect the biocompatibility of biomaterials are discussed below and shown in figure 7.1 (Wang *et al* 2004):

(i) Wettability of material surface: a hydrophobic surface adsorbs more proteins, hence researchers suggest making the surface of the biomaterial hydrophilic. There should be a balance between hydrophobic and hydrophilic properties at the biomaterial surface.
(ii) Presence of a functional group: the presence of a functional group on the polymer also helps in determination of biological activity of biomaterials.
(iii) Charge density in the biomaterial: the anionic surface of polymers absorbs less protein than the cationic surface as protein possesses a negative charge, hence poly-anions and neutral polymer are less cytotoxic as compared to poly-cation polymer.
(iv) Molecular mass of the biomaterial: a polymer having larger molecular mass shows less protein absorption and platelet adhesion.

As all types of vascular biomaterials come in contact with blood, one most important factor to be considered is the thrombosis formation by adsorption of proteins on the surface of a vascular biomaterial through intrinsic or extrinsic pathways. The intrinsic pathway starts when blood comes into touch with an

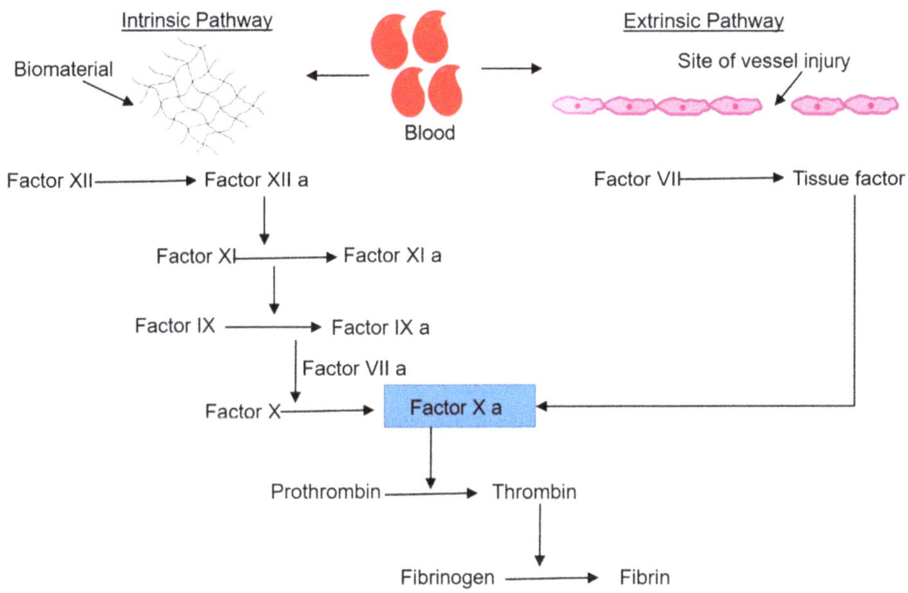

Figure 7.1. Intrinsic along with extrinsic pathways of blood coagulation followed during contact with biomaterials and vessel injury, respectively.

external material (say biomaterials) and extrinsic or exterior pathways start through a place of wound in a blood vessel. The intrinsic or interior pathway begins with converting factor XII (Hagemen factor) to factor XIIa. Factor XIIa changes factor XI (plasma thromboplastin) to Xia and makes factor IXa (Christmas factor). This IXa factor together with factor VIIa (stable) converts factor X (Stuart-power) to Xa. Later Xa transforms prothrombin to stimulated thrombin that acts with fibrinogen and fibrin is formed. This fibrin finally settles as red thrombin or coagulate. The extrinsic route begins from the discharge of tissue factor (collagen) with Von Willerbrand (Vwf) through the injured blood vessel. During this pathway, collagen acts as cofactor and activates factor X. In another way, factor VII will then stimulate factor IX, later this will stimulate factor X.

The non-thrombogenic surface on a vascular graft can be achieved by the formation of endothelial cell cladding upon the blood-contacting facet (i.e. lumen's surface) of the graft. As endothelial cells release the inhibitor of platelet aggregation (i.e. prostacyclin) and hence reduce platelet adhesion, this helps improve patency after endothelial layer establishment and also prevents thrombosis *in vivo* (Walluscheck *et al* 1996, Weinberg and Bell 1986). It is very crucial to take into account the compatibility of blood constituents with the biomaterials used. The biomaterials can be recognized as foreign by blood, resulting in the start of the thrombosis process followed by the thromboemboli formation. The factors such as surface structure (crystallinity/amorphous), surface composition (polar/acidic, acidic/basic), surface porosity, surface roughness/smoothness or surface mechanical compliance of biomaterials affects the blood material interaction. The researchers tried to enhance graft endothelialization and hence increase patency by the process of transplanting endothelial cells onto an artificially fabricated vascular graft. After seeding an endothelial cell upon the exterior layer of a vascular graft, an endothelial cell lining and associated basement membrane (neo-intima) will develop on the surface of the graft and overcome the factors responsible for graft failure. The endothelial lining produces a marker for vascular endothelium i.e. von Willebrand's factor and leads to formation of a permeability barrier (Weinberg and Bell 1986). This endothelial lining will also control leukocyte adhesion and platelet activation. As the endothelial cells possess an anti-thrombogenic property, hence proper technique must be used for the endothelization of a vascular graft. Some of the strategies are endothelial cells seeding on a prepared artificial vascular graft before implantation or by surface modification of the vascular graft to enhance attachment of the endothelial cell after implantation.

7.3 Synthetic polymer

The first arterial reconstruction using cylindrical arterial prostheses made up of fine mesh synthetic fiber fabrics (Vinyon-N) started the evolution of preparing vascular constructs using fabrics (Pourdeyhimi 1986, Voorhees *et al* 1952). Voorhees observed that when this fabric was implanted on the right ventricle of a dog, endothelial cells start proliferating on it. Later, the biological response (in terms of biodegradation, mechanical strength) of a variety of synthetic polymers was evaluated. Among them, polyethlene terephthalate (Dacron) and polytetrafluoroethylene (Teflon) were found to be suitable for the preparation of the vascular graft. The reason behind the use of this synthetic polymer is its capability to withstand mechanical load during the *in vivo* condition, porosity, flexibility, and degree of its biodegradation (Pourdeyhimi 1986). These two polymers were successfully used as a graft in the case of large (>10 mm internal diameter) and middle-size (6–10 mm internal diameter) blood vessels and were incompatible in a small-sized (<6 mm internal diameter) blood vessel due to compliance mismatch with a native blood vessel or thrombosis formation.

Dacron: Polyethylene terephthalate (PET) also known as Dacron was implanted as a vascular graft in 1957 by Julain. A Dacron graft can be bifurcated knitted Dacron graft, bifurcated woven Dacron graft or segmental composite bifurcated woven Dacron graft. Researchers evaluated the long-term performance of Dacron prosthesis as an arterial substitute in terms of late patency rate, histologic behavior in tissue specimen, incidence of healing complications and histologic behavior in serial angiograms. They also stated that factors responsible for improper arteriogenesis leading to abnormal incorporation of prostheses are infection in the adjacent tissue and uninfected tissue environment (due to exudation of blood, serum or lymph) (Szilagyi *et al* 1965). Stump and co-authors observed growth of endothelial cell with nuclei in implantation of knitted, crimped Dacron prostheses sutured with polyethylene suture (Stump *et al* 1963). Schmidt *et al* found that Dacron graft seeded with endothelial cell achieved highest patency as compared to non-seeded Dacron graft and even with cell seeded and non-seeded polytetrafluoroethylene(PTFE) graft (figure 7.2).

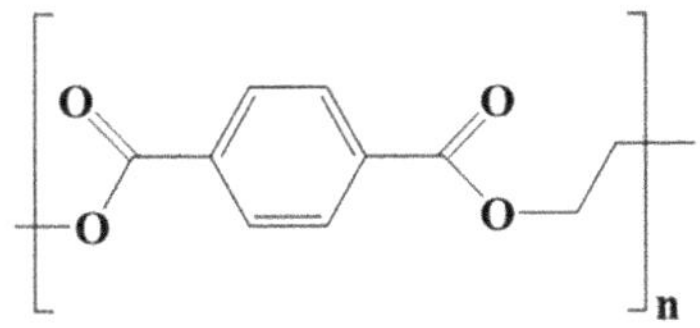

Figure 7.2. Chemical structure of Dacron.

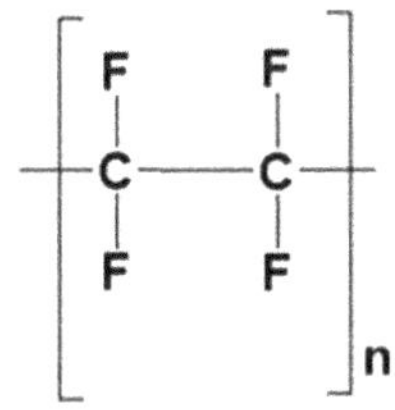

Figure 7.3. Chemical structure of polytetrafluoroethylene.

Expanded polytetrafluoroethylene (ePTFE): Expanded polytetrafluoroethylene is the most inert polymer and has a non-thrombogenic electro-negative surface which is not degradable. But the tissue in-growth and neointima growth was poor when implanted in humans (Ariyoshi *et al* 1997). The graft can clog after long-term usage as it possesses limited graft maturation by endothelialization. The intrinsic hydrophobicity of ePTFE limits endothelium formation. Also, the synthetic ePTFE graft is more prone to bacterial infection, increasing the sepsis risk and even risk of intimal hyperplasia and graft failure due to stenosis (Kirkton *et al* 2018). As the adhesion of endothelial cells on ePTFE is very poor, different pretreatment methods are taken into consideration to inhibit protein and platelet adsorption (Kim *et al* 2019). Walluscheck *et al* pre-coated ePTFE graft with RGD-tri-peptide sequence (arginine-glycine-aspartate). This graft was evaluated for shear stress resistance and endothelial cell retention using an artificial pulsatile flow circuit (Walluscheck *et al* 1996). Hoshi *et al* incorporated heparin (anticoagulant) over-poly(1,8-octanediol-co-citrate) (POC) modified ePTFE vascular substitute. This graft allows endothelial cell adhesion, shows expression of endothelial cell marker von Willebrand's factor along with vascular endothelial cadherin (Hoshi *et al* 2013). Bastijanic coated luminal surface of ePTFE with cell adhesive fluoro-surfactant polymer (FSP). They then degassed pores by hydration with ethanol and finally incorporated poly(ethylene glycol) supported hydrogel in graft intervals to allow integration of smooth muscle cells with walls of graft. The prepared graft solves the problem of endothelialization and intimal hyperplasia allowing fabrication of an off-the-shelf vascular graft (Bastijanic *et al* 2016). Substance P (SP) being a neuro-peptide has an immune-modulatory effect and induces different growth factors and cytokines. The SP and heparin were covalently conjoined through an elastic biocompatible polymer PLCL and were vacuum coated on a ePTFE vascular graft (figure 7.3). This vascular graft when implanted shows enhanced angiogenesis and long-term patency (Kim *et al* 2019). Even though ePTFE served as a vascular prosthesis, it does not lead to any immediate harm to the patient's body but the stenosis by thromboembolic and intimal hyperplasia restricted their use as a large diameter vascular graft in clinical applications.

Polyurethane: Polyurethane (PU) is an elastic biocompatible polymer with excellent tensile strength, appreciable elongation and consists of soft and hard segments. Guo *et al* fabricated a small diameter vascular graft by electrospinning and pulling-out method using PCL/PU as biomaterial. This graft proved to be appropriate to be utilized like a small diameter vascular graft after evaluation of its cell compatibility, mechanical properties, porosity and cyclic mechanical durability (Guo *et al* 2015). Bergmeister prepared a thermoplastic polyurethane polymer with a biodegradable chain extender and having properties much like PU (figure 7.4). This graft possesses enhanced compliance with mechanical stability and long term patency, low intimal hyperplasia formation (Bergmeister *et al* 2015).

Polydioxanone: Polydioxanone (PDO) turns out to be crystalline and also this polymer is absorbable by bio substance. It is nowadays also used as a suture to close an open cut or wound as it is very elastic and also retains strength brilliantly and possess shape memory (figure 7.5). It demonstrates a slow absorption rate as well as lower inflammatory reaction. Sell and co-authors in their research, prepared a non-woven scaffold using electrospun-blended nano-fibrous PDO and elastin to evaluate its applicability as a vascular graft. The fabricated graft shows sufficient mechanical integrity to overcome an aneurysm and has compliance similar to a native artery (Sell *et al* 2006).

Polycaprolactone (PCL): PCL is an aliphatic polyester with high tensile strength. It is widely used as a vascular graft due to its biocompatibility, cell affinity, slow degradation and low mechanical strength (Zhang *et al* 2012) (figure 7.6). Valence *et al* prepared a vascular prosthesis by electrospinning PCL into micro and nano-fiber serving as a shelf-ready vascular graft with appreciable patency, mechanical properties and endothelialization (de Valence *et al* 2012, Nottelet *et al* 2009). Zhang *et al* enhanced performance about hydrophobic PCL by incorporating natural lecithin into cholesterol-poly(ε-caprolactone) (Chol-PCL). The electrospunfibrous Chol-PCL/lecithin membranes demonstrated better hemocompatibility and cyto-compatibility (Zhang *et al* 2012).

The above-mentioned polymers can provide adequate mechanical strength but lack the reactive site for bio-functionalization and are hydrophobic. The reason behind the failure of a synthetic polymer as the vascular graft is poor endotheliza-tion, occlusion, inflammation, thrombosis, infection, and hyperplasia (Piterina *et al* 2009, Sell *et al* 2006).

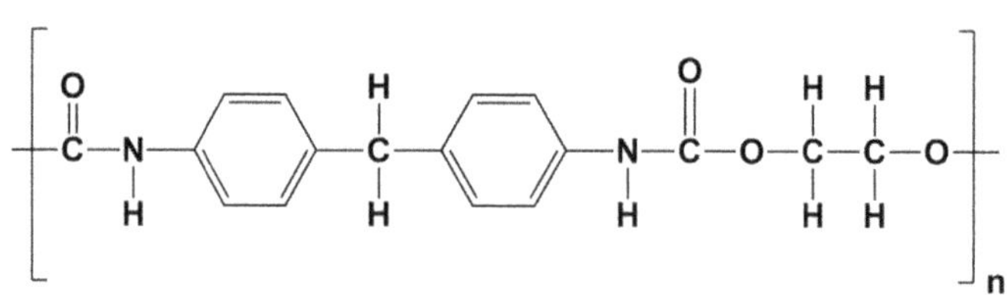

Figure 7.4. Chemical structure of polyurethane.

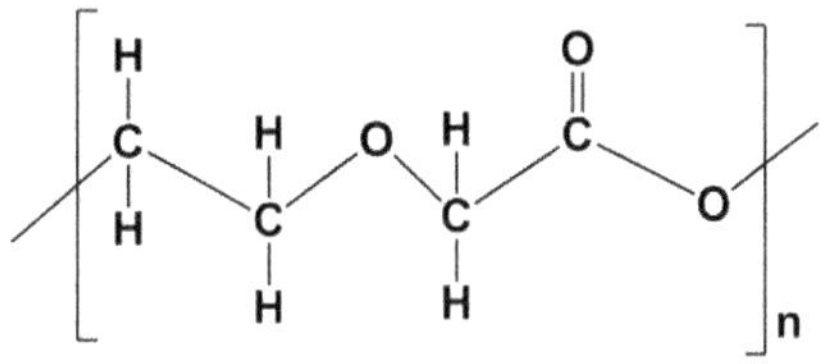

Figure 7.5. Chemical structure of polydioxanone.

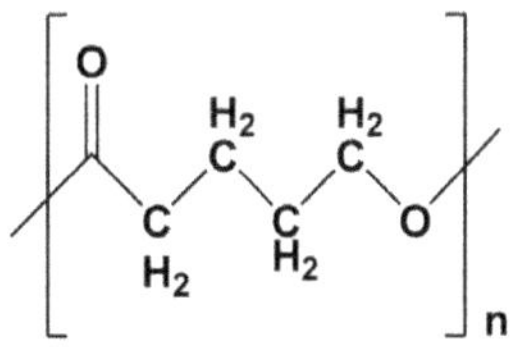

Figure 7.6. Chemical structure of polycaprolactone.

The main concern while using synthetic polymers is to make their surface non-thrombogenic, particularly in the case of repair of small and middle diameter damaged blood vessels. Even though the synthetic polymers provide adequate mechanical strength, they lack cellular compatibility and their degradation products may draw out unfavorable immune response *in vivo*.

7.4 Natural polymer

Blood vessel tissue engineering aims to develop scaffolds which mimic that of involuntary as well as biological attributes of an indigenous blood vessel. As vascular grafts need to be implanted in a mechanically changing environment, biomaterials must be able to sustain deformations without failure. Natural polymer leads to the preparation of scaffold with cellular affinity, possesses hydrophilicity but is often degradable creating empty voids for tissue regeneration and cellular migration.

Chitosan: Chitosan is a natural amino-polysaccharide, highly hydrophilic, and with sufficient mechanical strength but it lacks hemocompatibility due to its cationic profile causing negatively charged platelet adhesion. Its structure contains glycosaminoglycan (GAG) in ECM.

Fibrin: Fibrin is obtained from its precursor i.e. polymerized fibrinogen. The fibrinogen can be obtained from patients' own blood plasma, hence showing enhanced collagen deposition and ECM molecules (Ahmed *et al* 2008). It allows cell attachment and proliferation but has poor biomechanical strength. However, its disadvantage of biomechanical stability can be overcome by *in vitro* maturation process. Aper *et al* used high-speed rotation molding technique to allow centrifugal force to be applied inside the tubular shaped fibrin segment (Aper *et al* 2016). Wolf *et al* developed a digester or bioreactor for dynamic conditioning of a fibrin based tissue engineered vascular graft containing human umbilical arteries and smooth muscle cells. The lumen of the prepared vascular graft has been endothelialized with endothelial cells, and the graft has been exposed through surge and pressure and it proved to be a mechanically stable graft (Wolf *et al* 2018). Gui and colleagues prepared a small diameter vascular graft inside a pulsatile bioreactor by seeding smooth muscle cells and fibroblast cells in fibrin gel (Gui *et al* 2014).

Silk: Researcher have shown interest in using silk as a biomaterial for a vascular substitute due to its controlled degradability, excellent mechanical strength and low immune response. Filipe and researchers created a cell-free silk-based vascular construct by electrospinning and evaluated its involuntary/mechanical as well as biological attributes. Hence, the result came as formation of a graft with appropriate mechanical properties, negligible fibrin network formation, rapid endothelialization, graft patency and excellent biocompatibility with endothelial cells and blood (Filipe *et al* 2018).

Collagen: Weinberg and Berg fabricated the first vascular substitute by fabricating multiple collagen layers integrated with Dacron mesh into which endothelial cells, fibroblasts, as well as smooth muscle cells were added (Weinberg and Bell 1986).

7.5 Decellularized matrix

Decellularization is a way to create a vascular graft after complete removal of cellular materials from the tissue and preserving the essential proteins of the extracellular matrix (collagen, elastin, fibronectin). The decellularization process removes the cellular antigens by chemical, physical and enzymatic means without influencing the mechanical integrity and composition of the native extracellular matrix (ECM). The extracellular matrix will be decellularized using detergent, enzyme or high-hydrostatic pressure (HHP) method. The extent of decellularization of natural extracellular matrix can be calculated using histological staining or by quantification of residual DNA.

The decellularized biomaterials can be converted into a functional blood vessel by seeding cardiovascular cells like endothelial cells as well as smooth muscle cells. This recellularization of a decellularized vascular graft with endothelial cells allows its usage as a long-term vascular graft (figure 7.7). This decellularized tissue reduces immunogenicity and can withstand physiological blood pressure, making it suitable for allogenic application. These tissues have the capability to repair a damaged blood vessel *in vivo*. The small diameter vascular grafts can be fabricated using decellularized biomaterials.

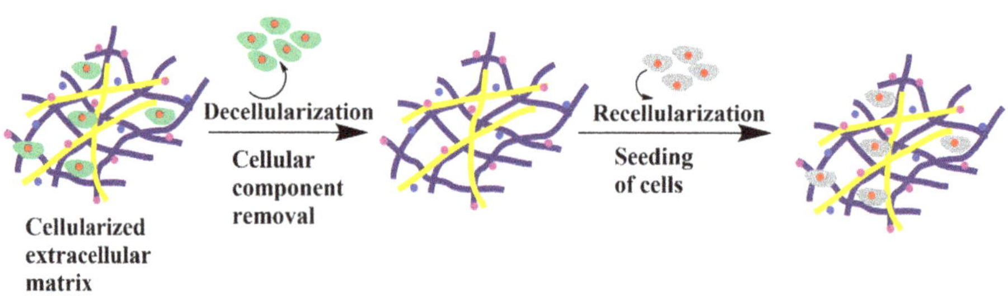

Figure 7.7. Schematic diagram of the process of cell decellularization.

7.6 Hybrid material

The main problem of an artificial vascular graft is its poor blood compatibility leading to platelet adhesion and plasma protein adhesion followed by thrombus formation (Yao *et al* 2014). Hence, the vascular graft must have a hemocompatible surface or combined with anti-thrombogenic properties. The vascular graft must have endothelium regeneration (to allow changes in circulating blood flow dynamics) to allow high patency and prevent intimal hyperplasia and thrombogenesis. The synthetic polymer has excellent mechanical strength but is bio-inert, hydrophobic, maintains structural integrity, has poor patency rate and lacks sites for bio-functionalization, whereas natural polymer is hydrophilic, possesses cellular affinity, contains reactive groups, is biocompatible with fast degradation creating space for migration of cells. To date many scaffolds with characteristics suitable for vascular graft have been prepared but lack these persistent properties when exposed to the physiological vascular microenvironment. The low resistance towards rapid degradation or effect of shear stress may lead to failure of the vascular graft. Therefore, there is a demand to prepare a hybrid vascular biomaterial which can withstand flow environment and high pressure without any collapse or degradation until it gets converted into the matured vascular construct. Lee and co-authors fabricated a scaffold with sufficient biomechanical properties and favorable conditions for vascular cell growth (Lee *et al* 2008). Yao *et al* prepared a hybrid vascular graft using electrospinning of PCL and chitosan followed by its functionalization by immobilization of heparin. The resulting vascular graft shows sufficient mechanical properties, enhanced growth of endothelial cells and good endothelization and better graft patency (Yao *et al* 2014). Heo and co-authors immobilized collagen type IV (COL-IV) on PLLA electrospun nanofibers. The prepared nanofibers were grafted with acrylic acid (AAc) using gamma-ray irradiation to enhance endothelium formation on the lumina of the vascular graft (Heo *et al* 2015). Tan and colleagues prepared scaffold by co-electrospinning sacrificial PVA with PCL/gelatin solution and immobilized by heparin to develop a scaffold with better mechanical properties, improved biocompatibility and reduced thrombosis (Tan *et al* 2016). As cellulose and chitosan possess the same $\beta - (1 \rightarrow 4)D -$ glucopyranose unit, researchers prepared a hollow tube with different ratio of cellulose and chitosan. They compared it with Dacron, ePTFE, human saphenous vein and human coronary artery and they found that mechanical and cell compatibility properties of a prepared tube show that it possesses the capability to be utilized as a coronary artery bypass graft (Azevedo *et al* 2013).

7.7 Assessment of practical use of a vascular graft

For checking the stability and potential of a vascular graft as a replacement for a damaged blood vessel in clinical applications, *in vitro* as well as *in vivo,* an estimate about the prepared graft is necessary.

***In vitro* assessment**

The *in vitro* biocompatibility of a vascular graft with endothelial cells and blood cells.

(i) Hemocompatibility: the efficiency of a vascular graft for *in vivo* applicability can be calculated by this most important parameter. The grafts must be analyzed as per ISO 10 993–4:2017 which specifies the evaluation of medical devices that come in contact with blood. The test vascular graft will be incubated with fresh blood and the evaluations are made regarding blood cell attachment, thrombus formation, fibrin network generation and protein adsorption. This interaction between vascular biomaterial and blood helps in analysis.

In vivo assessment

The *in vivo* biocompatibility, compatibility with blood, stability.

For *in vivo* evaluation of a vascular graft, it is implanted in an animal model with prior permission from the Animal Ethical Committee. The animal model will be anesthetized with proper calculated dosage of anesthesia. The restrained animal will be disinfected with disinfectant. The prepared vascular graft will be implanted in the damaged blood vessel and sutured. Post-surgical care will be taken and animals will be examined every day. After a specific time point, the sample implant will be harvested along with the host blood vessel for evaluating its potential to be used as a vascular graft.

(i) Histological examinations: this examination helps in determining compliance, aneurysm formation, patency, stenosis, presence of neointimal hyperplasia and evaluation of thrombus formation.
(ii) Scanning electron microscopy: for the study of structure about endothelial cells, presence of thrombus and neointima formation, the samples will be observed under scanning electron microscope.

7.8 Conclusion

The vascular graft used for treatment of vascular diseases must be thromboresistant and must be suitable for long-term usage. There is still an unmet demand for a novel biomaterial with desirable mechanical properties, vascular cell compatibility, blood compatibility, off-the-shelf availability that has clinical translatability. Advancements

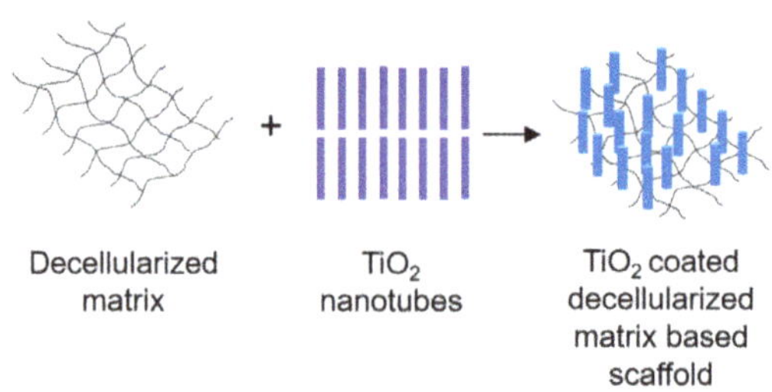

Figure 7.8. Schematic diagram showing the proposed technique to prepare a vascular graft by combining the benefit of mechanical properties of nanotubes with biological properties of a decellularized matrix. The obtained TiO_2 nanotubes based decellularized matrix will be seeded with vascular cells to prepare a tissue engineered vascular graft.

must be made in biomaterials so that they have sufficient mechanical properties, allow cell growth and prevent thrombosis.

An optimized scaffold which serves as a vascular graft with mechanical, architectural, physio-chemical and biomechanical properties can be fabricated by dispersing nanotubes chemically into a decellularized matrix, as shown in figure 7.8. As per the literature, titanium oxide (TiO_2) nanotubes support vascular cells.

References

Ahmed T A E, Dare E V and Hincke M 2008 Fibrin: a versatile scaffold for tissue engineering applications *Tissue Eng. Part B, Rev.* **14** 199–215

Aper T, Wilhelmi M and Gebhardt C *et al* 2016 Novel method for the generation of tissue-engineered vascular grafts based on a highly compacted fibrin matrix *Acta Biomater.* **29** 21–32

Aper T, Wilhelmi M, Hoeffler K, Benecke N and Haverich A 2016 Engineering of small diameter tubular compacted fibrin matrices *Bionanomaterials* **17** 13–7

Ariyoshi H, Okuyama M and Okahara K *et al* 1997 Expanded polytetrafluoroethylene (ePTFE) vascular graft loses its thrombogenicity six months after implantation *Thromb. Res.* **88** 427–33

Azevedo E P, Retarekar R, Raghavan M L and Kumar V 2013 Mechanical properties of cellulose: chitosan blends for potential use as a coronary artery bypass graft *J. Biomater. Sci. Polym. Ed.* **24** 239–52

Bastijanic J M, Kligman F L, Marchant R E and Kottke-Marchant K 2016 Dual biofunctional polymer modifications to address endothelialization and smooth muscle cell integration of ePTFE vascular grafts *J. Biomed. Mater. Res.* A **104** 71–81

Bergmeister H, Seyidova N and Schreiber C *et al* 2015 Biodegradable, thermoplastic polyurethane grafts for small diameter vascular replacements *Acta Biomater.* **11** 104–13

Crosby C O and Zoldan J 2019 Mimicking the physical cues of the ECM in angiogenic biomaterials *Regen. Biomater.* **6** 61–73

de Valence S, Tille J-C and Mugnai D *et al* 2012 Long term performance of polycaprolactone vascular grafts in a rat abdominal aorta replacement model *Biomaterials* **33** 38–47

Filipe E C, Santos M and Hung J *et al* 2018 Rapid endothelialization of off-the-shelf small diameter silk vascular grafts *JACC Basic Transl. Sci.* **3** 38–53

Gui L, Boyle M J and Kamin Y M *et al* 2014 Construction of tissue-engineered small-diameter vascular grafts in fibrin scaffolds in 30 days *Tissue Eng. Part A* **20** 1499–507

Guo F, Wang N and Wang L *et al* 2015 An electrospun strong PCL/PU composite vascular graft with mechanical anisotropy and cyclic stability *J. Mater. Chem.* A **3** 4782–7

Heo Y, Shin Y M, Lee Y B, Lim Y M and Shin H 2015 Effect of immobilized collagen type IV on biological properties of endothelial cells for the enhanced endothelialization of synthetic vascular graft materials *Colloids Surf.* B **134** 196–203

Hoshi R A, Van Lith R, Jen M C, Allen J B, Lapidos K A and Ameer G 2013 The blood and vascular cell compatibility of heparin-modified ePTFE vascular grafts *Biomaterials* **34** 30–41

Kim D, Chung J J, Jung Y and Kim S H 2019 The effect of substance P/heparin conjugated PLCL polymer coating of bioinertePTFE vascular grafts on the recruitment of both ECs and SMCs for accelerated regeneration *Sci. Rep.* **9** 17083

Kirkton R D, Prichard H L, Santiago-Maysonet M, Niklason L E, Lawson J H and Dahl S L M 2018 Susceptibility of ePTFE vascular grafts and bioengineered human acellular vessels to infection *J. Surg. Res.* **221** 143–51

Lee S J, Liu J, Oh S H, Soker S, Atala A and Yoo J J 2008 Development of a composite vascular scaffolding system that withstands physiological vascular conditions *Biomaterials* **29** 2891–8

Nottelet B, Pektok E and Mandracchia D *et al* 2009 Factorial design optimization and *in vivo* feasibility of poly(epsilon-caprolactone)-micro- and nanofiber-based small diameter vascular grafts *J. Biomed. Mater. Res.* A **89** 865–75

Piterina A V, Cloonan A J and Meaney C L *et al* 2009 ECM-based materials in cardiovascular applications: inherent healing potential and augmentation of native regenerative processes *Int. J. Mol. Sci.* **10** 4375–417

Pourdeyhimi B 1986 Vascular grafts: textile structures and their performance *Text. Prog.* **15** 1–30

Sell S A, McClure M J and Barnes C P *et al* 2006 Electrospunpolydioxanone–elastin blends: potential for bioresorbable vascular grafts *Biomed. Mater.* **1** 72–80

Stump M M, Jordan G L, Debakey M E and Halpert B 1963 Endothelium grown from circulating blood on isolated intravascular dacron hub *Am. J. Pathol.* **43** 361–7

Szilagyi D E, Smith R F, Elliott J P and Allen H M 1965 Long-term behavior of a dacron arterial substitute *Ann. Surg.* **162** 453–77

Tai N R, Giudiceandrea A, Salacinski H J, Seifalian A M and Hamilton G 1999 *In vivo* femoropopliteal arterial wall compliance in subjects with and without lower limb vascular disease *J. Vasc. Surg.* **30** 936–45

Tang D, Chen S and Hou D *et al* 2018 Regulation of macrophage polarization and promotion of endothelialization by NO generating and PEG-YIGSR modified vascular graft *Mater. Sci. Eng.* C **84** 1–11

Tan Z, Wang H, Gao X, Liu T and Tan Y 2016 Composite vascular grafts with high cell infiltration by co-electrospinning *Mater. Sci. Eng.* C **67** 369–77

Voorhees A B, Jaretzki A and Blakemore A H 1952 The use of tubes constructed from vinyon ‘N’ cloth in bridging arterial defects *Ann. Surg.* **135** 332–6

Walluscheck K P, Steinhoff G, Kelm S and Haverich A 1996 Improved endothelial cell attachment on ePTFE vascular grafts pretreated with synthetic RGD-containing peptides *Eur. J. Vasc. Endovasc. Surg.* **12** 321–30

Wang Y-X, Robertson J L, Spillman W B Jr and Claus R O 2004 Effects of the chemical structure and the surface properties of polymeric biomaterials on their biocompatibility *Pharm. Res.* **21** 1362–73

Weinberg C B and Bell E 1986 A blood vessel model constructed from collagen and cultured vascular cells *Science* **231** 397–400

Wissing T B, Bonito V and van Haaften E E *et al* 2019 Macrophage-driven biomaterial degradation depends on scaffold microarchitecture *Front. Bioeng. Biotechnol.* **7** 87

Wolf F, Rojas González D M and Steinseifer U *et al* 2018 Vascutrainer: a mobile and disposable bioreactor system for the conditioning of tissue-engineered vascular Grafts *Ann. Biomed. Eng.* **46** 616–26

Yao Y, Wang J and Cui Y *et al* 2014 Effect of sustained heparin release from PCL/chitosan hybrid small-diameter vascular grafts on anti-thrombogenic property and endothelialization *Acta Biomater.* **10** 2739–49

Zhang M, Wang K, Wang Z, Xing B, Zhao Q and Kong D 2012 Small-diameter tissue engineered vascular graft made of electrospun PCL/lecithin blend *J. Mater. Sci. – Mater. Med.* **23** 2639–48

Chapter 8

3D printing technology, bioink, fabrication technique of blood vessel and system used for cell culturing

Arindam Bit, Jasjit S Suri and Khemraj Desmukh

3D printing is newly emerging as a powerful tool for tissue engineering by implementing a 3D cell culture within complex 3D biomimetic architectures. In 3D printing, with the help of some external sources like light image pressure and so on, the materials are drafted into the required structure. The main conditions for 3D based printing are that the materials should be used for formation of scaffolds, be biocompatible, biodegradable and have some mechanical properties. 3D printing technology is capable of converting the 2D images collected from computed tomography (CT) or a magnetic resonance imaging (MRI) scan and some other non-penetrated scanning method into 3D images of required geometry structure. First, we give an introduction to various kinds of 3D printer methods used in recent times. After that we move to the recent biomaterial/bioink used in 3D printing technologies today for a realistic approach to tissue engineering (TE). We also emphasise the basic conditions for the choice of suitable bioinks for 3D printing. Recently, the use of combinational bio ink has been more popular. Examples of combinations of bioinks are cells, biomedical polymers, biomaterials. Different kinds of bioink material properties used in 3D printing are discussed in detail. This review also highlights the main improvement of composite bioinks and bioprinting methods to fabricate various tissues. Today, the fabrication of a blood vessel is challenging task. The formation of a blood vessel using a 3d printer brings organ engineering to a tremendous level. This chapter shows the different techniques used for formation of a blood vessel such as extrusion method, droplet method, laser method and other special methods, and the future anticipation of bioprinting of synthetic blood vessels. Bioreactors are a critical imminent part of TE strategy as they combine the fabrication of 3D functional tissues. The fundamental task of a bioreactor is to culture the cell seeding on the 3D printed scaffold. So we also

doi:10.1088/978-0-7503-2088-7ch8

emphasise the importance of the bioreactor and its types with properties available for cell culture.

8.1 Introduction

The uses of a 3D printer provide a dramatic revolution in medical health. A lot of scientists have made various achievements in biomaterials by use of 3D printing. This improvement in biomaterials encourages the engineer to use the 3D printer for medical research (Gupta *et al* 2018). In the current scenario different 3D printing technology/methods are available for generation of 3D printed products (Williams 2015). Various types of geometry/structure based on size, shape, structure, direction, cross section, area and mechanical properties are produced by the 3D printer (Christ *et al* 2015). A 3D bio-printer is the very important element of future organ bio-fabrication. The successful version of organ printing technologies depends on development of a 3D bio-printer capable of printing mortal/human tissue (Hesuani *et al* 2016).

The scaffold printed with the help of a 3D printer is providing a path and direction to the cell to make the organs. That the structure printed from the 3d printer represents the original structure of organs is owed to the feasibility of the biomaterials. The biomaterials available for TE must have some biological and general properties like being cost-effective, degradable, bio-compactable (Murdock and Badylak 2017, Gopinathan and Noh 2018). With the help of a microfluidic device we can perform scaffold-free 3D cell construction—it gives the artificial microsurrounding for cell production. Non-scaffolding methods consist of spheroid making and cell layer engineering (Kani *et al* 2016). The microfluidic device shows the strength possibility of a cell TE system (Vadivelu *et al* 2017, Li *et al* 2014). The nature of a scaffold alters the seeding of cells and the growth, proliferation and age of the cells. With the help of a scaffold, cells, and growth factors we are able to repair damaged tissue. The generation of tissue with the traditional method lacks control of orifice size, porosity, has limited design, development time, and navigation of architecture (Albrecht *et al* 2016). Convectional techniques used for scaffold generation lack good mechanical properties as compared to the original organ (Lu *et al* 2013). But 3D printing techniques have the ability to provide precise control over the scaffold shape and also help to perform the same operation in a repeated manner. For the fabrication of required tissues, the 3D structures must be able to support cell adhesion, growth, and separation.

One of the recent focuses of TE is on fabricating artificial biological organ substitutes either for transplantations or as models for drug screening. In such cases, one of the essential challenges is to fabricate or create viable and functional vasculatures to help cell growth within the fabricated scaffold (Kolesky *et al* 2014). Thus, current plans are focused on fabrication of vascular grafts which show biological properties the same as native vessels for implantations. In vascular TE we use arterial wall cells, endothelial cells (ECs) and smooth muscle cells (SMCs) with or without a biodegradable and bio-compactable scaffold (Pei *et al* (2002).

This chapter gives a general idea of bio-materials used for different 3D printing techniques for TE, in addition to their respective limitations. We highlight the different types of mechanism used for blood vessel formation. We also describe the various types of bioreactor used for blood vessel culture. We also provide an explanation of which bioreactor and 3D printer technology is used for a blood vessel.

8.2 3D printing technology for tissue engineering

All 3D printers are operated with different principles but all methods have some common steps that are:

(i) Geometry design;
(ii) Changing of geometry into standard tessellation language (STL) file format;
(iii) Choice of appropriate bio-material for printing;
(iv) As per governing phenomena of method, structure is printed coat-by-coat;
(v) Post-processing is changed according to technology of printer.

In the 3D printing process, first we create the structure with the help of CAD software. Then it is encoded into a standard tessellation language (.STL) file type because the printer understands only this format. The STL is converted into g code with the help of printer driver software. After reading the g code, the printer starts printing layer by layer according to g code. The 3D structure is printed by combining the image obtained from MRI and CT devices (figure 8.1).

8.2.1 Fused deposition modelling (FDM)

The 3D printer whose operation is based on FDM (figure 8.2) technology has a bed on which the printing is done, a nozzle (also called printer head) and printing material. The printer platform or the bed is made of a different kind of material. The printer head is attached in such a way that it can move, for which reason we attach the printer head to a belt. The whole assembly is designed in such a way that that it

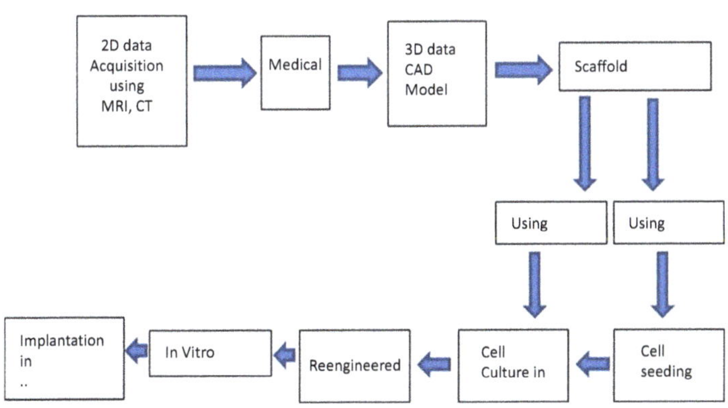

Figure 8.1. Printing step of the 3D bio-printer.

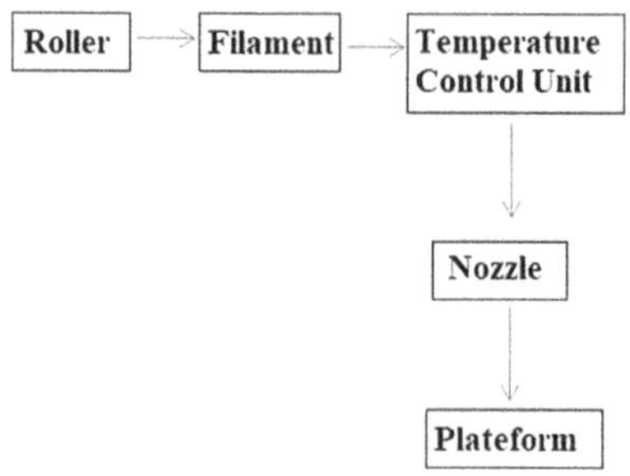

Figure 8.2. Schematic of FDM technology.

moves in three directions by use of a motor. All the activities of the material and head are adjusted by computer. The core material used for printing is some kind of plastic, whose melting is controlled by temperature, called thermoplastic. When thermoplastic material is heated and becomes melted repeatedly and then it hardens when the heat is lower or removed. At the start of printing, the raw material is dispended from fine filament via the warm head. It settles on the bed. The next layer that is dispended from the nozzle fuses with the layer below, making the structure from the bottom up, coat by coat. The temperature range for FDM is 180 °C and 350 °C as per the type of material used (Glatzel *et al* 2016). Habib *et al* (2016) use FDM technology for a 3D bio-printer. They use PLA material for printing and then do porosity, pore size calculation on some structure.

8.2.2 Selective laser sintering: the laser based

The laser based technology used in a 3D printer is called selective laser sintering (SLS). In SLS, the laser is used as an energy source for fusing the material used for making a 3D object. When we form any 3D structure of things from very fine crumbs using atomic diffusion then it is called sintering. The raw material in the form of powder used for making the structure is most commonly a plastic, such as nylon. When laser heats the powder then the particle present in the powder is fused together and solidified. When the first layer is formed, the bed of the SLS printer moves down and forms a new layer that is fused to the first layer. This step is continued until the whole object is created. When the entire object is fully created, it is left to cool down in the machine before being removed.

8.2.3 Stereolithography

Stereolithography is more commonly known as SLA (figure 8.3). 3D printing is one of the most popular and widespread techniques in the world of additive manufacturing. In an SLA system we use enough plastic in the form of a liquid. SLAs have some important body parts: a tank which has polymer, a perforated platform, an ultraviolet (UV) laser and a platform which is fully controlled. Firstly a fine layer of photopolymer is exposed above the perforated bed. The heating of the bed is done by UV laser according to the impression of the object. The places where UV light strikes are hardened rapidly and create the first coat. Once the initial layer is formed

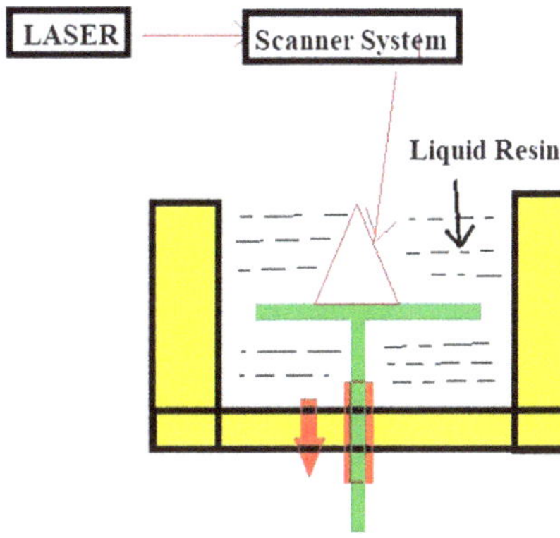

Figure 8.3. Schematic diagram of SLA.

then the platform is shifted down, and these are called the new surface layer of polymer. This task is repeated automatically until the entire structure has been formed. Objects created using SLA generally have very fine surfaces; one of the limitations is that the standard of printing depends upon the quality of SLA system used. The amount of time required to print the object is directly proportional the size of the SLA system used for printing. Due to its high resolution, SLA is the most adequate technique as compared to other techniques. But the resin used for the printing has very poor biodegradable properties (Gupta *et al* 2018).

8.2.4 Laser metal deposition

Laser metal deposition (LMD) is a new technology being adopted to augment, and in some cases replace, traditional methods for the remanufacturing, laser cladding, and 3D printing of metallic parts. If it is applied in the correct conditions then it has potential to save manufacturers time and money. LMD is also known as directed energy deposition (DED); it uses a system in which metallic powders held in one or more hoppers are passed through a deposition nozzle and heated with the help of a laser to produce a metallic bead. The motion system lays the metallic beads down, layer upon layer, to build up a part or to add layers of material to an existing part.

8.2.5 Digital laser processing

3D printing DLP technology is the same as SLA but use of the light source provides a difference between these two technologies. In DLP we use a conventional light source. Light generated from the source is directed using micro mirrors. The liquid crystal display panel is used as a photo mask. The surface where the light is incident hardens rapidly. The main difference between SLA and DLP is the light source. An arc lamp with a CDP are the sources of light for DLP. DLP is faster than SLA because in DLP the light is passed to the entire surface rather than a single pass in SLA. The resolution of DLP is higher than SLA. One advantage of DLP over SLA is that to execute the process only a shallow vat of resin is required, so that it produces low waste and operates at minimal expense.

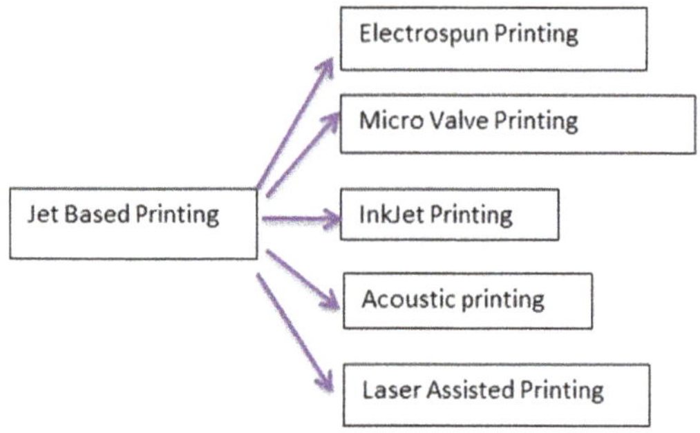

Figure 8.4. Types of jet-based printing technology.

8.2.6 Jet-based bioprinting technology

Various types of technology are used for 3D printing such as extrusion bio-printing, jet-based bioprinting and stereolithography. As compared to all current technology, jet based printing has many more advantages such as agility, maximum thoughput and single-cell printing. In this portion, we will highlight the various types of jet-based technology used for printing (figure 8.4).

8.2.7 Inkjet printing

There are three types of inkjet printing, named as electrostatic, thermal and piezoelectric forces (Li *et al* 2018). In electrostatic inkjet printing, we can develop droplets with the help of the volume of the fluid chamber (Kamisuki *et al* 1998). If we increase the volume of the ink chamber then the voltage between the pressure plate and the electrode are also increased.

In the absence of charge the pressure plate retains its actual shape. Increased pressure is used for withdrawing the droplet. As its name indicates, in thermal inkjet printing, the actuator is heat thermally, and then fine air bubbles are formulated which drain the droplet from the printer head; a short current pulse generates a temperature up to 300 °C (Cui and Boland 2009). Evaporation of ink is done by the thermal actuator which causes bubble formulation to drain out of the ink by force. With the help of the current pulse and ink viscosity, droplet formation can be adjusted (Li *et al* 2018). In piezoelectric inkjet printing, for changing the shape of a piezoelectric actuator to skew the fluid chamber we use the voltage pulse. Pressure change is highly dependent upon the fluid chamber volume, and fluid chamber volume is dependent upon the applied voltage. So the formation of the droplet is done due to reduction of surface tension at the printing head orifice (Li *et al* 2018). Most of the time thermal and piezoelectric printing methods are preferred. The thermal printer is more state-of-the-art compared to other printers.

8.2.8 Micro-valve printing

As the name indicates, the droplet is generated with the help of a valve. Here a solenoid coil and a plunger are used as a valve. This valve is used to provide the obstacle to the orifice (Faulkner-Jones *et al* 2013). A voltage is used to form a magnet. With the help of these magnets we can move the plunger. If the plunger is moved then the valve is moved up and down or moved left and right depending upon the position of the valve. Then the back pressure generated by the valve pushes the ink for draining (Hospodiuk *et al* 2017). Shear stress can be minimised by increasing the diameter of the nozzle (Faulkner-Jones *et al* 2013).

8.2.9 Acoustic printing

In this method an acoustic wave is used to generate droplets. When the acoustic radiation is more than the surface tension, the droplets are drained out from the nozzle. The droplet size is altered with the help of time; it can be manipulated by adjusting the time duration between two pulses with the cell concentration. The stability of the acoustic signal is directly proportional to the quality of droplet used for printing. Acoustic printing can be achieved with the help of high viscosity ink (Zhou *et al* 2019).

8.2.10 Laser-assisted printing

In laser-assisted printing, the removal of the droplet is dependent upon the laser (Zhou *et al* 2019). The laser source and a special substrate are essential for this system. This substrate is coated with some kind of special absorbing material. When a laser beam is incident on the substrate, due to heat the upper absorbing material is evaporated and causes bubble formation. The size of the droplets and the resolution can be changed by factors such as the laser source vitality, the width of substrate thickness and the characteristics of the ink.

8.2.11 Electrospun

The electrospun system has the potential to generate nanofibers and nanostructures (Zhou *et al* 2019). The workings of electrospun is the continuous stream. Any popular electrospun system must have the following: a high voltage power supply; an ink store device called a syringe and a syringe pump; a metallic needle; and a collector which acts as a negative terminal. First, the solution whose sheet is prepared is loaded into the ink sample storage device. When a positive terminal of high voltage is connected to the metallic needle it is treated as the anode and negative voltage is applied to the collector. When the applied voltage is more than the critical value, then the jet is removed from the point of the Taylor cone becoming thin and long, and quickly collects on the grounded collector. Electrospun parameters, solution and environmental parameters, are responsible for alter the functioning and output of the electrospun system (Zhou *et al* 2019).

8.2.12 Electrohydrodynamic jet printing

There are two types of electrohydrodynamic jet printing: the first one is pulsating droplets; and the second is continuous stream (Zhou *et al* 2019). It is also operated in high voltage pressure. This pressure is applied between the metallic needle and the collector for formation of droplets. Dripping can be achieved with the minimum electric field and a lower flow rate. So we can say there is a direct relationship between the electric field, flow rate and speed of the dripping. Bio-ink properties and applied back pressure are the main reasons for forming a critical voltage. Critical voltage is responsible for ejecting a regular stream from the print head (Zhou *et al* 2019). Because of its high pressure, electrohydrodynamic jet printing can create a droplet. The necessary condition for single-cell printing is that the size of the droplet is small compared to the orifice. In general, there is an inverse relationship between the droplet volume and the applied voltage.

For creating a droplet two printers, namely inkjet and micro-valve, need an orifice. Due to high viscosity as well as concentration, these droplets suffer from obstacles and clogging, so we need high pressure to withdraw the material from the nozzle. But when pressure is increased the cell viability decreases. So inkjet and micro-valve printing are mostly not used for printing with high cell density. The clogging problem does not arise with acoustic and laser-assisted printing because an orifice is not used in these two methods. The printer having no orifice can provide minimum shear stress during the material flow via the nozzle. Orifice-free printers can avoid the shear stress phenomena provoked by the material passing through a nozzle. So the orifice-free method can be used for large density bio-ink, but both these require surface tension. The viscous ink constraint is eliminated in electro-hydrodynamic jet printing (Zhou *et al* 2019).

8.3 Bio-ink/Biomaterial

The main ingredient of 3D bio-printing is the bio-ink, because it plays a crucial role for the development of the functional body part. The bio-inks used in 3D printing technology need different properties. Those properties are critical and must be kept in mind during the consideration as a bio-ink. Combining two or more materials enhances the properties of bio-ink.

Biomaterial is divided into two parts. The first is called cell-based scaffold and the second is called scaffold-free cell-based approach (Gopinathan and Noh 2018). In the first method, the bio-ink is nothing but the combination of biomaterial and live cells. The live cells eliminate the scaffold biomaterial, and all live cells proliferate and differentiate slowly. Cells are used as the ink in the second method. These cells are responsible for the generation of new tissues and with the help of these tissues we form functional tissue by use of deposition. The ideal bio-ink fulfils some biomaterial and biological criteria. Biological criteria mainly include biocompatibility, biodegradability, cytocompatibility. The output of a 3D printer is altered by different parameters such as viscosity of the bio-ink, surface tension of the bio-ink, the crosslinker and surface properties of the printer nozzle/head. The hydrophilicity and the viscosity of the bio-ink change the printing accuracy and doping of the live cell. There is a proportional relationship between viscosity of ink and extrusion pressure. If the bio-ink formulation is highly viscous, then more pressure is required for the extrusion. If we need high extrusion pressure then the flow of the solution from the print head through the fine head orifice may get overwhelmed.

8.3.1 Criteria for selection of biomaterial

Bioprintability, biocompactibility, shear thinning and bio-degradation are essential criteria for choosing a bio-ink for a 3D printer. The properties required are shear stress, biodegradability, biocompactibility and harmlessness to cells (Gopinathan and Noh 2018). Selection of bioink depends upon the requirements of the desired tissues and organs.

8.3.2 Types of biomaterials

Natural polymer: frequently, natural polymers are used as raw material for a 3D printer. Natural polymer permits printing in respect of temperature. One of the limitations of natural polymer is its limited availability. Collagen, chitosan, alginate etc, come under the category of natural polymer. The applications of these types of polymers are repairing of bone, cartilage, nerves, and skin. One of the deficiencies of natural polymer is its limited physical and chemical properties (Ji and Guvendiren 2017). We can also use a combination of natural polymers to increase the physical and chemical properties. Bio-inks having a combination of natural materials are alginate with gelatin/fibrin, silk fibroin with gelatine, agarose with collagen, chitosan with gelatine, cellulose with alginate and hyaluronan with cellulose.

Synthetic polymers: these are manufactured inside the workshop using chemical processing, they are mainly used in 3D printing (Gupta *et al* 2018). Natural polymers are beneficial, but most of the time synthetic polymers are used because of their characteristics. Good synthetic polymers have some important characteristics like high mechanical strength, proper degradation rate, and good treating ability. Examples of synthetic polymers are polycaprolactone, acrylonitrile butadiene styrene, poly-lactic acid (PLA), polyamide, poly-theretherketone, poly(lactic-*co*-glycolic acid) (PLGA), PVC, PEG (polyethylene glycol) etc.

Hybrid bio-inks: by merging natural and synthetic polymers we can increase the properties of biomaterial such as biosuitability, machine driven performance, properties dependent on temperature, and cross-link ability. In many applications researchers use heterogeneous materials getting the required physical and chemical properties that control the shear-thinning properties. One example of this type of combination is GelMA.

Bio-inks based on hydrogels and particles: with the help of adding specific nanomaterials we can improve mechanical properties of hydrogels in a dramatic manner. Some nanomaterials can be used as crosslinkers to enhance the mechanical strength of hydrogels (Ashammakhi *et al* 2019). Much previous work shows the use of biomaterial comprised of hydrogel, for example poly(-lactide-co-glycolide)-PEG (PLGA-PEG), being added to cell-laden carboxymethyl cellulose (CMC) for enhancing the characteristics. Some examples of nanoparticles are silicates, hydroxyapatite, tricalcium phosphate, bioactive glass, carbon nanomaterials, etc.

Biomolecule-contained bio-inks: various kinds of biomolecules are added to the bio-ink to increase the properties of biomaterials (Ashammakhi *et al* 2019). Growth factor can be used in biomolecules for a hydrogel. For example, one research group, designed ink called GelMA; in this bio-ink VEGF was added, acting as a growth factor for bone tissue. Biomolecules such as bone morphogenetic proteins, vascular endothelial growth factor, fibroblast growth factor, transforming growth factor, peptide motifs, extracellular matrix, stromal cell–derived factor, platelet-rich plasma, and some other molecules can be added in bio-ink.

Apart from the above, other materials that can be used as bio-inks are fibrin, cell-u-lose, silk, ECM bio-inks, cell aggregates, cell spheroids, etc.

8.4 Blood vessel formation using a 3D printer

One of the states-of-the-art of TE is developing a synthetic biological construct either for transplantations or repair. Current technology makes accessible the control of deposition of cells, with growth factor to enhance vascular growth. Formation of a blood vessel structure is a challenging task in the current era. Fabrication of vascularized products forces organ construction to the next level. The better solution of the problem is bioprinting. This section gives the idea for biofabrication of blood vessel constructs. extrusion-based, droplet-based and laser-based bioprinting techniques are some useful methods used for printing of a blood vessel. The following sections provide details of the method available for fabrication of a blood vessel and possible future work (Ng *et al* 2018).

8.4.1 Anatomy of a blood vessel

Three layers are used for the formation of an *in vivo* blood vessel. The innermost layer is the tunica intima, the middle layer is called tunica media and the outermost layer is called tunica adventitia, sometimes also called tunica externa. There are three major factors considered in the innermost layer, the first of which is the endothelium. The endothelium is fabricated as a thin single layer of squamous epithelium. The second factor is basal lamina that has type IV collagen and laminin. The last factor is the subendothelial layer, which has some loose combinative tissue. Among the three factors, endothelium is the thinnest layer, and it plays a very important role for blood vessel formation. The main work of endothelium is forming a non-thrombogenic barrier, which provides protection to tissue from infection. SMCs are present in tunica media, and these cells are arranged in circular form about the tunica. SMCs containing a muscular coat are arranged circularly around the tunica media along with type I and III collagen. This middle layer is thick in arteries as compare to veins. The last layer, the tunica adventitia is the outermost layer of the blood vessel having connective tissue of embedded fibroblasts. This layer protects the blood vessel from enclosing tissue and it also provides some additional support to the blood vessel. The outermost layer has vasa vasorum and nervivaso-rum. It is autonomic nervous. The vascular network is mainly a vasa vasorum. The anatomies of arteries and veins are very close to each other but the thickness structure of tunica and their cellular and protein compositions are different. Most of the time arteries have a broader tunica media and a slim tunica adventitia. Both veins and arteries are classified into three categories. These are classified or categorised according to parameters such as size and category of tunica media. In all types of blood vessel capillaries have very small diameter. In all types of veins tunica adventitia is thicker than the arteries. Veins are generally classified into four types. This classification is based on size (Ng *et al* 2018).

8.4.2 Self-assembly approach

This unique method is proposed by Ng *et al* (2018). Here the gelatin is used as biomaterial and coated in a tissue culture plate. SMCs along with fibroblasts are used as cells for the self-assembly approach. Then the sheet is rolled with the help of a tubular structure to make a blood vessel construct. This blood vessel structure contains a smooth muscle layer in the inner layer and a fibroblast layer in the outer layer. These inner and outer layers are equivalent to the tunica media and tunica adventitia, respectively. The output product of the self-assembly approach has good mechanical properties. These mechanical properties depend upon the types of fibroblasts used. There is some basic difference between the proposed method and the TE approach. In the self-assembly approach, cells generally leave a trail to make the blood vessel structure. In the TE approach, first a tubular channel is formed with the help of suitable biomaterial, then the cell is cultured on it. So in the self-assembly approach we can directly control the cells for tubular channels which is much easier than other methods available for forming tubular structure.

8.4.3 Extrusion-based 3D printing system with rotatory printing device

Rotatory printing devices have a motion control system, multinozzle distribution system, rotary system, and control system for temperature. The biomaterials can be extruded with the help of a pressure control system, the motor is used to control the rotating rod, and the nozzle speed is handled by a linear motor. v_1 shows the nozzle speed, v_2 shows the speed of material extrusion and v_3 shows the motorspeed. If we can set the speed of the nozzle higher than v_1, then the dispersion phenomenon occurs on the material. Otherwise, the biomaterials will illustrate the accumulation phenomenon. An alginate–gelatin (AlGe) solution is used as a biomaterial for making the vessel structures. We print the blood vessel structure with the help of alginate–gelatin (AlGe) solution under all considered conditions. By changing some parameters we can change the thickness of the printed structure (Liu *et al* 2017). Those authors have developed a formula for the thickness of the vessel but that formula is not accurate, so we modified the formula for matching the experimental value close to the mathematical value.

The thickness can be derived by equation (8.1)

$$H = \alpha\sqrt{\left(\frac{v_3 c}{2v_2}\right)^2 - L^2} - r \tag{8.1}$$

where,

H is the thickness;
α is invariant constant;
L is displacement of nozzle;
r is rotating rod radius.

The relationship between viscosity and thickness of printing product is as follows.

1. **Bio-ink is Newtonian flow.**
   We know that

$$H = \alpha\sqrt{\left(\frac{v_3 c}{2v_2}\right)^2 - L^2} - r \tag{8.2}$$

   and the relationship between v_2 and viscosity is

$$F_{vis} = \eta\frac{Av_2}{l} \tag{8.3}$$

   where
   η is viscosity;
   A is cross-sectional area of the pipe;
   v_2 is viscosity of bio-ink;
   l is the length of pipe.

From equations (8.2) and (8.3), we solve for the relationship between H to obtain

$$H = \sqrt{\left(\frac{v_3 c\eta A}{2F_{vis}l}\right)^2 - L^2} - r^2 \tag{8.4}$$

   Equation (8.4) is the required expression.

2. **Bio ink is non-Newtonian flow.**
   A 'power-law' is used for the non-Newtonian fluid model. It gives a general relationship between viscosity ν, and the strain rate γ.
   The relation is given as

$$\nu = k\dot{\gamma}^{n-1} \tag{8.5}$$

   where,
   k is the flow consistency index ($m^2\ s^{-1}$);
   γ is the strain-rate (SI units s^{-1});
   n is the flow behaviour index.

Based on the flow behaviour index η:

1. If $0 < \eta < 1$: the fluid shows 'pseudoplastic or shear thinning' beahaviour. Here a smaller value of η, means a greater degree of shear thinning.
2. If $\eta = 1$: the fluid shows Newtonian behaviour.
3. If $1 < \eta$: the fluid shows 'dilatant or shear thickening' behaviour with higher value resulting in greater thickening.

We know that

$$H = \alpha\sqrt{\left(\frac{v_3c}{2v_2}\right)^2 - L^2} - r \tag{8.6}$$

and we also know that (for non-Newtonian fluid)

$$F_{\text{vis}} = \eta A\frac{dv_2}{dx} \tag{8.7}$$

η is viscosity;
A is area of the pipe;
$\frac{dv}{dx}$ is velocity gradient.

Solving equation (8.7) for velocity v_2

$$v_2 = \int\frac{F_{\text{vis}}}{\eta A}$$

Substituting the value of v_2 in equation (8.6), we get

$$H = \alpha\sqrt{\left(\frac{v_3c}{2\int\frac{F_{\text{vis}}}{\eta A}}\right)^2 - L^2} - r \tag{8.8}$$

But the value of viscosity of a non-Newtonian fluid according to power law is

$$\eta = \nu = k\gamma^{\cdot n-1} \tag{8.9}$$

so putting the value of η into equation (8.8), we get

$$H = \alpha\sqrt{\left(\frac{v_3c}{2\int\frac{F_{\text{vis}}}{Ak\,\gamma^{n-1}}}\right)^2 - L^2} - r \tag{8.10}$$

8.4.4 Drop and demand method

Drop and demand are a new emerging trend for fabrication of vascular structure. These methods have a heating bed. The vessel-like structure is printed with sacrificial core and gelatine in the reactor, and temperature of the reactor is maintained at 5 °C. The minimum distance between two droplets is 0.5 mm. When the distance is less than the minimum distance then droplets are fused together, and if the distance is greater than minimum distance then droplets create an irregular pattern. We can print the first layer of ECs by using hydrogels which are very close to the sacrificial material. SMCs are spread over the scaffold before filling the outer hydrogel. We can maintain the maximum width of the SMC layer up to 425 μm (Schöneberg *et al* 2018).

8.4.5 Hydrogel bio-printed micro-channel

Bertassoni *et al* used agarose as the material for blood vessel 3d printing. In the proposed paper, agarose fibers are printed with a precise and controlled environment. After that, a hydrogel precursor is incident on the entire printed object and after some time it is photo-polymerized. After the spray of gelation, the printed structure is easily removed from the surrounding photo-cross-linked hydrogels. For making an adequate perfusable network without any need for template dissolution for this, printed templates can be removed efficiently. To prove that the presented model has working ability, cell-laden methacrylatedgelatin (GelMA) hydrogels are used in a standard platform (Bertassoni *et al* 2014).

8.4.6 Laser-based 3D bioprinting of a blood vessel

Digital light projection (DLP) and laser-based stereolithography are also know as laser-based bioprinting techniques. A combination of small mirrors is called a digital mirror, which improves the performance of the DLP. The digital mirror is operated according to a digital signal. The laser beams are guided by a small mirror. When the laser beam is incident on a photo-curable/sensitive biomaterial the material gets hardened. In DLP GelMa is used as cross-linker and UV is used for the energy source. HUVECs are used as the cells for this type of printing. DLP has limited application and experiments done with DLP are very costly. DLP has a lack of ability to print large structures (Sarker *et al* 2019) (figure 8.5).

8.4.7 Embedded bioprinting for vascular engineering

This technique is proposed by Burdick and colleagues. They used supramolecular hydrogels based on addition of adamantane (Ad) and β-cyclodextrin (CD). Standalone HA shows good biocompatibility, but when we mix Ad and CD to HA than biocompatibility increases. This is a minimum requirement for embedded printing. With the help of a support matrix via self-heating we can stablize the

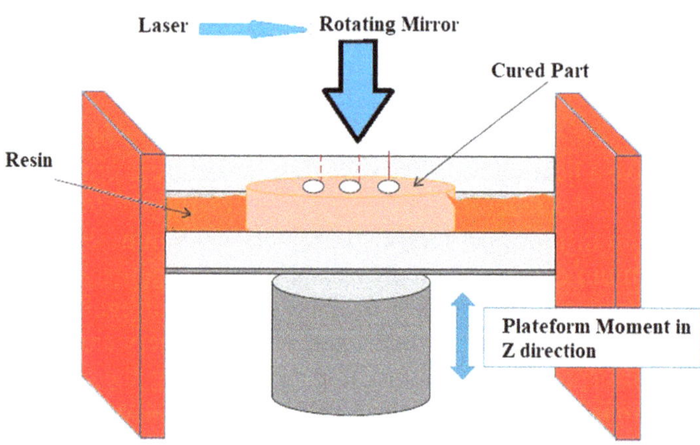

Figure 8.5. Laser-based bioprinting technique.

biomaterial in the structure. 3D filaments are incorporated inside a support hydrogel. With the help of size and movement of needles we can control the diameter of the filament. FRESH-enabled bioprinting of complex structures is done using CAD software. Sometimes microparticles-based soft granular hydrogels are an alternative material for embedded bioprinting. Microparticles-based granular gels easily move between solid and fluid state. The Angelini Group reported bioprinting of a vessel structure using hydrogel, colloids, silicones and living cells (Hu and Zhang 2018).

8.5 Importance of bioreactors

When we deal with tissue fabricated by a 3D printer then a bioreactor plays an important role. So the main task of a bioreactor is to control the technical parameter cell culture which affects the biological output. The technical parameters are temperature, pressure, flow, pH value, viscosity, velocity, shear stress etc. Bioreactors help the production of tissue as follows: (1) balancing the steady state cell deposition on the printed structure; (2) handling the technical parameters which affect the cell culture, controlling substantial parameters; (3) providing mass transfer with high facility; (4) providing automatic steps. In respect of tissue culture or for any kind of tissue application a bioreactor fulfils both biological and technical parameters for improving the performance.

Lei Song *et al* (2012) proposed a fully automated bioreactor of small diameter. This computer-operated bioreactor system can provide the same environment required for a culture graft. This computer controlled bioreactor maintains the concentration of oxygen and carbon dioxide in the media and required amount of nutrients. This bioreactor has an automatic feedback control system, with various sensors, an advanced controller, a gas operated magnetic valve, and a gas/liquid and liquid/liquid exchanger. ECs and SMCs are cultured in both a computer-controlled and non-computer-controlled system. Then they are examined after 14 days with the help of MTT assay and we find that the output of the computer-controlled perfusion system is higher than the pulsatile bioreactors because the computer-controlled bioreactor has a very tight degree for fabricating small diameter vessels.

Juan Wang *et al* (2016) proposed a bioreactor (figure 8.6), where, with the help of certain parameters, we can observe Mg degradation in *in vivo* and *in vitro* conditions. The authors reported that the *ex vivo* bioreactor permits culture with and without a blood vessel for study of Mg degradation in comparison with *in vitro* conditions and within *in vivo* conditions. They found that the degradation rates on the diffusion models were not affected. On the other hand, degradation of Mg is greater *ex vivo* at the fluid flow conditions as compared to *in vivo*. The degradation rates on the diffusion models are almost the same, although the loss of Mg is greater in flow conditions *ex vivo* as compared to *in vivo*.

Barak Zohar *et al* (2019) designed the multi-flow view (MFV) bioreactor. This allows the user to observe the cell culture in real time along with flow. The MFV bioreactor is used for long-period culture and has four logical constructs and these are observed in diverse flow conditions. The MFV bioreactor provides a facility to the user to examine cellular dynamics under a disciplined and controlled flow environment.

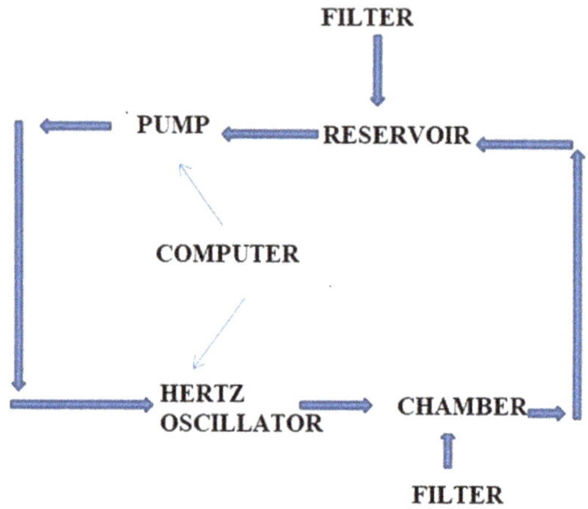

Figure 8.6. Sketch of a bioreactor system.

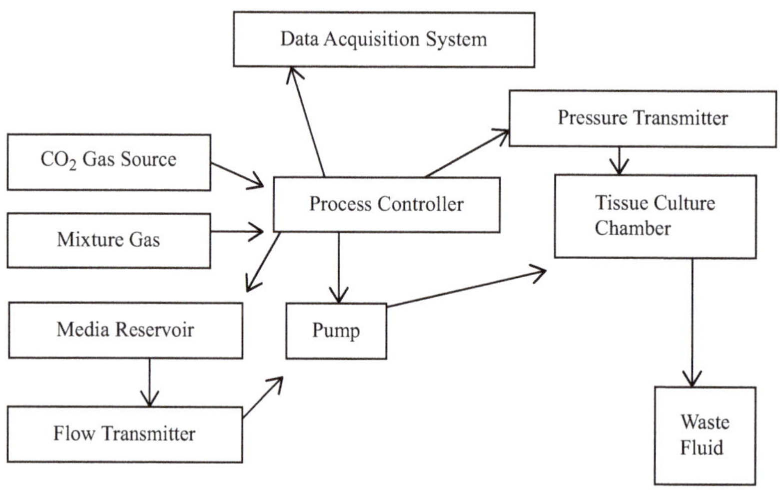

Figure 8.7. Schematic of perfusable blood vessels.

Hidekazu Sekine *et al* (2013) discussed perfusable blood vessels for use *in vitro* (figure 8.7). ECs are used in a bioreactor for formation of a triple layer blood vessel cell culture. The biological product is cultured in a customised bioreactor. The cell culture and proliferation are done inside the bioreactor. The important component of the bioreactor is an apulsatile pump, a closed encloser, pH transmitter, flow and pressure transmitter, carbon dioxide gas source, process controller for providing precise control and data acquisition system to store and record measured data. For making a bed or sheet, the printed structure is located inside the bioreactor. The container holds the media distributed with the help of the pump. Pressure generated by the pump is sensed by the pressure sensor. All the process related parameters such as pH, temperature, pressure, concentration of carbon dioxide gas, are controlled by a process controller.

Korin *et al* (2009) examined the effect of automated periodic 'flow-stop' perfusion systems, which were used for culture of mammalian cells. In a micro- channel type bioreactor we use a 'flow-stop' mechanism because these techniques provide the most convenient cell culture. The expansion of human foreskin fibroblasts represent that the cell expansion rate is increased in the micro-channel as compared to traditional culture plates. Human embryonic stem cells (HESCs) are highly unstable to shear stress, for a small change in flow rate, even at very minimal flow rates colonies will detach. So the adoption of a perfusion bioreactor is prohibited for HESCs proliferation. In the 'flow stop' mechanism a very small oscillation is applied for flow. A small oscillation in flow withdraws some of the short flow pulse pull out stash factors; stash factors and cell functioning is refreshed by deep static incubation. The flow-stop method is crucial for culture.

In TE there are four types of bioreactors used: (1) static, and (2) mixed flasks, (3) rotating wall, and (4) perfusion bioreactors. Bioreactors have three types of flow: static, turbulent and laminar, and they transport nutrients from the chamber to organs. Bioreactors are also divided on the bases of mass transfer and shear stress rate used for cell culture. Table 8.1 provides a comparison between the technical parameters considered for various bioreactors.

Adams *et al* (2011) proposed a mechanical bioreactor. Cyclic/pumping mechanical loading increases the health and ECM synthesis in TE blood vessels (TEBVs). The main aim of the project was to design an apparatus to provide pumping enclosing a span on cell-containing TEBV rings proliferating inside a silicone cylindrical vessel. Currently, mechanical bioreactors have a limitation in sample size, and handling. The proposed device holds 30 samples at a time, and these 30 samples are located in five different chambers, meaning six samples per chamber. This gives the permission to the operator to alter the media environment, time for culturing and frequency.

Lauren Ferrechio *et al* (2009) designed a project, the aim of which was to draw a bioreactor whose output is small diameter tubular and completely cellular vascular grafts. A cartridge was designed as a housing for five hollow, porous fibers, all used to provide support for expansion of cells seeded into the bioreactor. A pump was used for flow of media to form a dynamic culture system. The design was verified by examining the fluid flow and forces generated inside the tube. The result obtained from flow testing, indicates that the bioreactor is convenient for cell seeding and vascular graft generation.

8.6 Conclusion

In this chapter, we described different types of 3D bio-printing technology used for printing, and what are the materials used for 3D printing. In all the methods jet-based bioprinting offers immense hope in medicine and TE, due to its very well defined property characteristics such as sharpness, resolution, and accurate handle on deposition structure. Currently developed multicomponent bio-inks are the most popular biomaterial for 3d printing. This is because multicomponent bio-ink fulfils the requirement for 3D printing material used in TE applications. Common,

Table 8.1. Comparison between technical parameters considered in various bioreactors.

Bioreactor type	General descriptions	Mass transfer mechanism	Shear stress	Special usage	Tissue	Considerations
Static culture	Non-continous culture with different nutrients	High diffusion	Minimum	Cell growth		Uniform structure of cell constructs and nutrient diffusion limitations
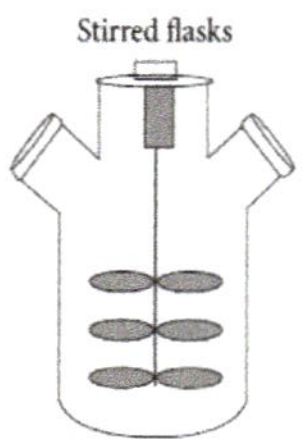	Magnetically stirred	High convection	High	Proper seeding of scaffolds	Cartilage	Appropriate scaffold and balance between increasing mass transfer and modulating shear stresses
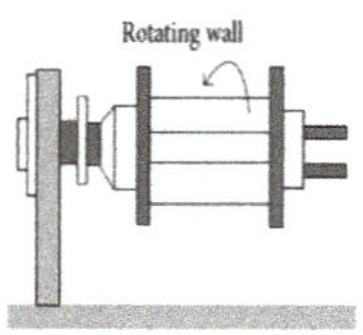	Rotating at a speed so the constructs in the reactor are maintained 'stationary' in a state of continuous free fall	High convection	Minimum	Tissue product required laminar flow	Cartilage, bone and skin	Operating conditions (e.g., speed of rotating) especially for growing large tissue mass
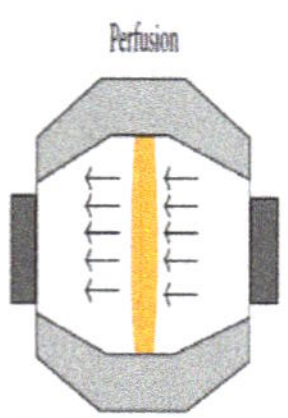	Flow of medium over cell sheet	High diffusion and medium convection	Medium	Related to human tissue	Epithelial cells, intestinal, bone, cartilage, and arteries	Inserting the cell inside the human body

artificial, or natural–synthetic raw materials and various kinds of cells, are used in poly- or multicomponent bio-ink. Some nanobiomaterials can be added to improve the quality of biomaterial. The fundamental task of this chapter is to provide a brief review of diversified approaches or technology used for the formation of blood vessels commonly used in vascular engineering. There are various types of bioreactor available for cell culturing among which the rotatory bioreactor give the best performance for blood vessel cell culture. Microfluidic bioreactors have great potential for cell biology studies and tissue engineering applications.

References

Adams K, Bishop K, Casey E, Gwyther T and Rolle M 2011 Design of a bioreactor to cyclically strain tissue engineered blood vessel rings *2011 IEEE 37th Annual Northeast Bioengineering Conference (NEBEC)* pp 1–2 IEEE

Albrecht L D, Sawyer S W and Soman P 2016 Developing 3D scaffolds in the field of tissue engineering to treat complex bone defects *3D Print. Add. Manuf.* **3** 106–12

Ashammakhi N, Ahadian S, Xu C, Montazerian H, Ko H, Nasiri R, Barros N and Khademhosseini A 2019 Bioinks and bioprinting technologies to make heterogeneous and biomimetic tissue constructs *Mater. Today Bio* **1** 100008

Bertassoni L E *et al* 2014 Hydrogel bioprinted microchannel networks for vascularization of tissue engineering constructs *Lab Chip* **14** 2202–11

Christ S, Schnabel M, Vorndran E, Groll J and Gbureck U 2015 Fiber reinforcement during 3D printing *Mater. Lett.* **139** 165–8

Cui X and Boland T J B 2009 Human microvasculature fabrication using thermal inkjet printing technology**30** 6221–7

Faulkner-Jones A, Greenhough S, Jason A K, Gardner J, Courtney A and Shu W 2013 Development of a valve-based cell printer for the formation of human embryonic stem cell spheroid aggregates *Biofabrication* **5** 015013

Ferrechio L, Lopez A, Skelly J and Thompson J 2009 *Bioreactor Design for Vascular Tissue Engineering* (Worcester, MA: Worcester Polytechnic Institute)

Glatzel S *et al* 2016 A portable 3D printer system for the diagnosis and treatment of multidrug-resistant bacteria *Chem* **1** 494–504

Gopinathan J and Noh I 2018 Recent trends in bioinks for 3D printing *Biomater. Res.* **22** 11

Gupta S, Bissoyi A and Bit A 2018 A review on 3D printable techniques for tissue engineering *BioNanoScience* **8** 868–83

Habib F N, Nikzad M, Masood S H and Saifullah A B M 2016 Design and development of scaffolds for tissue engineering using three-dimensional printing for bio-based applications *3D Print. Additive Manufac.* **3**

Hesuani Y D, Pereira F D A S, Parfenov V, Koudan E, Mitryashikin A, Replyanski N, Kasyanov V, Knyazeva A, Bulanova E and Mironov V *et al* 2016 Design and implementation of novel multifunctional 3D bioprinter *3D Print. Additive Manufact.* **3**

Hospodiuk M, Dey M, Sosnoski D and Ozbolat I T 2017 The bioink: a comprehensive review on bioprintable materials *Biotechnol. Adv.* **35** 217–39

Hu N and Zhang Y S 2018 3D bioprinting blood vessels *3D Bioprinting for Reconstructive Surgery* (Amsterdam: Elsevier) pp 377–91

Ji S and Guvendiren M 2017 Recent advances in bioink design for 3D bioprinting of tissues and organs *Front. Bioeng. Biotechnol.* **5** 23

Kamisuki S, Hagata T, Tezuka C, Nose Y, Fujii M and Atobe M 1998 A low power, small, electrostatically-driven commercial inkjet head *Proc. MEMS 98. IEEE. Eleventh Annual Int. Workshop on Micro Electro Mechanical Systems. An Investigation of Micro Structures, Sensors, Actuators, Machines and Systems (Cat. No.98CH36176)* https://doi.org/10.1109/MEMSYS.1998.659730

Kani M H, Chan E C, Young R C, Butler T, Smith R and Paul J W 2016 3D cell culturing and possibilities for myometrial tissue engineering *Ann. Biomed. Eng.* **45** 1746–57

Kolesky D B, Truby R L, Gladman A S, Busbee T A, Homan K A and Lewis J A 2014 3D bioprinting of vascularized, heterogeneous cell-laden tissue constructs *Adv. Mater.* **26** 3124–30

Korin N, Bransky A, Dinnar U and Levenberg S 2009 Periodic 'flow-stop' perfusion microchannel bioreactors for mammalian and human embryonic stem cell long-term culture *Biomed Microdevices.* **11** 87–94

Li P, Lai D, Zhang F and Feng T 2014 Numerical modeling and simulation based on finite element method for internal cardiac defibrillation in a human torso 2014 *7th Int. Conf. on Biomedical Engineering and Informatics* pp 456–60

Li X, Chen J, Liu B, Wang X, Ren D and Xu T 2018 Inkjet printing for biofabrication *3D Printing and Biofabrication* (Cham: Springer) pp 283–301

Liu H, Zhou H, Lan H, Liu T, Liu X and Yu H 2017 3D printing of artificial blood vessel: study on multi-parameter optimization design for vascular molding effect in alginate and gelatin *Micromachines* **2017** 237

Lu T, Li Y and Chen T 2013 Techniques for fabrication and construction of three-dimensional scaffolds for tissue engineering *Int. J. Nanomed.* **8** 337–50

Murdock M H and Badylak S F 2017 Biomaterials-based *in situ* tissue engineering *Curr. Opin. Biomed. Eng.* **1** 4–7

Ng H Y, Lee K-X A, Kuo C-N and Shen Y-F 2018 Bioprinting of artificial blood vessels *Int. J. Bioprinting* **4** 140

Pei M, Solchaga L A, Seidel J, Zeng L, Vunjak-Novakovic G, Caplan A I and Freed L E 2002 Bioreactors mediate the effectiveness of tissue engineering scaffolds *FASEB J.* **16** 1691–4

Sarker M D, Naghieh S, Sharma N K, Ning L and Chen X 2019 Bioprinting of vascularized tissue scaffolds: influence of biopolymer, cells, growth factors, and gene delivery *Hindawi J. Healthcare Eng.*

Schöneberg J, Lorenzi F D, Theek B, Blaeser A, Rommel D, Kuehne A J C, Kiessling F and Fischer H 2018 Engineering biofunctional *in vitro* vessel models using a multilayer bioprinting technique *Sci. Rep.* **8** 10430

Sekine H, Shimizu T, Sakaguchi K, Dobashi I, Wada M, Yamato M, Umezo M and Okano T 2013 *In vitro* fabrication of functional three-dimensional tissues with perfusable blood vessels *Nat. Commun.* **4** 1399

Song L, Zhou Q, Duan P, Guo P, Li D, Xu Y, Li S and Zhang Z 2012 Successful development of small diameter tissue—engineering vascular vessels by our novel integrally designed pulsatile perfusion-based bioreactor *PLoS One* **7** e42569

Vadivelu R K, Kamble H, Shiddiky M J A and Nguyen N 2017 Microfluidic technology for the generation of cell spheroids and their applications *Micromachines* **8** 94

Wang J 2016 Ex vivo blood vessel bioreactor for analysis of the biodegradation of magnesium stent models with and without vessel wall integration *Acta Biomater.* **1** 546–55

Williams C B 2015 The status, challenges, and future of additive manufacturing in engineering *Comput. Des.* **69** 65–89

Zhou D, Chen J, Liu B, Zhang X and Xua X L T 2019 Bioinks for jet-based bioprinting *Bioprinting* **16** e00060

Zohar B, Blinder Y, Ephstein M, Szklanny A A, Kaplan B, Korin N, Mooney D J and Levenburg S 2019 Multi-flow channel bioreactor enables real-time monitoring of cellular dynamics in 3D engineered tissue *Commun. Biol.* **2** 158

Chapter 9

Blood flow evaluation in different circulatory systems

Arindam Bit and Jasjit S Suri

Pathogenesis of a diseased blood vessel is a critical dynamic phenomenon which intervenes in most of the immune disorders in the human body. Hemodynamic investigation of these blood vessels is of prime importance, because it helps in predicting acute consequences at the early stage of a diseased vessel. An extensive survey of literature is made on flow analysis of blood in diseased vessels of different anatomical positions.

9.1 Introduction

An extensive literature survey has been made on the numerical studies of a stenosed artery. Significant contributions by different authors have been observed. A 3D flow model was considered to evaluate the effects of wall shear stress (WSS) and vortex on the distribution of monocytes in the channel (Pritchard *et al* 1995). U937 cells in suspension form were perfused through this model which also contains a stenosis feature. The suspended particles were re-circulated. WSS, separation and re-integration in vortices were simulated in this model. A video micrograph was used to evaluate the velocity of rolling speed of cell and its adherence to the wall of the tube. Adhesion patterns of cells were also evaluated in the presence of chemotactic peptide gradients. A linear relationship was obtained between the rolling velocity of the cell and WSS, which is inversely proportional. Radial velocity and distribution of vortices are also influenced by cell-adhesion tendency. An increase in cell-adhesion to a wall coated with chemotactins was also observed.

Bluestein *et al* (1999) had evaluated the phenomenon of vortex-shedding in a laminar flow of simulated blood through a stenosed synthetic vessel. An axi-symmetric stenosed geometry (with 84% depth of stenosis) was considered for quantitative and qualitative evaluation of fluid flow in a streamline flow regime. Evaluation was modelled at different Reynolds (Re) numbers based on variable

doi:10.1088/978-0-7503-2088-7ch9

upstream velocity and channel width. Vortices generated at a localized spot were found to have a linear relationship with Re (size and duration of vortices). This observation helps us in understanding the formulation of aggregation in platelets and its deposition nearer to the region of stenosis. Carotid stenosis was also evaluated by using multiple grading methods. ECST and NASCET were used for grading of stenoses. The outcome of these two methods was compared with the common carotid method, and the inter-observer recurrence was evaluated. Results obtained were cross-verified with ultrasound output. The ultrasound method demonstrated superior accuracy over the above mentioned method.

Mittal *et al* (2001) had investigated pulsating flow of blood in a modelled arterial stenosis using large eddy simulation (LES) method. LES facilitates the evaluation of flow parameters at large Re of 2000. It evaluates progressive attenuation of the vorticity of the shear layer and transition states of the flow regime. Vortex shedding has been found to occur at a defined large frequency, and the feasible implication of this for minimal invasive diagnosis of the stenosed region was criticized. Further, Varghese and Frankel (2003) performed numerical modelling of pulsatile flow in a stenotic vessel in a turbulent regime. Turbulent flow had been modelled using the Reynolds-averaged-Navier–Stokes (RANS) evaluation approach. Multiple turbulence models were used to evaluate their influence on the results. The $\kappa-\omega$ turbulence model was found to be the best fit model at low as well as high Re. Peak WSS was found at the throat of the stenosis, while least WSS was found at the distal end of the stenosis, leading to flow-separation.

Lee *et al* (2003) also investigated turbulent flow through series stenosis using a numerical technique. Depth of stenosis was kept from 20%–60% of the vessel diameter. Flow regime of laminar flow was found to fit $\kappa-\omega$ laminar flow modelling. Inter-stenosis length, depth of stenosis, and Re of blood flow was found to influence the length of recirculation at the post-stenosis downstream region. Maximum value of centreline velocity and turbulence intensity was found progressively larger along the downstream secondary stenosis points. Growth of Re was found to have a positive effect on vorticity and centreline disturbance intensity. Re attenuates with an increase in number of secondary stenoses. In another turbulent study conducted by Ryval *et al* (2004), a two-equation model was used for solving pulsatile flow in a stenosed tube. Wilcox's standard $\kappa-\omega$ model and its transient variant were used here. Optimum grid resolution for the model was achieved by considering steady flow through the stenosed region. Inlet boundary condition was fixed by reviewing velocity magnitude and turbulence-field from experimental studies. Womersley flow profile was implemented at the inlet boundary and the outcome of numerical study was compared with experimental study. A two-equations turbulence model was found to replicate physiological flow condition in this numerical study.

Influence of carotid bifurcation on the risk assessment of stenosis in young and older populations was evaluated (Thomas *et al* 2005). Geometry was derived from a stack of MRI images of carotid bifurcation of 25 diseased subjects, and one control subject of a younger age group. Techniques with adequate robustness were employed to characterize 3D carotid arteries (of internal carotid artery, external carotid artery and common carotid artery). A higher similarity index of geometrical parameters was observed among the MRI images of young individuals, suggesting the existence of an inter-relationship between vascular geometry, its ageing, athelosclerosis and local hemodynamics. In another study, Suo *et al* (2007) conducted hemodynamic study of shear stresses in the mouse aorta in order to investigate the implications of atherosclerosis. Micro-CT methodology was used for reproducing numerical studies of the WSS of the aorta of mice that was found to be of higher magnitude than that of human beings. A quantum-dots based method was employed to evaluate cell adhesion to the aortic wall based on intercellular molecular protein-1 activation. A region with lower WSS was found promoting such phenomena. Atherogenic molecule distribution within the aorta of mice was analogous to stenosis distribution in the human aorta. It was found that WSS vector determines the trajectory of flow mediated inflammatory responses.

Kock *et al* (2007) determined stresses in carotid plaque vessels incorporating the fluid-structure-interaction (FSI) effect. Plaques may become vulnerable in the presence of mechanical stress on the protective fibrous cap due to flow of fluid. In order to evaluate it further, stenoses of different length, depth, thickness (of fibrous) and lipid core size were employed in the study. The maximal stresses generated thereby differed significantly. Evaluation of fibrous cap stresses might be effective in assessment of plaque severity, and thus enhance risk management in patients with carotid atherosclerosis. In another study, Morbiducci *et al* (2007) investigated the risk of atherogenesis in aorto-coronary bypass, and found helical flow as a fluid dynamic signature for such consequences. A correlation between helical flow structures and vascular wall indices for atherosclerosis was found. The study includes numerical modelling of bypass geometries which hinges to ascending aorta at variable angulation relative to aortic outflux. k-ϵ eddy viscosity model was employed for fluid dynamic simulation. Helical flow at the vicinity of bypass models was evaluated qualitatively. Oscillatory-shear-index (OSI) was found inversely proportional to helical flow index. It indicates the inhibitory role of helical flow structure to the formation of plaque by regulating the mechano-transduction pathway of cells.

Dehlaghi *et al* (2008) analyzed WSS in a stented artery using CFD techniques. The influence of stent geometry on hemodynamics in a stented carotid artery model with different stent geometries was reported. It was concluded that shear stress was influenced by three stent design parameters: strut spacing, strut profile, and number of struts. Strut design parameters such as strut spacing, strut profiles, and number of struts influences the WSS. Designing optimum and efficient struts requires the analysis of WSS at pre-, post- and stent-strut regions. Inclusion of flow dividers in the designing of stent enhances the magnitude of WSS. A carotid artery with stenosis and bifurcation also promotes disturbance in flow (Tan *et al* 2008). Analysis of flow was performed by transition and turbulence modelling. A patient-specific carotid bifurcated model was created by segmenting, followed by reconstructing of MRI images of the corresponding patient with 70% stenosis. A laminar flow model with a Menter's shear stress transport turbulence (SST) model and its scale adaptive simulation (SAS) variant were used in pulsatile simulations for analysis of velocity profiles. WSS and turbulent intensity were evaluated. A variation in value of WSS was found between the laminar flow model and the SST model, whereas the SAS model shows a similar WSS response to the SST model.

Transition to turbulence modelling of blood flow in an eccentric stenosed vessel was investigated by Varghese *et al* (2008). Results of their study suggested that distinct transitional flows were essential prior to effective utility of such models at low Re value, under critical hemodynamic conditions where turbulence might develop. In another study, Ro and Ryou (2010) conducted numerical simulation of the turbulent blood flow in a bifurcated artery with stenoses under recursive body acceleration, with non-Newtonian properties of blood. This is a validation of the $\kappa-\varepsilon$ model and Malin's turbulence model with a power law viscosity model of blood. The laminar flow model exhibited under-prediction of blood viscosity and WSS, whereas the standard $\kappa-\varepsilon$ model showed over-prediction. The influence of body acceleration in a 2D flow regime of a stenosed blood vessel was also computed (Shaw *et al* 2010). In that case, the Casson viscosity model of blood was considered. An elastic vessel was considered analogous to the arterial wall. Periodic body acceleration imposes unsteady flow through the arterial wall, and it was analysed in the presence of an external magnetic field.

WSS in coronary artery is found to be influenced by progression and translation of atherosclerotic plaque and arterial remodelling (Samady *et al* 2011). Twenty patients of coronary artery disease were consider with their radio-frequency intravascular ultrasound. These signals were used for boundary condition fixing in numerical modelling for evaluation of WSS. A low WSS coronary segment induced greater plaque and progression of necrosis and constructive remodelling. A high WSS coronary segment induced a higher necrotic core and calcium deposition. It also induces aggressive fibrous and fatty acid tissue and rapid expansion modelling, resulting in regeneration of an acute phenotype.

Biochemical studies of atherosclerosis in a coronary artery were also made (Ohayon *et al* 2014). Conjugated mechanical and hemodynamic factors were found regulating the growth of plaque. Biomechanical factors associated with rupture of plaque were also evaluated. Evaluation of elasticity of the wall, local strain and mechanical instabilities were made using the IVUS technique in order to access the progress of complex plaque geometry. The contour of the plaque geometry in dynamic form reveals the estimation of elastic modulus as a parameter for determining the stability of plaque. An improvement of the method was achieved by successful correlation of this estimation with histological estimation. X-ray angiography and IVUS methods were combined to evaluate follow-up studies of disease progression and hemodynamics in a stented artery (Vukicevic *et al* 2014). 3D reconstruction of a stenosed vessel was obtained using x-ray angiography and IVUS technique in a fusion manner. The Navier–Stokes flow equation was coupled with the reaction diffusion equation in the presence of a realistic wall boundary condition to evaluate flow phenomena of blood in such a vessel. IVUS–ECG gated frames were aligned transiently for extraction of baseline and periodic recording of the flow phenomena. THe right coronary artery underwent half-yearly follow-ups three times for assessment of hemodynamics. Results when compared with numerical results, indicate a higher confidence limit of numerical results.

A comprehensive survey of numerical investigations of stenosed blood vessela are made here. A few uncovered areas in these investigations include comparative assessment of different rheological models of blood in diseased conditions, proper assignment of flow rates with respect to diameter of blood vessel, and identification of the most typical state of stenosis, which might also be treated as a healthy state for blood flows through the diseased blood vessel.

9.2 Methodology

The geometrical model of an aortic arch is considered from a database of patient-specific CT-scan data. It uses stacking of images (CT scan) of the upper thorax. Differentiation in spatial resolution was used as a tool of segmentation for segregating ascending aorta (AA) and descending aorta (DA). Inlet diameter and length of the ascending part are 28 mm and 34 mm, respectively, whereas, length and diameter of the descending limb are 96 mm and 21 mm, respectively. The brachiocephalic artery, left common carotid artery, and left sub-clavian artery originate from the arch of the aorta with inlet diameters of 10 mm, 6 mm, and 5 mm, respectively. Outer and inner curvature of aortic arch are of length 102 mm and 73 mm, respectively. Developed geometry is given in figure 9.1.

Stenosis is introduced into the descending limb, and it is varied from 25% to 75% in order to observe the flow disturbances in the entire segment of vessel.

Different imaging modalities are used to obtain anatomical information inside deeper layers of an individual non-invasively. Several imaging modalities include

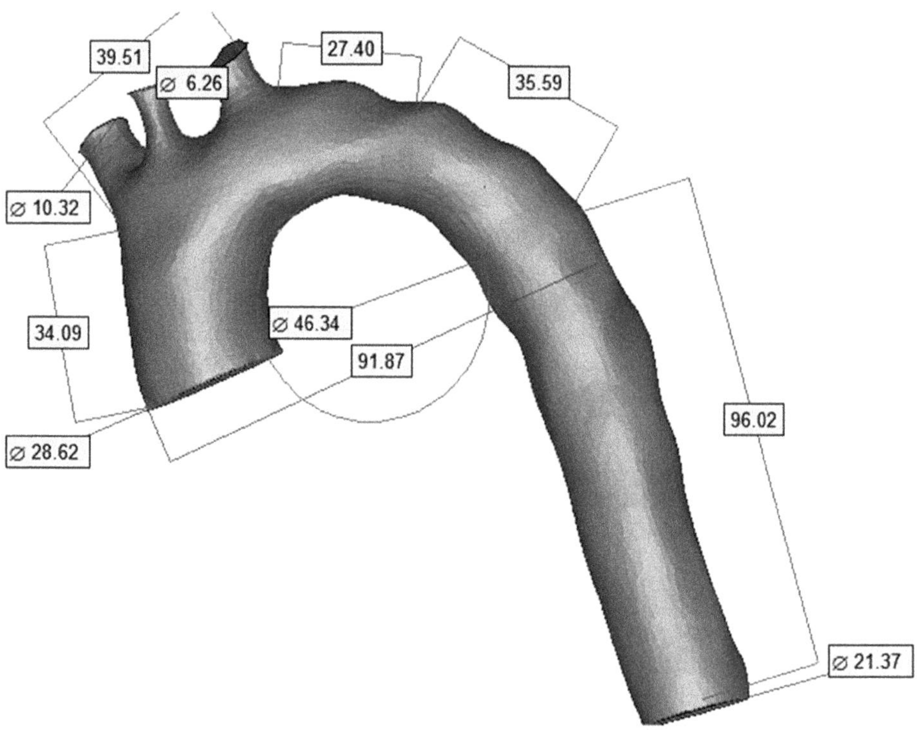

Figure 9.1. 3-D geometry of the aortic arch with ascending and descending limbs.

computerised tomography (CT) technique, magnetic resonance imaging (MRI) technique, ultra sonography (USG) technique, etc. Among these three techniques, CT scan modality uses radiation causing ionized particles for its operation. Thus it is harmful for any individual being exposed to such modality for a longer duration of time. But at the same time, it produces higher resolution images than the other two imaging modalities. This technique is more susceptible for higher density cell-information (like bone, ligaments) in the human body. An MRI scan on the other hand uses the water-density information of any cell in the body, and produces images with resolution function of water density. Soft tissues (like muscles, blood vessels, tendons) are best visualised by this imaging modality. The MRI scan can also be made time dependent with introduction of a phase-contrast reagent in order to view flow information in any annular region, this technique is known as four-dimensional phase contrast MRI imaging.

Information regarding volume and shape of the major aorta are obtained by scanning the thoracic region of the patient along three principle axes. Discrete scanning is done along each principle axis with a sample interval of 1 mm in the spatial domain (i.e. thickness of each slice). Spatial information thus obtained at every sample space has been represented in three forms, namely axial view (top to bottom), coronal view (left to right), and saggital view (anterior to posterior) as represented in figure 9.2.

The first task for generating a solid model is to create a segmentation known as threshold operation. Each slice of the CT scan is a resultant matrix of x-ray attenuation coefficients of scanned tissue. A region of interest is selected by defining a range of grey values. Setting a range of grey values for blood flowing through the aorta is difficult to determine. Thus, the region of interest is marked on the axial view in order to create profile lines.

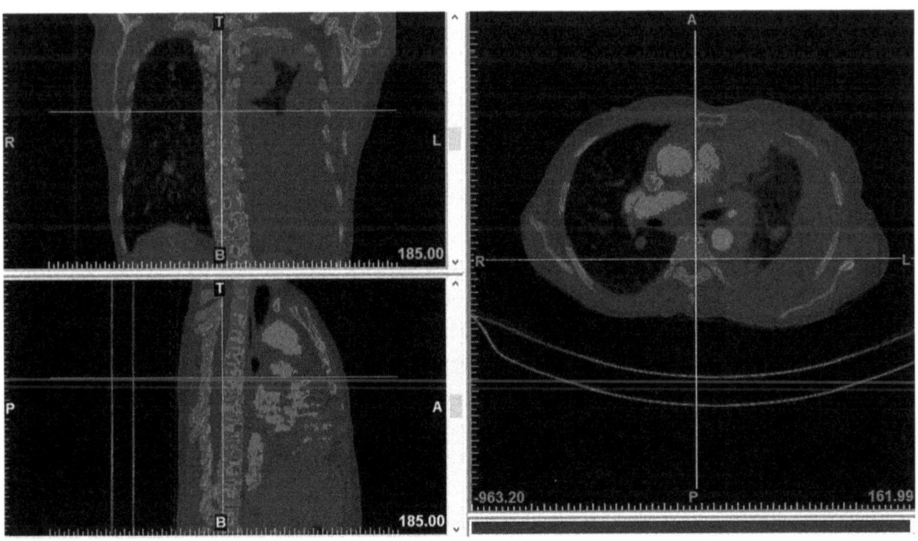

Figure 9.2. Different views of discrete slices (CT thorax) at a sample point.

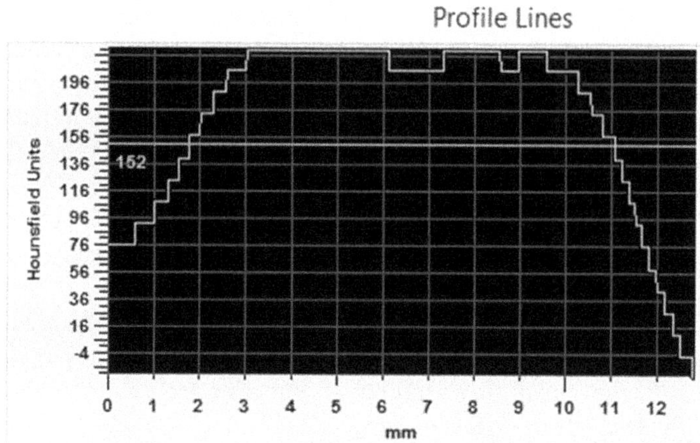

Figure 9.3. Profile lines correspond to region of interest.

Profile lines thus obtained (as shown in figure 9.3) reflect the range of grey value for blood flowing through the artery. Threshold values are indicated by upper and lower boundaries. Thr mask is an association of pixels belonging to this range of grey values.

The Hounsfield number for blood vessels is obtained by a 130 kVp, 96 mAs Siemens (Spirit) machine in the range of 152–229 (Technical User Manual, Spirit, Siemens). Average Hounsfield number is found to be 187.6854, with standard deviation of 20.9436. A slice increment of 0.999 mm is used during the procedure, while slice thickness is confined to 1.000 mm. The algorithm used for image interpolation is coded as B41s. Orientation of the gantry is made at RAB. A transformed image in the image intensifier is sampled corresponding to a pixel size of 0.723 mm. Exposure (aperture) of the x-ray tube (emitter) is manipulated in order to create a field of view of 37.00 cm. A histogram plot of the segmented image after threshold operation is given in figure 9.4.

Properties of the segmented image after threshold operations include minimum and maximum Hounsfield value of 158 and 222, respectively, average Hounsfield value of 185.5679, and its standard deviation as 21.7095. Total number of pixels included in the segmented volume is 1146 644 mm^3, whereas mask volume becomes 598 777.7045 mm^3.

The first step towards geometry-reconstruction is region growing. It mainly eliminates noise and separates structures that are not connected. Region growing allows segmenting a feature based on similarity of grey values within a fixed grey value range. It provides a simplified form of blood-vessel segmentation from CT images. The creation of this new mask is started by selecting a pixel. The algorithm starts comparing the grey values of the neighbouring pixels. Pixels with grey value which fit the pre-defined threshold according to equation (9.1) will be added to the new mask.

$$|\hat{\imath} - i| < d \tag{9.1}$$

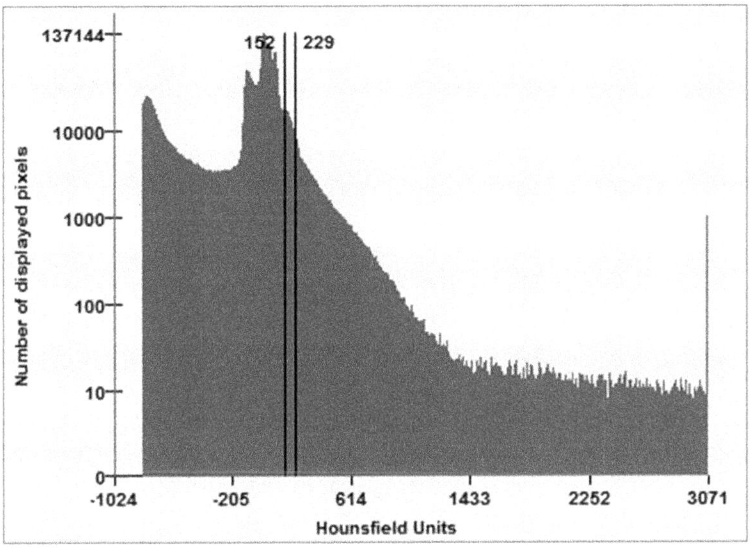

Figure 9.4. 3D histogram of Hounsfield number corresponds to region of interest.

where $\hat{i}$: average grey value, i: new grey value, and d: deviation. However, in order to eliminate local artefacts and to separate distinct blood vessels at the region of bifurcation, manual editing is done with local mean threshold value.

Region growing is followed by segmentation of enmarked pixel distribution in the spatial domain. On segmentation of the geometry (of all stacks of slices), dimensions of the segmented geometry are as follows:

X: 153.19 mm (minimum), 217.65 mm (maximum), 64.46 mm (delta);

Y: 88.87 mm (minimum), 178.95 mm (maximum), 90.09 mm (delta);

X: −1112.21 mm (minimum), −894.19 mm (maximum), 218.02 mm (delta).

Total volume of the segmented geometry is found to be 115 336.05 mm^3, whereas surface area becomes 22 903.18 mm^2. The total number of triangles and nodes found in segmented geometry are 24 098 and 12 053, respectively. Discredited spatial statistics of segmented geometry includes average Hounsfield value of 179.8855, standard deviation 19.2466, and total number of pixels as 221 892.

The contour interpolation method is used during segmentation procedure. However, continuity is preferred, keeping shell reduction details largest by 1 unit, slices positions are made from −1125.2 mm to −862.2 mm. In order to limit real-time execution load, a matrix reduction is made by 2 × 0.7227 mm (XY resolution) and 2 × 0.9999 mm (Z resolution). Eight iterations with smooth factor of 0.98 are used for smoothening of re-constructed geometry. Compensated shrinkage is activated during this process. Triangle reduction is made by the advanced edge reducing mode tolerance of 0.0903 mm, edge angle of 15.00°, and 20 iterations. A newly formed 3D aorta (containing ascending limb, descending limb, CCA, brachiocephalic artery, left carotid artery) is already shown in figure 9.1.

Generation of centerline is necessary to define cross-sections and associated boundary conditions. The MedCAD operation is used to generate the centreline.

Resolving resolution is kept at 0.100 mm, and distance between controls points is made to be 0.100 mm. Twenty iterations are used for creating the centreline for the entire geometry. Various parametric measurements are used in order to obtain 3D information of the segmented blood vessel. The centerline properties for the segmented geometry include branches, namely one inlet and four outlets. The centreline with dimensions for cross-sectional diameters at various segments of the vessel are given in figure 9.5.

Similar methodology is used to obtain a 3D reconstructed geometry of a bifurcated carotid artery with severe stenoses in between the onset of the right carotid artery (RCA) and inter carotid artery. Developed reconstruction from the above CT scan data is done by the formation of three dimensions of the RCA from the stack of CT scan slices.

Finite element modelling was done by initiating the process of re-meshing. An inspection scene is created to examine the quality of meshes while constructing the geometry. The number of mesh elements is reduced by fixing the mesh geometry. Then, by automatic re-meshing, most of the bad segments of mesh elements are eliminated. During the procedure, shape quality threshold is kept at 0.400, and maximum geometry error permissible is 0.100.

Total numbers of triangles in the re-constructed geometry are 21 316, whereas the number of points is 10 662. Volume and surface area of this geometry are 108 411 mm^3 and 21 204 mm^2, respectively. The inspection scene is created for inspecting the qualities of triangles, as shown in figure 9.6. Mesh quality can be defined in various methods. All of them are dealing with formation of triangles,

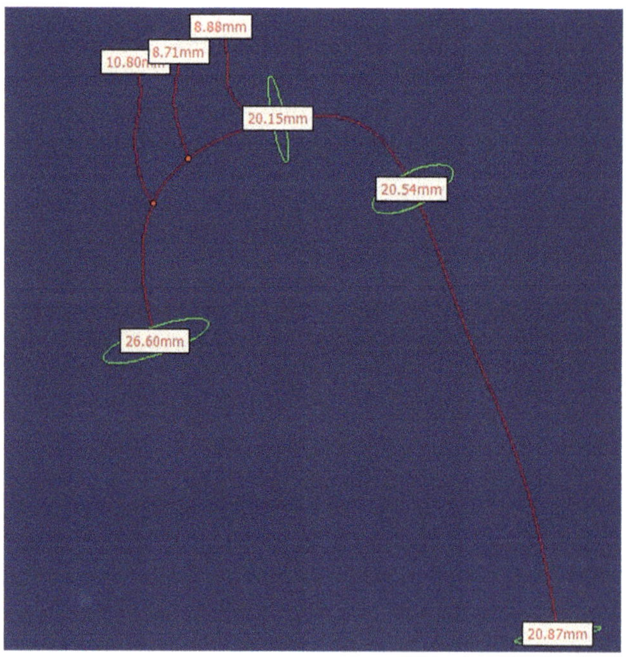

Figure 9.5. Centerline of the arch of an aorta with radial dimensions.

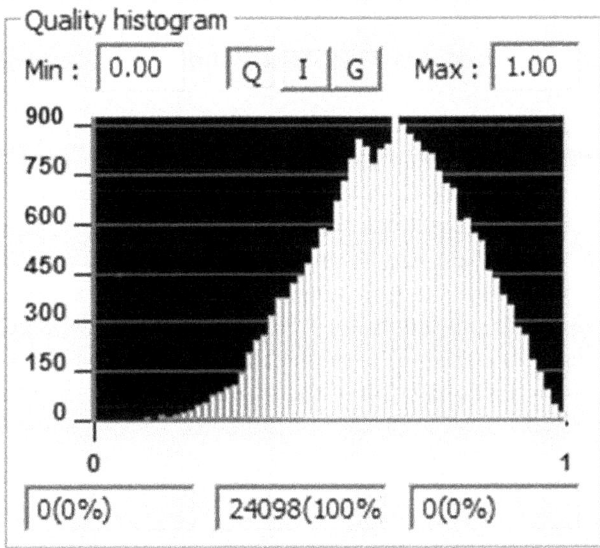

Figure 9.6. Quality histogram.

which leads to mesh formation. Quality of mesh is governed by quality of mesh-triangles such as length, height, base, ascribed and inscribed circles, and angle. The highest quality mesh consists of a maximum number of equilateral triangles.

Various quality parameters used for automatic remesh operations, and are described in detail as follows:

1. R-in/R-out (normalised): It defines the ratio of radius of inscribed to ascribed circles of an equilateral triangle. Its value is normalized to the factor of 2, and lies between 0 and 1.
2. Height/base (normalised): It is defined by the ratio of the height of the triangle and the length of its base (multiplied by sine 60°).
3. Height/base (absolute): It is defined by the ratio of height of triangle to that of the base. Magnitude of the ratio varies from 0° to sine (60°).
4. Skewness (normalized): It is defined as the ratio of actual area of the triangle to that of an equilateral triangle with the unique ascribed circle.
5. Smallest angle (normalized): It is equal to the ratio of smallest angle of a triangle and 60°.
6. Largest angle (normalized): It is defined by the ratio of difference of 180° and largest angle to 120°.
7. Smallest angle (absolute): It categorizes triangles based on their smallest angle. Its magnitude lies between 0° and 60°.
8. Angle ration (normalized): It is defined as the ratio of smallest to largest angle.
9. Edge ratio (normalized): It is defined by the ratio of smallest to the largest edge.

10. Equi-angle skewness (normalized): It is defined as the minimum of value (smallest angle/60°) or (180°—the largest angle)/(120°).
11. Stretch (normalized): It is defined as (inscribed circle radius/longest edge length)$^{1/2}$.

Various inspection parameters are used for quantifying quality of the mesh. Largest angle (absolute) should be equal to 60°. The bad edge (normalized) is found to be equal to 1 (i.e., the triangle has no bad edges). The comparison is done in two steps. First of all, the smallest neighbour is selected from all the neighbours of the actual triangle. Next, the size of this triangle is compared with the actual triangle. If the smallest neighbour is still larger than the actual triangle, the growth value is 1. This means that the actual triangle is smaller than all of its neighbours. In the other case, the growth factor is the fraction of the small neighbour and the actual triangle. We can also express this mathematically as equation (9.2):

$$\text{Growth factor} = \min\left[\frac{\text{Min}A_i}{A_*},\ 1\right] \tag{9.2}$$

where A = size definition (Area or largest edge length);

A_*= size of current triangle;

A_i = size of neighboring triangles.

To determine the neighbours, the ring definition can be created as follows.

Once the split based re-mesh is done, to smooth the transitions from small to large triangles, the re-mesh based on growth factor is used. Minimum and maximum edge length parameters during the process are kept at 0.005 mm and 2.00 mm, respectively. Maximum geometrical error allowed in the process is 0.02 mm.

Triangle-reduction methodology leads to localised shrinking of the geometry. Thus it requires an overall smoothing operation. Global smoothing is done for the entire geometry. Position of the center point is altered according to the positions of eight other points of eight triangles which are originally connected. The ratio-parameter is kept as high as 5. Now, the new position of the center point is mainly determined by the position of the other points of the triangles. A smoothing algorithm has the side effect that the part is shrinking. A shrinkage compensation factor will compensate shrinking. Negative compensation of −0.005 is used to move the center point in the opposite direction (out of 'smooth' position), so points oscillate forwards and backwards. But the oscillation of center points will be aborted when the angle between the normal of the neighbouring triangles is bigger than the given triangle. Throughout the entire smoothing process, the point will not be moved when it is at the bad edge. Curvature smoothing is done for confirming the new position of points based solely on geometry.

Maximum edge length for this locality is kept at 2 mm, and the influence area is kept at 3 mm. A volume mesh is created thereafter following the procedure for surface mesh. Table 9.1 summarizes the mesh statistics of the entire process of 3D re-construction and the same post-enhanced smoothening.

Table 9.1. Mesh statistics for the arch of an aorta.

Parameters	Pre-smoothing	Post smoothing
Number of degrees of freedom	50 746	11 256
Number of mesh points	2 232 497	924 546
Number of elements	846 249	453 402
Tetrahedral	639 462	282 908
Number of boundary elements	213 256	161 506
Triangular	83 256	52 796
Number of edge elements	5524	2536
Number of vertex elements	8784	3859
Minimum element quality	0.359	0.122
Element volume ratio	0.0001	0.0025

Table 9.2. Mesh element statistics of diseased and healthy vessel.

Carotid artery	Types of mesh element	Total no. of elements	Total area (mm^2)	Total volume (mm^3)	Inlet diameter (mm)	ECA diameter (mm)	ICA diameter (mm)	Height (mm)
Right CA (Diseased)	Tetrahedral: 178 808; Triangular: 22 516; Pentagonal: 99 903; Quadrilateral: 291	312 615	3438.76	4921.87	9.09	2.39	2.73	161.75
Left CA (Healthy)	Tetrahedral: 178 968; Triangular: 32 590	211 642	2942.26	3293.18	5.02	3.23	3.59	156.91

Finite element method studies were conducted by discretizing the geometry in the first step. Grid-optimization was achieved by grid-independent study. A later section briefs about the grid-convergence study. The mesh statistics of optimized mesh configuration is shown in table 9.2.

Most researchers assumed blood to be Newtonian while transporting through arteries. But in the case of a diseased vessel, antigen–antibody particulates presence makes blood heterogenous in nature. Thus, in such a situation blood very much behaves as non-Newtonian. A power law viscosity model (Walburn and Schneck 1976) was implemented in the current study. Equations (9.3) and (9.4) show stress and viscosity expression of blood using power law implementation.

$$\bar{\tau} = k|\bar{\gamma}|^{n-1}\bar{\gamma} \tag{9.3}$$

and

$$\mu(|\bar{\gamma}|^*) = k\frac{U_\infty^{n-1}}{l^{n-1}}|\bar{\gamma}|^{*n-1} \tag{9.4}$$

Chien *et al* (1967) calculated healthy blood sample parameters as $k = 14.67 \times 10^{-3}$ Pa-s, and $n = 0.7755$. Fluid becomes shear thickening for $n > 1$, and otherwise shear thinning. The characteristic parameter for a power-law-model-based flow is Re_{PL}.

The power law model was unable to realise Newtonian expression of shear-thinning fluid either at very low or very high shear stress. Thus, the Cross model (Cross 1965) was employed for making a viscosity model shear rate dependent, as given by equation (9.5).

$$\mu(|\bar{\gamma}|) = \mu_\infty + \frac{(\mu_0 - \mu_\infty)}{\left[1 + \left(\frac{\bar{\gamma}}{\bar{\gamma}_c}\right)^n\right]} \tag{9.5}$$

where $\mu_0 = 0.0364$ Pa$-s$ represents viscosity of blood at a ultra low shear rate, while $\bar{\gamma}_c = 2.63\ s^{-1}$ is considered to be reference shear rate, and $n = 1.45$ is the model constant.

The equation for mass conservation for cylindrical, axi-symmetric geometries is given by equation (9.6) as,

$$\frac{\partial\rho}{\partial t} + \frac{\partial(\rho v_z)}{\partial z} + \frac{1}{r}\frac{\partial(\rho r v_r)}{\partial r} = 0 \tag{9.6}$$

where r is the radial coordinate and z is the axial coordinate. v_r and v_z are the respective velocity components.

In cylindrical coordinates, for axi-symmetric problems the axial (z) momentum equation is given as,

$$\frac{\partial(\rho v_z)}{\partial t} + \frac{\partial(\rho v_z v_z)}{\partial z} + \frac{1}{r}\frac{\partial(\rho r v_r v_z)}{\partial r} = -\frac{\partial p}{\partial z} + \frac{\partial(\tau_{zz})}{\partial z} + \frac{1}{r}\frac{\partial(r\tau_{zr})}{\partial r} \tag{9.7}$$

Similarly, the radial (r) momentum equation is given as:

$$\frac{\partial(\rho v_r)}{\partial t} + \frac{\partial(\rho v_z v_r)}{\partial z} + \frac{1}{r}\frac{\partial(\rho r v_r v_r)}{\partial r} - \frac{\rho v_\theta^2}{r} = -\frac{\partial p}{\partial r} + \frac{\partial(\tau_{zr})}{\partial z} + \frac{1}{r}\frac{\partial(r\tau_{rr})}{\partial r} - \frac{\tau_{\theta\theta}}{r} \tag{9.8}$$

Clearly, for axi-symmetric problems, two additional terms appear in the radial momentum equation. These terms are $\rho v_\theta^2/r$ in the left hand side of equation (9.8), which is essentially an acceleration term (or convection term) and $\tau_{\theta\theta}/r$ in the right hand side of equation (9.8), which is a viscous force, arises from the tangential stresses due to curvature effect. The viscous stresses appear explicitly, and for Newtonian fluids they are given by the set of equations (9.9) as follows;

$$\tau_{xx} = \mu\left[2\frac{\partial u}{\partial x} - \frac{2}{3}(\nabla . v)\right] \tag{9.9a}$$

$$\tau_{yy} = \mu\left[2\frac{\partial v}{\partial y} - \frac{2}{3}(\nabla.\ v)\right] \tag{9.9b}$$

$$\tau_{yx} = \tau_{xy} = \mu\left[\frac{\partial u}{\partial y} + \frac{\partial v}{\partial x}\right] \tag{9.9c}$$

It may be noted here that for incompressible flows, the divergence of velocities, $\nabla.\ v = 0$ and hence, the expressions for stresses.

In the case of a 3D patient-specific blood vessel, emphasis is given on outlet boundary conditions by varying pressure profiles at the exit of different blood vessels. Inlet flow rates were measured at the centreline of an ascending section of an ascending aorta very much away from the arch of the aorta. The flow profile thus obtained was then transformed into Womersley's flow profile by implementing the Womersley formula (Womersley 1955). It provides a velocity profile at the cross section instead of at a point or line. The unsteady laminar flow of an incompressible fluid with a constant density and viscosity through a pipe can be realized by Womersley's equation. In the presence of a transient response of pressure gradients, velocity in a function is defined by the coefficient of a complex Fourier series of the former. By interpolation of transient response of flow-rate, the inlet velocity $u(t)$ was approximated by a Fourier series function. The period 'T' of the cardiac pulse was discretized in equal interval of length dt (d$t = T/M$), and the regular sampling interval point is given by $ts = s\mathrm{d}t$, $s = 0,1....\mathrm{M} - 1$. Therefore, the sampled flow rate values are stacked in a manner of $u_0 = u(t_0)$, $u_1 = u(t_1)$,......$u_{m-1} = u(t_{m-1})$, respectively. $M - 1$ uniformly sampled values of inlet flow rate were employed with respect to time pulse of maxima $N = (M - 2)/2$ harmonics of the Fourier series that is computed. For the pulse, the values of corresponding Fourier series coefficients are given in table 9.3.

$$\begin{aligned} u(t) = 0.1223 &+ a_1 \times \cos(\omega t) + b_1 \times \sin(\omega t) \\ &+ a_2 \times \cos(2\omega t) + b_2 \times \sin(2\omega t) + a_3 \times \cos(3\omega t) \\ &+ b_3 \times \sin(3\omega t) + a_4 \times \cos(4\omega t) + b_4 \times \sin(4\omega t) \\ &+ a_5 \times \cos(5\omega t) + b_5 \times \sin(5\omega t) + a_6 \times \cos(6\omega t) \\ &+ b_6 \times \sin(6\omega t) + a_7 \times \cos(7\omega t) + b_7 \times \sin(7\omega t) \\ &+ a_8 \times \cos(8\omega t) + b_8 \times \sin(8\omega t) \end{aligned} \tag{9.10}$$

Table 9.3. Values of coefficients of equation (9.10).

Type-II waveform							
a_1	−0.0708	a_5	0.000 3287	b_1	0.050 06	b_5	0.012 44
a_2	−0.010 54	a_6	0.031 45	b_2	0.0267	b_6	0.011 78
a_3	0.017 74	a_7	0.002 938	b_3	0.059 51	b_7	0.017 59
a_4	−0.010 43	a_8	0.003 075	b_4	0.0504	b_8	0.009 723

where $\omega = 4.436$, and values of coefficients a's, and b's of equation (9.10) are shown in table 9.3.

Womersley's velocity profile thus transformed from equation (9.10) is given in equation (9.11) as $w(t)$. And it is applied to the inlet of the vessel at aortic arch origin cross section.

$$w(t) = 2 \times \left(1 - \frac{z^2 + y^2}{R^2}\right) \times u(t) \tag{9.11}$$

Largely varying outlet pressures (at descending aorta (DA), left common carotid artery (LCCA), brachiocephalic artery (BCA) and left subclavian artery (LSA)) are employed as outlet pressure boundary conditions for the aortic geometry. The 4D phase-contrast MRI technique revealed translational changes in neighbourhood outlet boundaries. Outlet pressure profiles are expressed as milli Pa. A set of combinations of implementation of boundary conditions (as given in table 9.4) are framed in order to interpolate the effects of different outlet boundary conditions for influencing flow structures. The first column of the table indicates inlet boundary conditions for the AAO cross section, and the remaining four columns indicate outlet boundary conditions on all outlets of the geometry. The ticks indicate implementation of corresponding outlet boundary condition at respective cross-sections of the 3D model.

The wall of the blood vessel has several special features, which makes it ideal for flow of blood through its lumen. Vital components for attenuating clotting in the vascular system are: (1) smoothness on the surface of the endothelial layer, preventing intrinsic clotting system activation; (2) layer of glycocalyx on the endothelium surface in order to repel clotting factors and platelets; and (3) thrombomodulin protein binding to the endothelial membrane. A conjugate of the thrombomodulin–thrombin complex helps draining out of thrombin, triggers plasma protein and cascade anticoagulation process by triggered factors V and VIII. Smoothness and the glycocalyx protective layer degrade on destruction of the endothelial wall. This triggers factor XII and platelets which in turn initiates the intrinsic pathway of clotting. On interaction of these factors with sub-endothelial collagen, the clotting process becomes more powerful.

The wall boundary condition is used to bind the fluid and the solid region. In the present investigation, a no-slip boundary condition is enforced at walls (endothelial

Table 9.4. Configuration of inlet-outlet boundary conditions on a case-by-case basis.

Case no.	AAO	DAO	BCA	LSA	LCCA
Case 1	✓				
Case 2	✓	✓			
Case 3	✓	✓	✓		
Case 4	✓	✓	✓	✓	
Case 5	✓	✓	✓	✓	✓
Case 6	✓		✓	✓	✓

lining, also known as tunica interna) during fluid flow through the diseased blood vessel. It is given by equation (9.12).

$$u = v = w = 0 \tag{9.12}$$

9.3 Results and discussions

A 3D model of the aortic arch is obtained from a patient-specific aortic arch. Synthetic stenosis is created at the descending limb of an aorta. The heights of plaques range between 25% and 75%. The effects of outlet boundary conditions at LCCA, BCA, LSA and DA on OSI and relative residual time (RRT) at different zones (descending limb, inferior and superior curvature of the aortic arch) of the geometry are observed with an objective of identification of dead zones along the aortic arch.

Figure 9.7 shows the variations of OSI at different section of a stenosed arch of an aorta at 25% stenosis in the descending limb for all six cases of boundary conditions. Case 5 shows the highest response of OSI of 0.3 and 0.5 at the descending part of the aorta and at the superior curvature wall of the aorta (as in case 2).

A descending aorta with 25% blockage shows minimum response to OSI distributions at inferior curvature of the aortic arch. Similarly, it has been found that case 6 shows minimum response of OSI value at all three regions of the aortic arch. This indicates that non-zero pressure boundary conditions at outlets (especially at the distal end of a descending aorta, downstream of stenosis) are greatly influencing the OSI distribution along the entire length of the aortic arch.

Figure 9.8 shows the variations of OSI at different sections of a stenosed arch of an aorta at 75% stenoses in the descending limb for all six cases of boundary conditions. Case 5 shows the highest response of OSI of 0.22 and 0.50 at the descending aorta and inferior curvature of the aortic arch, respectively. However, a higher value of OSI is observed of 0.14 at the region of the superior curvature of the aortic arch while considering case 5. A descending aorta with 75% blockage shows minimum response to OSI distribution at the superior curvature of the aortic arch. In the case of 75% stenoses, numerical investigation using case 6 combinations of

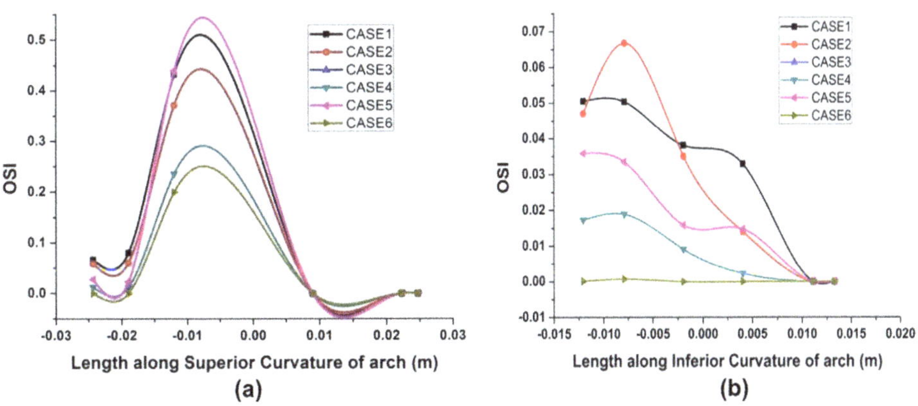

Figure 9.7. OSI plot at six different boundary conditions in a vessel with 25% blockage in the descending limb at (a) inferior and (b) superior curvature of the aortic arch.

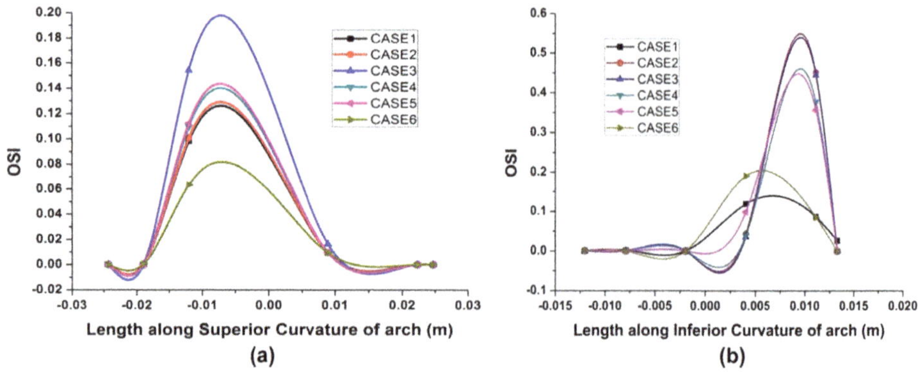

Figure 9.8. OSI plot at six different boundary conditions in a blood vessel having 75% blockage in the descending part at (a) inferior and (b) superior curvature of the aortic arch.

boundary conditions shows minimum response of OSI value at all three regions of the aortic arch, similar to that reflected in the aortic arch with stenoses of 25% and 50% severity. Again, it confirmes that non-zero pressure boundary conditions at outlets (especially at the distal end of the descending aorta, downstream of stenosis) greatly influence the OSI distribution along the entire length of the aortic arch. The response of OSI for both cases 1 and 6 are minimal in all three segments of the aortic arch.

Figure 9.9 shows the variations of RRT at different sections of a stenosed arch of an aorta at 25% stenosis in the descending limb for all six cases of boundary conditions. Case 5 shows highest response of RRT of 7000, 33 and 130 at the descending aorta, inferior curvature and superior curvature of the aortic arch, respectively. Descending aorta with 25% blockage shows minimum response to RRT distributions at the inferior curvature of the aortic arch, while the descending limb itself shows abruptly maximum response to RRT distributions. Similarly, it has been found that case 6 shows minimum response of RRT value at superior curvature of the aortic arch. This indicates that non-zero pressure boundary conditions at outlets (especially at the distal end of the descending aorta, downstream of stenosis) are greatly influencing the RRT distribution along the superior curvature of the aortic arch.

In the remaining two sections of the vessel (i.e., descending limb and inferior curvature of the aortic arch) cases 1 and 5 show significantly dampened responses to RRT distribution in comparison to that of rest of the cases. This confirms the significance of non-zero outlet boundary conditions in the descending limb of geometry like an aortic arch.

Figure 9.10 shows the variations of RRT at different sections of the stenosed arch of an aorta at 75% stenosis in the descending limb for all six cases of boundary conditions.

In the case of a carotid artery, bifurcation followed by stenosis at the ICA of the left branch of coronary arteries induces anomalies in blood flowing through the RCA at the upper neck and brain regions. The effective structure of blood flow can

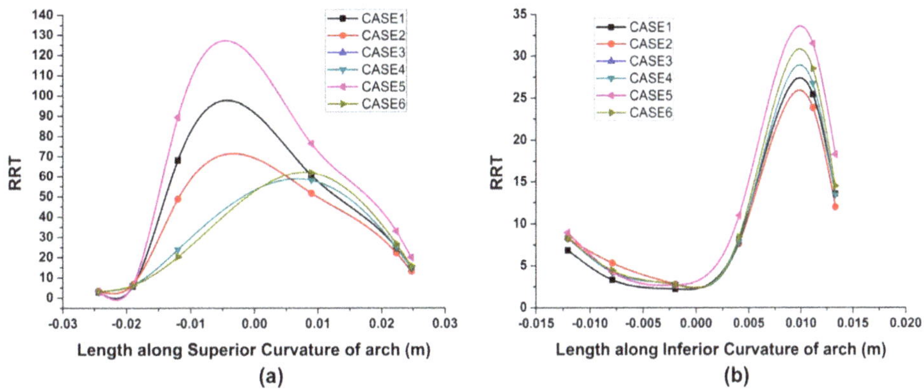

Figure 9.9. RRT plot at six different boundary conditions in a vessel with 25% blockage in the descending limb at (a) inferior and (b) superior curvature of the aortic arch.

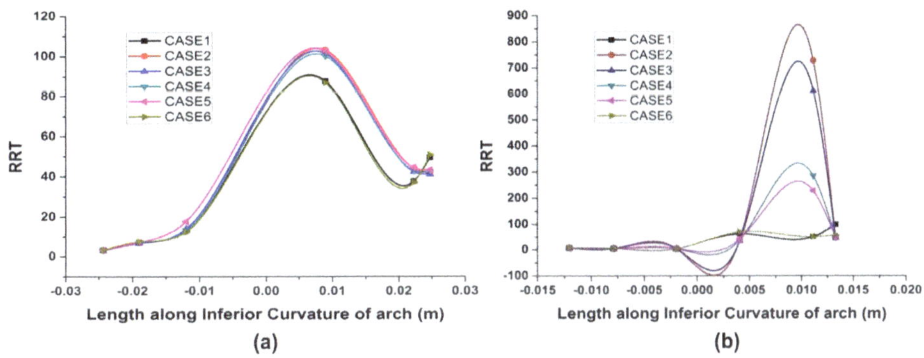

Figure 9.10. RRT plot at six different boundary conditions in a blood vessel having 75% stenosis in the descending limb at (a) inferior and (b) superior curvature of the aortic arch.

be seen in a diseased RCA. A transient response can be recorded by obtaining a streamline velocity plot at regular time steps in line with cardiac cycles. WSS can also be observed from its contour plot on the vessel wall. Hemodynamic studies were also conducted at higher Re (equivalent to stressful activities of brain and sudden neck-impact) for assessment of point identification of a vessel for its rupture. Quantitative assessment was done by evaluating WSS on the wall at various points.

Variations in streamline representation of the velocity plot at the region of bifurcation of the carotid artery at Re II, while implementing the power law model for substituting viscosity of diseased blood were made. Velocity streamline at every time step of the cardiac cycle shows development of flow. An elevated flow regime is observed developing at the bifurcated region between the internal carotid artery and the subsequent-LCA over time. These kinds of localized anomalies at the region of bifurcation might occur with due influences from the stenosed the RCA.

Higher flow rate in this region simultaneously influences higher risk of secondary stenosis. Vortices are initially observed downstream of the flow from bifurcated regions near walls of the daughter-left carotid branch and internal-left carotid artery. However, at the end of one complete cardiac cycle, vortices are found even in the upstream flow, in regions of pre-bifurcation of the left carotid artery. Also, it is observed that maximum flow rate is greatly reduced from 0.6 m s^{-1} to 0.1 m s^{-1} due to development of vortices in the flow regime.

On further increased flow rate, it is observed that an elevated flow regime is developed at the bifurcated region between the external and internal wall of the left carotid artery, with a tendency of higher magnitude of velocity at the beginning of the internal side of the left carotid artery over time. This kind of recurrence of localized anomalies at the region of bifurcation confirms the occurrence of the trend with due influences from the stenosed RCA. Higher flow rate in this region simultaneously influences higher risk of secondary stenosis in the internal left carotid artery.

Surprisingly, vortices are initially observed downstream of the flow from the bifurcated regions near the walls of the external left carotid branch only as compared with the velocity streamline observed at lower flow rate. Spatial distribution of vortices remains unchanged for the entire cardiac cycle in the downstream region of bifurcation, however, newly developed vortices are also observed in the upstream of bifurcation at the end of one complete cardiac cycle. It is also observed that maximum flow rate reduces from 0.9 m s^{-1} to 0.5 m s^{-1} due to development of vortices both upstream and downstream of the flow regime.

The region of stenosis gives rise to the development of high shear rate in the RCA and it is found consistent over the entire cardiac cycle. However, it is observed that the presence of stenosis at the region of bifurcation induces development of vortices in external branches of the RCA. Size of vortices increased with progression of time steps. Presence of stenosis at the bifurcation induces larger vortices in the main branch of the RCA over the entire cardiac cycle. Critically elevated shear rate is found far downstream of the external limb of the RCA. Higher flow rate in this region simultaneously influences higher risk of secondary stenoses in the external limb of the left carotid artery. Spatial distributions of vortices remain unchanged for the entire cardiac cycle in the downstream region of bifurcation.

On further increased flow rate, it is observed that an elevated flow regime is developed downstream of the stenosis in the internal branch of the RCA. Structures of vortices are found mostly irregular in shape on increased flow rates in both branches downstream of stenosis. However, influence of stenosis in producing vortices is mostly observed in the opposite branch (external branch) of the RCA, whereas, vortices vanish from the affected branch of the vessel.

This signifies that chances of occurrence of secondary stenoses are greater in the downstream flow of the opposite branch than the branch of origin of stenosis. Developments of vortices are also observed upstream of the bifurcation, in the main branch of the RCA with reduced sizes of vortices. Spatial distribution of vortices remains unchanged for the entire cardiac cycle in the downstream region of bifurcation. It is also observed that maximum flow rate reduces from 6.275 m s^{-1} to 1.9 m s^{-1} due to development of vortices both upstream and downstream of the flow regime, and sustainability of stenosis in the region of bifurcation.

Variety in physical features of the endothelial layer of the vessel wall was evaluated using WSS spatial and temporal distribution in the region. Both gradient and harmonic based descriptors are evaluated in this study. Endothelial cell tension was evaluated by WSS spatial gradient (WSSG), which is evaluated from tensor components of WSS parallel and perpendicular to the TAWSS vector (*m* and *n*, respectively), and it is given by equation (9.13).

$$WSSG = \frac{1}{T}\int_0^T \sqrt{\left(\frac{\partial \tau_{w,\,m}}{\partial m}\right)^2 + \left(\frac{\partial \tau_{w,\,n}}{\partial n}\right)^2}\,dt \tag{9.13}$$

Response of WSSG to different rheological model was found and mentioned. The Cross model has been found to exhibit higher response to flow parameters like WSSG. This indicates the higher efficiency of the Cross model for high-shear regime flow in non-deterministic geometrical models with bifurcated features. It is observed that WSSG is also influenced by flow rate. At higher flow rate, WSSG is found to be higher over the integral of time.

Maximum absolute gradient in the magnitude of WSS over a cardiac cycle is also known as temporal gradient of WSS (WSST), and it is calculated by equation (9.14) as:

$$WSST = \max\left(\left|\frac{\partial|\tau_w|}{\partial t}\right|\right) \tag{9.14}$$

Behavior of WSST does not follow a pattern similar to WSSG in the case of the Cross model. Behavior of WSST is influenced by both rheological model and flow rate (for the latter, WSST is inversely proportional) in a healthy blood vessel. For a stenosed vessel when exposed to higher flow rate of blood, WSST oscillates with increase in flow rate. WSST magnitude varies more profoundly along the axial direction than the radial direction. This difference in magnitude of WSST can be found at a local minima at the primary steps of the cardiac cycle, and it enhances with progress of the cardiac cycle.

A tentative metric of flow disturbances can be measured from harmonic components of WSS. Further, it was confirmed that the responses to the flow disturbances (induced by WSS) interact by trans-sensory behavior of endothelial cells. Amplitudes of transient WSS can be quantified by the Fourier decomposition method. Amplitude of highest frequency at the end of decomposition is known as dominant harmonics (DH). DH is calculated by equation (9.15) as:

$$DH = \max(F_w(nw_0)),\ F_w \equiv FFT(|\tau_w|),\ \omega_0 = 2\pi/T \tag{9.15}$$

Observation indicates an increase in flow-rate leads to increase in DH in diseased vessel. However, in the case of a normal blood vessel, DH shows an oscillatory response. The Cross model also shows an influence on the behavior of DH distribution.

9.4 Conclusion

Patient-specific vessel-geometries, namely the aortic arch and bifurcated carotid artery, were selected for evaluation of flow parameters in a stenosed blood vessel. Stenosis was also introduced in the descending limb of an aortic arch with varying depth of stenosis from 25% to 75%. Influences of outlet boundary conditions on flow structures in several severities of blockages were investigated to obtain distributions of fluid mechanical parameters like WSS, OSI and RRT along the entire length of the aortic arch. It was found that these parameters were measured with maximal efficacy when patient-specific non-zero pressure profiles were implemented at the outlets of the vessel. Induction of secondary stenoses on the inferior wall of the aortic arch can be found in the presence of acute stenosis in the descending limb of the aorta. The superior wall of the aortic arch was also prone to secondary stenoses in the presence of mild stenosis in the descending limb of an aorta.

In the case of bifurcated a carotid artery model, stenosis was already found in the first-generation branching of the RCA, whereas the left carotid artery was found in a healthy condition. Hemodynamics of blood was investigated to calculate distribution of shear stress parameters like velocity streamline, WSS and gradient and harmonic markers of WSS to measure the influence of stenosis on the RCA to produce secondary stenoses in the rest of the bifurcated carotid artery at different flow rates. Secondary stenosis was vulnerable in the external branch of the RCA. A similar tendency to secondary stenoses can also be found in LCA in the presence of primary stenosis inthe internal branch of the RCA. It was also found that on increased flow rate, the external branch of the RCA was exposed to an abrupt value of shear stress.

References

Bluestein D, Gutierrez C, Londono M and Schoephoerster R T 1999 Vortex shedding in steady flow through a model of an arterial stenosis and its relevance to mural platelet deposition *Ann. Biomed. Eng.* **27** 763–73

Chien S, Usami S, Dellenback R and Gregersen M 1967 Blood viscosity: influence of erythrocyte deformation *Science* **157** 827–9

Cross M M 1965 Rheology of non-Newtonian fluids: a new flow equation for pseudoplastic system *J. Colloid Sci.* **20** 417–29.

Dehlaghi V, Shadpoor M T and Najarian S 2008 Analysis of wall shear stress in stented coronary artery using 3D computational fluid dynamics modelling *J. Mater. Process. Technol.* **197** 174–81

Lee T S, Liao W and Low H T 2003 Numerical simulation of turbulent flow through series stenoses *Int. J. Numer. Methods Fluids* **42** 717–40

Mittal R, Simmons S P and Udaykumar H S 2001 Application of large-eddy simulation to the study of pulsatile flow in a modelled arterial stenosis *J. Biomech. Eng.* **123** 325–34

Morbiducci U, Ponzini R, Grigioni M and Redaelli A 2007 Helical flow as fluid dynamic signature for atherogenesis risk in aortocoronary bypass: a nerical study *J. Biomech.* **40** 519–34

Ohayon J, Finet G, Floch S L, Cloutier G, Gharib A M, Heroux J and Pettigrew R I 2014 Biomechanics of atherosclerotic coronary plaque: site, stability and *in vivo* elasticity modeling *Ann. Biomed. Eng.* **42** 269–79

Pritchard W F, Davies P F, Derafshi Z, Polacek D C, Tsao R, Dull R O, Jones S A and Giddens D P 1995 Effects of wall shear stress and fluid recirculation on the localization of circulating monocytes in a three-dimensional flow model *J. Biomech.* **28** 1459–69

Ro K C and Ryou H S 2010 Numerical study on turbulent blood flow in a stenosed artery bifurcation under periodic body acceleration using a modified k-e model *Korea-Aust. Rheol. J.* **22** 129–39

Ryval J, Straatman A G and Steinman D A 2004 Two-equation turbulence modeling of pulsatile flow in a stenosed tube *J. Biomech. Eng.* **126** 625–35

Samady H, Eshtehardi P, McDaniel M C, Suo J, Dhawan S S, Maynard C, Timmins L H, Quyyumi A A and Giddens D P 2011 Coronary artery wall shear stress is associated with progression and transformation of atherosclerotic plaque and arterial remodeling in patients with coronary artery disease *Circulation* **124** 779–88

Shaw S, Murthy P V S N and Pradhan S C 2010 The effect of body acceleration on two dimensional flow of Casson fluid through an artery with asymmetric stenosis *Open Transport Phenomena J.* **2** 55–68

Suo J, Ferrara D E, Sorescu D, Guldberg R E, Taylor W R and Gidden D P 2007 Hemodynamic shear stresses in mouse aortas: implications for atherogenesis *Arterioscler. Thromb. Vasc. Biol.* 346–51

Tan F P P, Soloperto G, Bashford S, Wood N B, Thom S, Hughes A and Xu X Y 2008 Analysis of flow disturbance in a stenosed carotid artery bifurcation using two-equation transitional and turbulence models *J. Biomech. Eng.* **130** 061008-1-11

Thomas J B, Antiga L, Che S L, Milner J S, Steinman D A H, Spence J D, Rutt B K and Steinman D A 2005 Variation in the carotid bifurcation geometry of young versus older adults: implications for geometric risk of atherosclerosis *Stroke* **36** 2450–6

Varghese S S and Frankel S H 2003 Numerical modelling of pulsatile turbulent flow in stenotic vessels *ASME J. Biomech. Eng.* **125** 445–60

Varghese S S, Frankel S H and Fischer P F 2008 Modeling transition to turbulence in eccentric stenotic flows *J. Biomech. Eng.* **130** 014503-1-7

Vukicevic A M, Stepanovic N M, Jovicic G R, Apostolovic S R and Fillipovic N D 2014 Computer methods for follow-up study of hemodynamic and disease progression in the stented coronary artery by fusing IVUS and X-ray angiography *Med. Biol. Eng. Comput.* **52** 539–56

Walburn F J and Schneck D J 1976 A constitutive equation for whole human blood *Biorheology* **13** 201–19

Womersley J R 1955 Method for the calculation of velocity, rate of flow and viscous drag in arteries when the pressure gradient is known *J. Physiol.* **127** 553–63

Chapter 10

Fabrication techniques of artificial blood vessels

Arindam Bit, Jasjit S Suri and Sharda Gupta

In the real world a large proportion of the mortality rate is due to cardiovascular disease. The reconstruction of a functional blood vessel substitute is still a demanding task to improve outcomes of cardiovascular surgery. Unfortunately, autologous blood vessels are insufficient in number and allogeneic grafts are prone to immunogenic rejection. The synthetic grafts available have contributed well in the case of giant width blood vessesl but in the case of limited width blood vessels, they have shown poor patency and an inability to grow. A tissue engineering (TE) blood vessel can perhaps be an alternative to address these shortcomings by utilizing vascular cells, biochemical signals and scaffolds. The combination of TE approach can help in fabrication of a tissue-engineered blood vessel with the desired mechanical properties, interconnected pores and elasticity. This chapter discusses different TE techniques for vascular graft fabrication.

10.1 Introduction

Peripheral arterial disorder, atherosclerosis, carotid stroke, cardiac infarction, arteritis and valvular disease are some of the dominant factors of disease in the blood vessel. The common disorders associated with blood vessels are stenosis or occlusion. Blood vessel transplantation is an important technique of vascular disease treatment. With the help of blood vessels we can expand all types of functional tissue, hence there is a high demand for a proper blood vessel fabrication technique.

The knowledge about complete biology of vascular biology will help in developing vascular tissue constructs which can be used to solve various clinical problems. Blood vessels (arteries and veins) are responsible for transportation of nutrients and oxygen to tissue and removal of waste products from the tissue. The mature healthy blood vessel walls have three layers. Each of the three layers has its own specific function and unique structure. The innermost layer, called tunica intima, has endothelial cells, tunica media has connective tissue and smooth muscle cells (SMCs) and the outermost layer, called tunica adventitia, has irregular connective

doi:10.1088/978-0-7503-2088-7ch10

tissue. The media layer has an aligned multilayered SMC extracellular matrix arranged in a helical configuration. This aligned SMC helps in maintaining blood pressure through relaxation/expansion by changing wall diameter, hence affecting the mechanical properties of the vessel wall (Rim *et al* 2018). The thickness of intima is variable and is expressed as a media thickness ratio (Taki *et al* 2017). The media has a thickness in the range of 125 and 350 μm.

Blood vessels are classified into two types: large diameter (inner diameter > 6 mm) and small diameter (inner diameter < 6 mm). The difference in diameter leads to different blood flow velocities, hence a different design specification in each case.

The conventional treatment for blood vessel regeneration is the surgical technique for revascularization of an unhealthy vessel using a graft. Bypass surgery refers to the surgical revision of blood flow through a blood vessel segment in the case of narrowing or blocking of the blood vessel. An incision is made across a blocked vessel and a new pathway for blood flow is made using a graft. The graft can be either from an individual's native vessel or a synthetic vessel.

The current options of vascular graft for blood vessel reconstruction for replacing damaged blood vessels are allografts, autologous graft, or synthetic vascular graft. The golden standard for bypass surgery is the patient's own vein or artery (Chlupác *et al* 2009). For a lesser width graft we use autologous vessels like saphenous vein and internal thoracic artery (Koschwanez and Reichert 2013). However, this has the limitation of non-availability due to amputation or vascular disease. Synthetic polyester materials such as polytetrafluoroethylene (PTFE), expanded polytetrafluoroethylene (ePTFE) or polyethylene terephthalate synthetic fiber (Dacron) can be used for the construction of prosthetic devices. For the replacement of a blood vessel using a synthetic graft, a mandrel is implanted into the patient which is then covered with a knitted synthetic graft. After a few days of implantation, a 'Sparks' mandrel' construct forms across the mandrel. Finally, the fibres are harvested and then implanted as a bypass graft. Synthetic prosthetic grafts show good results for a larger diameter vessel but face immune rejection in the case of a small diameter vessel. The reason behind this rejection is thrombosis, intimal hyperplasia or aneurysm leading to blood vessel occlusion. Its failure in the case of a small diameter vessel is due to lack of endothelial cells (leading to surface thrombogenicity) and hemodynamic disturbances (causing anastomotic intimal hyperplasia) (Chlupác *et al* 2009). Also, in the case of children, the blood vessel implant must have growth potential which is missing in the case of a synthetic graft.

In addition, low blood velocity in the case of a small diameter vessel is one of the causes of failure of synthetic material. The patency rate of synthetic material vessels is lower than native vessels especially in the case of a small diameter vessel. There are many limitations of vascular bypass grafting. One is that patients may not have healthy vessels acceptable to harvest for surgery. There can be development of degenerative graft disease in vein grafts or arterial graft such as atherosclerosis. Vascular graft infection is another devastating limitation of vascular grafts. Staphylococcal species are the predominant species causing graft infections, hence a careful clinical examination is essential (Sohail *et al* 2010, Settembrini and Settembrini 2017). To prevent degenerative graft disease, reoperation is probably the most appropriate option but it can be life threatening in the case of old aged patients or patients with vascular disease.

Due to some of the disadvantages of the above mentioned grafting technique, an artificial blood vessel must be manufactured using two ways depending upon the material available to us, called synthetic or natural biomaterial. Weinberg and Bell (1986) developed an artificial blood vessel with the help of collagen material and then cultured with the help of TE technique. For representing of blood vessel structure we can use endothelial cells, fibroblast or SMC. The prepared vascular scaffold will support the growth of vascular cells which creates the extracellular matrix and ultimately forms a functional bloodvessel.

Today TE is the most widely and suitable platform for blood vessel formation by utilizing the cell source (adult stem cell or embryonic stem cell), bio-reactive signals (growth factor, adhesion factor) and scaffolds (synthetic, natural or composite material). In TE, researchers explored the culturing of vascular cells or differentiated stem cells on the synthetic or natural scaffold of a tubular shape inside a bioreactor to experience a different type of mechanical or chemical stimulation similar to the vessel environment for the fabrication of a functional blood vessel replacement graft. This leads to the generation of the prosthetic graft of synthetic material.

The output of the TE approach is a multilayered blood vessel identical with permeability and cytocompatibility. The tissue engineered blood vessels are basically multilayered tubes created from extracellular based material and vascular cells. The ideal vascular graft should be: suturable, easy to handle, be non-thrombogenic, have improved patency, have anti-calcification ability, possess vasoactive properties, be mechanically durable and biocompatible, able to withstand long-term hemodynamic stress (pressure-resistant), its surface should allow adherence of endothelial cells (Karaman *et al* 2017, Funamoto *et al* 2010). It should be permeable to solutes, cells, and water. It must be elastic to avoid aneurysm formation. In fabricated structure the motive of an endothelial cell is to interrupt the formation of thrombosis during blood flow condition. The engineered vascular graft needs to be studied to check its suitability as an *in vivo* functional vessels. Some of the parameters which decide its utility as artificial blood vessels are burst pressure, tensile strength, elastic modulus, suture retention test, compliance test, and dynamic fatigue test.

This chapter summarizes the relevant techniques for the fabrication of an artificial blood vessel. Several techniques have emerged for the fabrication of a vascular graft. These include cell sheet technology, freeze-drying, stereolithography, electrospinning, and 3D printing techniques. This chapter will provide an overview of blood vessel reconstruction from stem cells, scaffolds and scaffold-free constructs.

10.2 Cell types used for blood vessel regeneration

The endothelial cells are located in the innermost layer of a blood vessel and are in contact with blood. These cell layers are responsible for prevention of platelet adhesion. As the surface of the endothelium is negatively charged, it repels negatively charged blood protein and platelets responsible for thrombosis. The thickest layer, tunica media, contains SMCs, which contract and relax in response to the endothelial and systemic factor during blood flow (Stocum 2012). The outer layer adventitia is made up of fibroblasts. The endothelial cell located in the lumen is used to escape thrombosis.

For the functional blood vessel morphogenesis, blood vessel formation is divided into two process: vasculogenesis and angiogenesis (sprouting, branching of pre-existing endothelial cells) (Geudens and Gerhardt 2011). In the process of vasculogenesis, as illustrated in figure 10.1(a), mesoderm-derived cells (angioblasts) differentiate into hemangioblasts. These progenitors of endothelial cells and blood cells then migrate to specific locations and aggregate to form primitive blood vessels (Karamysheva 2008). Then the existing vascular network expands and transforms into a vascular net (Potente *et al* 2011, Herbert and Stainier 2011, Betsholtz 2018). Through the process of angiogenesis, as shown in figure 10.1(b), pre-existing blood vessel walls are destroyed, endothelial cells get activated, sprout and start to migrate. The migrated endothelial cells are remodeled in the tubular structure as explained in figure 10.1.

For fabrication of synthetic blood vessel development, endothelial progenitor cells (EPCs) can be obtained from adult blood and can help in endothelium regeneration (Liu *et al* 2007). During the process of vascularization, the endothelial progenitors derived from bone marrow are migrated, differentiated and incorporated into the damaged vessel (Betsholtz 2018, Zhang *et al* 2014). The various cell sources for different blood vessel layers are discussed in table 10.1. SMCs are specialized cells which provide a response to vasoactive stimuli by contracting and expanding the diameter of the vessel and provide structural integrity (Metz *et al* 2012). Because of limited proliferation properties of SMCs, other sources of cells must be considered in blood vessel fabrication. But the adult blood derived EPC has less proliferation capability and there is difficulty in harvesting cells, hence bone

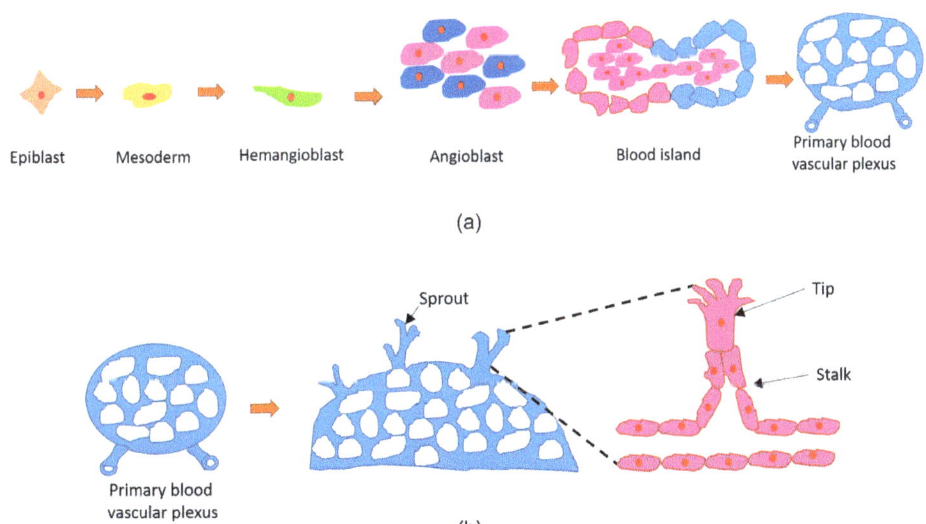

Figure 10.1. Formation of a blood vessel through (a) vasculogenesis and (b) angiogenesis.

Table 10.1. Cell source for a different layer of blood vessels.

Blood vessel layer	Cell type	Cell source	References
Tunica intima, lumen	Endothelial cells	Human cord blood derived endothelial progenitor cells (hEPCs), embryonic stem cells, human coronary artery endothelial cells	(Jung *et al* 2015, Zhang *et al* 2007)
Tunica media	Smooth muscle	Human mesenchymal stem cells (hMSCs), embryonic stem cells	(Jung *et al* 2015, Rim *et al* 2018)
Tunica adventitia	Fibroblasts	Human artery derived fibroblast	(Kelm *et al* 2010)

marrow derived EPCs are preferred. The vascular cells can be derived from the mesenchymal stem cells (MSCs) (endothelial cells (ECs) and SMCs) and these vascular cells can help in cell-therapy for blood vessel formation. By culturing mesenchymal stem cells in the presence of endothelial growth supplements, it can be converted into an endothelial phenotype. Oswald and co-authors were successful in establishing a protocol for differentiation of bone marrow stem cell into ECs by culturing in the presence of fetal calf serum and vascular endothelial growth factor (Oswald *et al* 2004). Suzuki *et al* (2016) obtained SMCs from the human bone marrow derived MSCs and then cultured via laminin.

10.3 Techniques for the regeneration of blood vessels

For cure of vascular diseases, most of the time we use vascular grafts. Many synthetic grafts are available but do not meet the specified requirements. The most preferable technique for manufacturing vascular grafts is TE. The approach for the fabrication of an artificial blood vessel consists of development of a three-layered structured: intima, media and adventitia representing the original structure of the blood vessel. An artificial tissue engineered blood vessel (TEBV) has resistance to infection and is non- thrombogenic. This section details the techniques employed for the fabrication of an artificial blood vessel, as explained in figure 10.2. Further, the pulsatile flow environment of blood *in vivo* can be produced by placing the artificial vessel inside a suitable bioreactor which will be highlighted in detail in the following segments.

10.3.1 Scaffold-free technology

Cell sheet technology is a scaffold-free technique that employs a cell source derived from somatic cells or stem cells. Each cell sheet type has different cell sheet preparation steps and cell sheet harvest method. This technology provides a non-

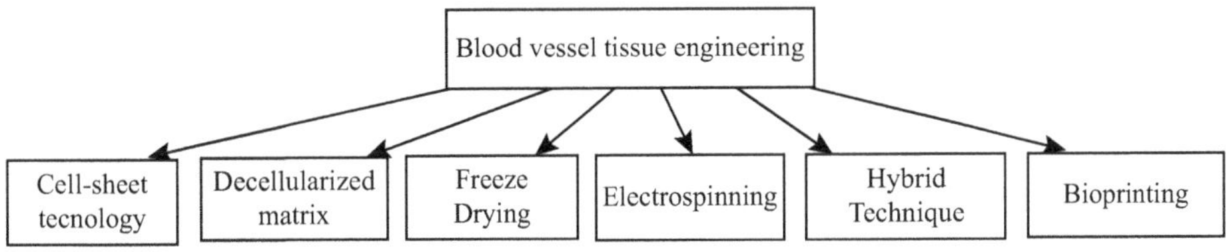

Figure 10.2. Sketch of the various vascular tissue engineering techniques.

invasive way for cell detachment and attachment by use of a temperature-responsive polymer. Thermo-responsive polymers are stimuli-responsive, which utilizes the phase separation behavior at high and/or low temperature. In the case of upper critical solution temperature (UCST) polymer, the polymer gets solubilized after heating and in lower critical solution temperature (LCST) it gets insolubilized on heating. In this technology, a temperature responsive polymer is coated at nanometer thickness on the surface of a culture dish. Every temperature dependent polymer has an LCST. At temperature greater than LCST, this polymer shows hydrophobic characteristics, and at a lower temperature it possess hydrophilic characteristics. Hence, at regular cell culture temperature i.e. 37 °C, polymer coating allows cell culture. But at lower temperature due to swelling and rapid hydration of the temperature responsive polymer, the cell detaches itself spontaneously from the surface of the polymer. This presence of extracellular matrix (ECM) and membrane protein (fibronectin and integrin) allows its easy adhesion with host tissue without sutures. There are different fabrication techniques to modify the thermoresponsive polymer surface on a culture dish such as plasma polymerization, UV irradiation, solvent cast or electron beam polymerization which shortens the cell sheet detachment time. Many thermoresponsive polymers are developed but only a few are suitable for cell adhesion as some are unable to adhere cells even above their LCST. The static water angle contact, swelling ratio, wettability and molecular mobility of the polymeric surface is responsible for cell adhesion property. At LCST, the thermo-responsive polymer undergoes a transition from coil to globular state in aqueous solution; the polymeric chain gets hydrated and expands its confirmations. This hydration of the polymeric chain initiates cell detachment, as shown in figures 10.3 and 10.4.

The most commonly used temperature responsive polymer used for this technique is poly(*N*-isopropylacrylamide) (PNIPAAAm) as its LCST is around 32 °C (approximately equal to human body temperature) when we manufacture a 3D tissue construct for that we arrange the cell sheet layer by layer without using any scaffold. This technology is most widely used for the reconstruction of diverse tissue (Haraguchi *et al* 2012, Labbé *et al* 2011). The ECM microenvironment guides neovascularization by promoting the growth of endothelial cell ECs (Moschouris *et al* 2016, Papavasiliou *et al* 2012).

For the calculation of LCST of any thermo-responsive polymer, to have knowledge of the transition behavior can be achieved by calculation of the radius of gyration (Rg) of the polymeric chain at different temperatures, the temperature alter the Flory–Huggins interaction parameter between chain and water molecule and

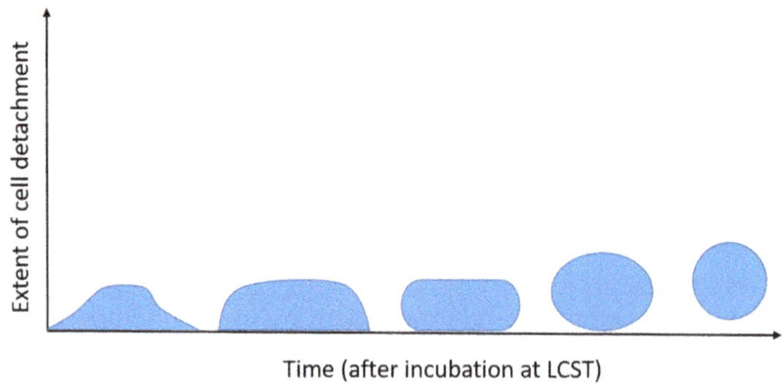

Figure 10.3. Diagrammatic representation of cellular shape change during cell detachment from a thermo-responsive polymer coated surface during incubation at temperature lower than LCST.

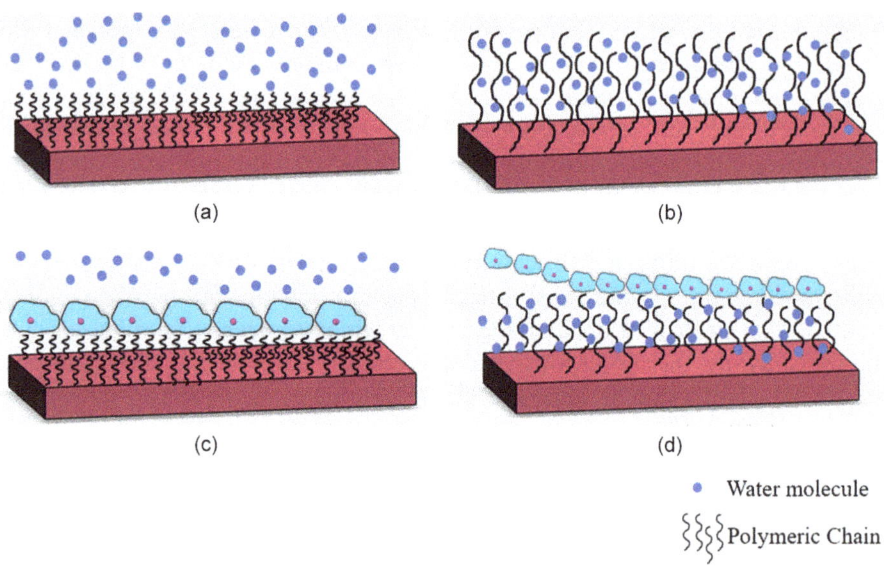

Figure 10.4. Figure representing cell detachment from a thermoresponsive polymer during cell sheet technology. At temperature $T >$ LCST, the thermoresponsive polymer chain with or without cells will be in aggregate form (a, c). The thermoresponsive polymer chain will hydrate and swell resulting in coil formation allowing cells to detach from its surface (b, d).

end-to-end distance of chains in a water molecule (Koochaki *et al* 2018). The radius of gyration (Rg) of the polymeric chain at different temperatures to evaluate chain structure compactness can be determined by equation (10.1):

$$R_g^2 = \frac{1}{N}\left\langle \sum_{k=1}^{N}(r_i - r_{\mathrm{com}})^2 \right\rangle \tag{10.1}$$

where r_i = position vector of each atom of polymer chain;

r_{com} = vector of center-of –mass (CM) of polymer chain;
$\left\langle \sum_{k=1}^{N}(r_i - r_{com})^2 \right\rangle$ = end-to-end distance of chain.

The calculation of interaction energy between thermo-responsive polymer chain and water molecule (E_{Int}) can be calculated as per equation (10.2):

$$E_{Int} = E_{(polymer/water)} - (E_{(polymer)} + E_{(water)}) \tag{10.2}$$

where $E_{(polymer/water)}$ = total potential energy of the equilibrated thermo-responsive polymer/water structure;
$E_{(polymer)}$ = total potential energy of thermo-responsive polymer/water structure;
$E_{(water)}$ = total potential energy of the water molecules (calculated by removing the thermo-responsive polymer chain from the equilibrated thermo-responsive polymer/ water structure).

The energy of mixing (ΔE_{mix}) of the polymeric chain and water molecules can be calculated as per equation (10.3).

$$\Delta E_{mix} = \varnothing_A\left(\frac{E_{coh}}{V}\right)_A + \varnothing_B\left(\frac{E_{coh}}{V}\right)_B - \left(\frac{E_{coh}}{V}\right)_{mix} \tag{10.3}$$

where $\varnothing_A$ = volume fraction of thermo-responsive polymer;
$\varnothing_B$ = volume fraction of water;
$\left(\frac{E_{coh}}{V}\right)_A$ = calculated cohesive energy density of thermo-responsive polymer;
$\left(\frac{E_{coh}}{V}\right)_B$ = calculated cohesive energy density of water;
$\left(\frac{E_{coh}}{V}\right)_{mix}$ = calculated cohesive energy density of mix.

The Flory–Huggins interaction parameter (χ) can be determined as per equation (10.4):

$$\chi = \left(\frac{\Delta E_{mix}}{RT}\right)V_m \tag{10.4}$$

where R = gas constant;
T = absolute temperature;
V_m = reference molar volume.

Cell sheet technology has rich and intact ECM, hence researchers have utilized this technique to form a prevascularized 3D cell sheet. With this technology, functional 3D tissue can be developed by varying the number of cell sheets as per thickness of tissue to be constructed. L'Heureux *et al* prepared a TEBV, which is manufactured with the help of sheet based technology for small-diameter arterial revascularization using fibroblasts. This artificial blood vessel has mechanical properties similar to a native blood vessel (L'Heureux *et al* 2006). Sekine *et al* (2013) prepared a triple layer cardiac blood vessel and then a culture using a

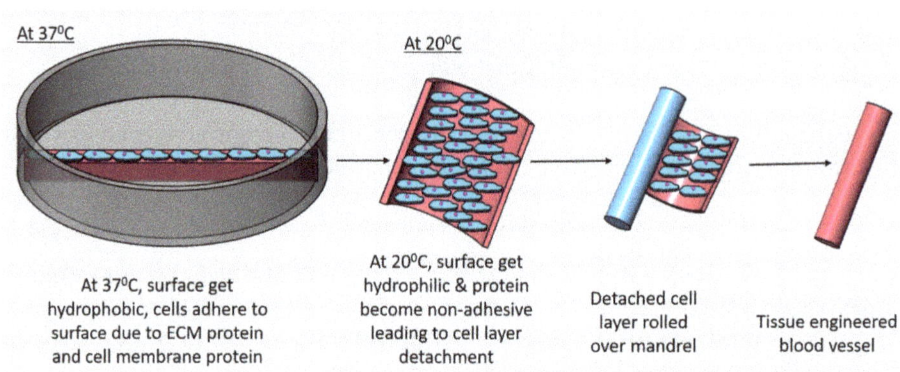

Figure 10.5. Image showing stepwise preparation of artificial blood vessel using cell-sheet technology.

pulsatile bioreactor. Ren *et al* prepared a cell sheet (figure 10.5). For fabrication of cell sheet they use a mesenchymal stem cell (hMSC) obtained from human bone marrow. After that this cell sheet is culture on a dish and cultured till it reaches confluence.

Jung and co-authors fabricated a poly(dimethyl siloxane) (PDMS) based nano-grating substrate using lithography, onto which hMSCs were seeded. The hMSCs layer was then detached from the PDMS surface and rolled across the mandrel to form a TEBV. They found that the cell sheet-based construct is deprived of the necessary mechanical support of a scaffold, which can be improved by adding more cell layers and more maturation time. However, the three-layered TEBV has a burst pressure similar to the human native vessel (Jung *et al* 2015).

10.3.2 Freeze drying

The freeze drying process can be summarized in two steps: freezing of solution at low temperature, removal of ice from the frozen sample by a process of sublimation and desorption. In the process of freeze drying, the polymeric solution dissolved in a suitable solvent is allowed to cool under freezing point. At the freezing point, due to the sublimation process, the solvent is evaporated leaving behind the porous scaffold. This method has been utilized in the fabrication of an artificial blood vessel with interconnected pores (figure 10.6). It has been shown that it is capable of forming a porous scaffold.

Firstly, Zhu *et al* prepared a novel human-like collagen solution and glutaraldehyde cross-linked chitosan solution, which is poured into a tubular mold to prepare an adventitia layer scaffold. The scaffold was biocompatible, demonstrated enhanced adhesion and proliferation of human venous fibroblast cells on it and interconnected pores (Zhu *et al* 2009a). They prepared a silk-fibroin blended human-like collagen based vascular scaffold using the freeze drying process to improve the mechanical (tensile) property of the scaffold. SMCs were incident in the scaffold and combining the elastic property of silk fibroin and the biocompatibility of collagen enabled them to fabricate a blood vessel media layer scaffold (Zhu *et al* 2009b). They further constructed the intima layer by the same freeze drying approach, but by

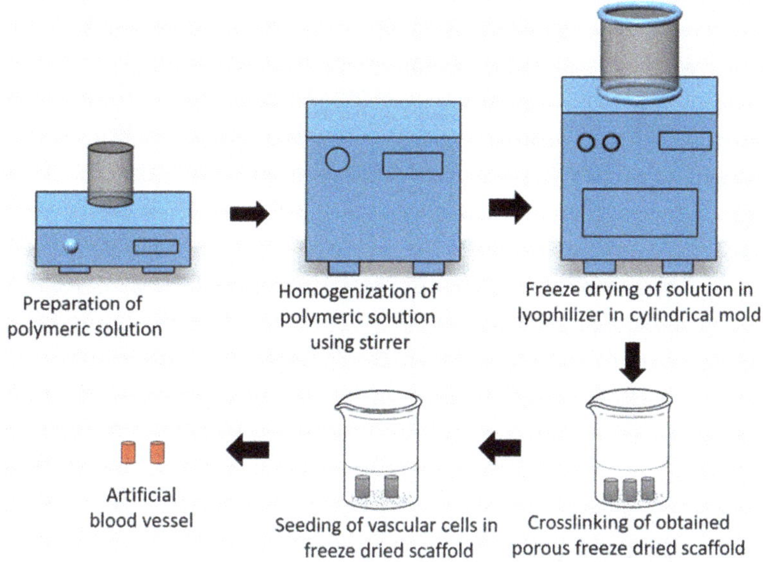

Figure 10.6. Schematic representation of the freeze drying technique for fabrication of an artificial blood vessel.

using hyaluronic acid and human-like collagen into which ECs were added. The addition of hyaluronic acid in the scaffold showed improved endothelialization, which is the requirement in the intima layer of a blood vessel (Zhu *et al* 2014).

10.3.3 Decellularized vascular graft

Decellularization is an effective technique for removal of cells and nuclear matter from the tissue while maintaining the extracellular matrix to represent a 3D scaffold. To facilitate patients with tissue-engineered vascular grafts, mechanically strong vessels are taken, and decellularized to serve as a scaffold (Dahl *et al* 2003). Decellularization of tissue can be done using: chemical method (acid or detergents or alkaline treatment); physical method (mechanical agitation); or enzymatic method. After decellularization, collagenous ECM with mechanical properties similar to native tissue is prepared, onto which autologous or genetically modified cells can be seeded (figure 10.7). With the help of these techniques we can fabricate tissue-engineered blood vessels with minimum time.

Many researchers have managed to upgrade the cell seeding technique on a porcine decellularized common carotid artery by using magnetic force and magnetite nanoparticles (Mag-seeding) to seed vascular cells on it. The result shows increased cell seeding efficiency on a tubular scaffold using Mag-seeding technique (Shimizu *et al* 2007).

These above-mentioned treatment techniques can change the mechanical properties of tissue. Funamoto and co-authors have tried to overcome this problem by using high-hydrostatic pressure (HHP) technique. They have produced an undamaged extracellular matrices based scaffold after decellularization of a porcine blood

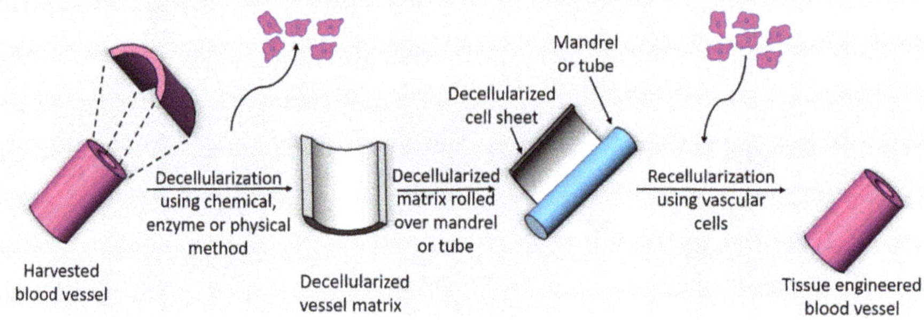

Figure 10.7. Schematic diagram showing artificial tissue-engineered blood vessel fabrication using decellularization technique.

vessel without using any chemicals. The allogeneic transplantation study determined its suitability to be used as a vascular graft as no clot forms on the luminal surface, and it also withstands high arterial blood pressure (Funamoto *et al* 2010).

In arterial TE, cells are isolated from patients, cultured and maintained to design a tissue-engineered blood vessel. The complete techniques take a long time for graft fabrication. porcine SMCs are incident on the polyglycolic acid based scaffold using a perfusion system. They concluded that a tissue-engineered vessel can be decellularized to quickly form connective tissue which otherwise takes months (Quint *et al* 2011). The research is ongoing for formation of a blood vessel from human SMCs. They found that the developed vessel supported neointima growth with ECs and SMCs with high patency (Quint *et al* 2012).

As discussed earlier, endothelium is essential for preventing thrombosis, intimal hyperplasia, and calcification, which can lead to occlusion of the blood vessel. To enhance endothelialization of the decellularized matrix, Lee and his team have done surface modification of a decellularized matrix by coating it with polydopamine and cell adhesion peptides (Lee *et al* 2014).

10.4 Bioprinting technique

For the fabrication of a tissue engineered vascular graft that mimics the function and structure of a native vessel, 3D printing technology is the preferred option. The step-by step workflow of 3D printing technique is as follows (Gupta and Bit 2018, Gupta *et al* 2018):

1. Geometry design;
2. Slicing the 3D model into 2D to make it printable;
3. Printing of a 3D model by 3D printing technology;
4. Post processing or surface modification of 3D model.

The 3D printing technology is sub-divided into three categories depending on the printing technique and ink: extrusion based, inkjet based and light based printing.

In extrusion based printing technique, material in the form of hydrogel, bioink or cell suspension is extruded from the nozzle and deposited on the substrate by the application of pneumatic, mechanical or thermal printing technique. In the extrusion based printer, the printer head or stage moves along the *z*-axis after successful completion of printing of one layer. Using extrusion, Kolesky *et al* produced a vascularized, heterogeneous construct. For developing such a construct, four independently controlled printer heads were used: one printing PDMS ink; the second depositing pluronic F-127 to develop embedded vasculature; the third containing gelatin methacrylate (GelMA) to create an extracellular matrix; and the last one containing cell-laden GelMA ink. The removal of pluronic by cooling the construct below 4 °C leads to formation of a vascular network into which endothelial growth media and human umbilical veinendothelial cells (HUVECs) were injected for endothelialization of the vascular network (Kolesky *et al* 2016).

The inkjet based technique employs the ejection of drops of material from a nozzle onto the substrate. In this case, material viscosity must be within printable range. Also, this type of printing is not preferred for printing of high-density cells as it can lead to clogging of the nozzle. The inkjet based printer utilizes piezoelectric, thermal and micro-valve printing techniques to print the desired model. The drops are generated by heat in the case of thermal inkjet, and a transient pressure wave in the case of piezoelectric bioprinting (Gupta *et al* 2018). Utilizing the thermal ink jet printing technique, Cui *et al* used human microvascular endothelial cells (HMVECs) and fibrin as bioink to fabricate a very fine vessel containing aligned cells. The cells proliferated inside the channel and microvasculature was formed (Cui and Boland 2009).

The light based printing technique utilizes energy of light to crosslink the photosensitive polymer. The various light based systems are stereolithography, digital light processing and laser-assisted bioprinting. Ringeisen *et al* constructed a lumen and lumen network using a biological laser bioprinting (BioLP) technique. The HUVECs and human umbilical vein smooth muscle cells (HUVSMCs) were incubated for 24 h for differentiation into lumina, then a scaffold was placed into a biological laser bioprinter to deposit HUVMSCs on top of it (Wu and Ringeisen 2010). Using laser assisted bioprinting, controlling of the microvascular network using printing of ECs on type I collagen biopaper in a specific pattern was investigated (Zhang *et al* 2007). The ECs spread out when printed alone, however, when co-cultured with mesenchymal stem cells, they remain in the printed area and formed capillaries (Bourget *et al* 2016). Zhu and co-authors fabricated prevascularized tissue using microscale continuous optical bioprinting of vascular cell encapsulated hydrogel. This polymer solution was exposed to an UV pattern with a vascular channel mask in the middle. This technique enhances vascularization without use of any sacrificial material, a new pre-vascularization technique was developed (Zhu *et al* 2017).

10.5 Electrospinning

The electrospinning technique is able to generate a multilayered blood vessel by multilayering electrospinning, and it is advantageous as it can help to prepare nano-sized dimensions of a natural extracellular matrix. With the help of high electric field we can generate micro-or nanofibers from a polymeric solution. This technique employs the electrostatic force to fabricate such a fibrous matrix. Fabrication of fibrous scaffolds by electrospinning of polymers with or without using solvent are widespread nowadays for vascular grafts (Wakuda *et al* 2018). The electrospinning unit consists of high voltage power supply, syringe pump and collector (rotating mandrel). The polymeric solution placed in a syringe pump is injected through the needle of the syringe after application of a high voltage (figure 10.8). In electrospinning technique, with the control over the electrospinning parameters: solution parameters (concentration, viscosity, surface tension, molecular weight), processing parameters (operating voltage, collector rotation speed, distance between collector and tip, flow rate), a micro-to-nanoscale fibrous matrix with controlled composition, porosity, pore size, fiber diameter, structure and mechanical properties can be fabricated. The printed blood vessel structure called a blood vessel scaffold shows the morphological and mechanical properties of an actual blood vessel located *in vivo* (Hasan *et al* 2015, Vaz *et al* 2005).

For the improvement of cell–scaffold interaction, surface modification can be done. Two of the surface modification techniques employed for modification are hydrophobic interaction (Wang *et al* 2017) and low vacuum plasma treatment (Yalcinkaya *et al* 2016). To mimic mechanical properties and morphology of the blood vessel, Vaz and co-authors fabricated bi-layered tubular scaffold using multilayering electrospinning. They first electrospun polycaprolactone (PCL), which is the innermost layer, onto which a layer of polylactic acid (PLA) was electrospun with each polymer having its own printing parameters. This multilayer electrospun bilayered scaffold supported cell growth and proliferation and possessed characteristics suitable for a blood vessel substitute (Vaz *et al* 2005). Now researchers are exploring this technology for fabrication of small diameter vascular grafts. These tubes demonstrate a burst pressure of 575 ± 17 mmHg, which is more than the physiological pressure range (80–120 mmHg), are more compliant than a synthetic graft and have good cytocompatibility for use as a blood vessel substitute (Marelli *et al* 2010). Researchers explored the use of piezoelectric polymer poly(vinylidene fluoride-trifluoroethylene) for the blood vessel formation by incorporating zinc

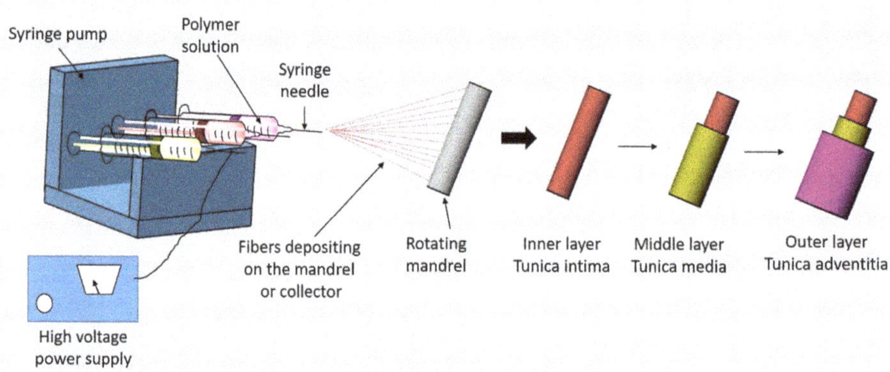

Figure 10.8. Fabrication of an artificial blood vessel using electrospinning technique.

oxide nanoparticles on it. They found that this electrospun polymeric scaffold proved to be biocompatible (Augustine *et al* 2017). Marcolin and group explored a chemical reaction to crosslink gelatin on electrospun silk fibroin tubular scaffold to enhance cell interaction (Marcolin *et al* 2017).

10.6 Hybrid scaffold fabrication technique

Surface topography, roughness and chemical composition are various characteristics that alter the attachment and proliferation of cells. Various methods are available for construction of organs. Each method has its own benefit and limitation. To overcome the problems of each technique, a hybrid scaffold fabrication technique can be utilized by combining the techniques to produce a 3D scaffold. For instance, blood vessels require a highly porous structure with predefined internal architecture. The combination of fabrication techniques will help to meet the specific requirements for the fabrication of blood vessels.

A decellularized matrix as a blood vessel graft has the advantages of non-immunogenicity and mechanical properties. As fabrication of an artificial blood vessel is a time consuming process, for the availability of such a scaffold for clinical trial it must be preserved long term with proper sterilization. Hence researchers freeze-dried the decellularized artery, which leads to the development of a decellularized matrix with intact ECM components. This dried decellularized scaffold can be sterilized easily using gamma radiation or ethylene oxide (Sheridan *et al* 2013). The extent of decellularization of tissue must be considered for the optimal production of a suitable scaffold. The under-decellularization may produce a scaffold with immunogenic cells and an over-decellularization process may damage blood vessel structure and can degrade or deform elastin protein (this is responsible for elasticity, compliance to vessels and allows smooth flow of blood) (Gong *et al* 2016). However, a decellularized blood vessel has properties like anti-thrombogenicity, but generation of a small diameter blood vessel from a porcine source is very difficult. Hence, researchers tried to fabricate a small diameter vessel off optimal shape and size using a large diameter porcine aorta. For this, they have removed the outer layer of the porcine aorta, decellularized the aortic-intima-media sheet, and rolled it into a mandrel of appropriate size. Finally, the coating of polyurethane was done over it using the electrospinning technique. Modulating the thickness of the coating can help to meet the required mechanical properties of the native blood vessel (Wu *et al* 2019).

10.7 Characterization of an artificial blood vessel

During implantation, the graft must have proper mechanical properties to withstand the blood flow velocity. Tests need to done to check the suitability of the vessel-like structured enveloped. Some of them are:

Burst pressure: This can be calculated in many ways:

(i) Using a pressure gauge: the vascular graft segments are cannulated and a pressure applied at the rate of 80 to 100 mmHg s^{-1} using phosphate buffer saline until graft failure. A sampling rate of 3 Hz pressure is recorded using a pressure gauge (Konig *et al* 2009).

(ii) Using a tensiometer: this techniques utilizes Laplace's law for a blood vessel (Melchiorri *et al* 2014). For this, grafts were cut into sections and kept on the metal hook of a tensiometer. The vascular grafts are stretched at 1 mm min^{-1} rate until graft failure. The burst pressure can be calculated from equation (10.5):

$$P = \frac{UTS \times t}{r} \tag{10.5}$$

where P = internal pressure;
UTS = ultimate tensile strength;
t = wall thickness;
r = radius of graft.

(iii) Tank gauging system: the blood vessel grafts are filled with hydrostatic pressure from a carbon dioxide tank until failure. The burst pressure is recorded from the tank gauge.

Compliance: The compliance test is done to check suitability of a graft as a blood vessel substitute. It can be evaluated using different standards:

(i) ANSI 7198: the vascular grafts are tensioned and pressurized with phosphate buffer saline (PBS). To measure the external diameter, high-resolution images are taken at low and high pressure, respectively. For inner diameter measurement, a small section from the same graft is fixed in formalin for histology. The wall thickness and internal diameter are analyzed using magnified images of histology (Konig *et al* 2009). The compliance is calculated as per equations (10.6) and (10.7):

$$\% \frac{\text{Compliance}}{100} mmHg = \frac{\frac{R_{ip2} - R_{ip1}}{R_{ip1}}}{p_2 - p_1} \times 10^4 \tag{10.6}$$

where p_2 = higher pressure;
p_1 = lower pressure;
R_{ipx} = internal diameter at pressure x.

$$R_{ipx} = \sqrt{R_{opx}^2 - (R_i + t_o)^2 + R_i^2} \tag{10.7}$$

where R_{opx} = measured external radius at pressure x;
R_i = measured internal radius at rest;
t_0 = measured wall thickness a at rest.

(ii) Varying pressure within the graft: compliance is evaluated by calculating the change in vessel diameter on application of pressure within the tube (Atlan *et al* 2018). The compliance can be calculated using equation (10.8):

$$C = \frac{d_{120} - d_{80}}{d_{80}} \quad (10.8)$$

where d_{80} = graft diameter at 80 mmHg pressure;
d_{120} = graft diameter at 120 mmHg pressure.

Suture retention test: This test is done as per ANSI 7198 specification. The native suture retention strength in the case of an internal mammary artery is 138 ± 50 gmf (Konig *et al* 2009). The suture retention strength of a tissue engineered blood vessel must be compatible with the native blood vessel value. For calculation of suture retention strength, a suture is passed through the graft at a gap of approximately 10–20 mm from the graft deadline. The outer side is tied by a clamp. The entire structure is dragged out from the deadline at a constant rate of 120 mm min^{-1} before failure (Melchiorri *et al* 2014, Konig *et al* 2009, Atlan *et al* 2018).

10.8 Mechanical properties of an artificially fabricated blood vessel by different techniques

The efficacy of the blood vessel depends upon its mechanical properties and it must have burst pressure, compliance and suture retention strength similar to a native blood vessel. The mechanical properties of different types of blood vessel structure by different methods are listed in table 10.2.

10.9 Conclusion

Large diameter blood vessels can be easily transplanted using synthetic grafts. Particular attention is paid in the case of a small diameter blood vessel that requires a graft that can avoid thrombosis formation. The replacement of casualty's blood vessel is still very difficult task, but it can be made easier via TE. Various kinds of tissue engineered blood vessel surrogates have been fabricated and experimented on in animals, and a few have undergone human clinical trials.

A native blood vessel extracellular matrix consists of a large amount of collagens, proteins and growth factors providing a microenvironment suitable for growth of vascular cells. The decellularized matrix can be obtained by removing cellular components of native ECM. The blood vessel decellularized matrix must be sterilized prior to freezing and finally lyophilized to prepare a fine powder, as explained in figure 10.9. This powdered decellularized matrix may be blended with natural biomaterials like hyaluronic acid, elastin or collagen and vascular cells to prepare a bioink for subsequent bioprinting of artificial blood vessels. The tissue engineered blood vessel developed from this bioink will have the capability to regenerate endothelium and will possess mechanical strength similar to a native blood vessel. This cell seeded construct developed without bioreactors will be cost effective and hence reduce the cost factor. Hopefully, this bioprinted tubular scaffold can be fabricated as an off-the-shelf vascular and patient specific graft which may be immunologically accepted by humans. This scaffold may provide a microenvironment and biochemical signals necessary for the growth of vascular cells.

Table 10.2. Table showing mechanical properties of different artificial blood vessels.

S. no.		Fabrication technique	Vessel internal diameter (mm)	Suture retention strength	Compliance (%)	Burst pressure	References
1.	L'Heureux *et al* 2006	Cell sheet	4.5	162 ±15 gf	1.5 ± 0.3	3468 ± 500 mmHg	(L'Heureux *et al* 2006)
			1.5	—	—	3668 ± 1865 mmHg	
2.	Konig *et al* 2009	Cell sheet	4.8	152 gf	3.4	3503 mmHg	(Konig *et al* 2009)
3.	Wang *et al* 2009	Electrospinning	4.5	1.07 ± 0.07 N	—	111.4 ± 2.6 kPa	(Wang *et al* 2009)
4.	Drilling *et al* 2009	Electrospinning	5	—	—	2000 mmHg	(Drilling *et al* 2009)
5.	Gong *et al* 2016	Decellularization	1.5	—	—	2060 ± 127 mmHg	(Gong *et al* 2016)

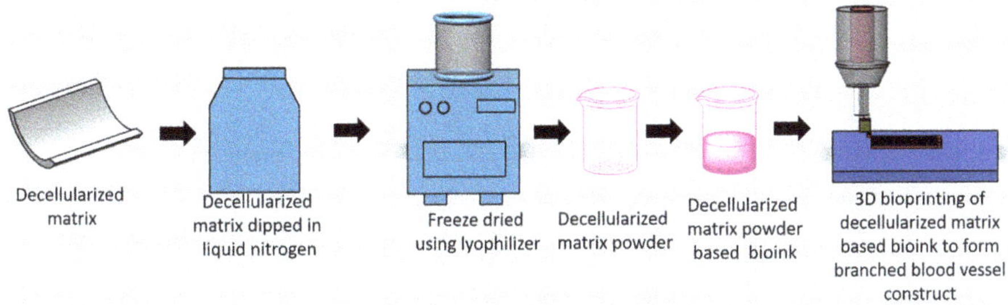

Figure 10.9. Schematic diagram of a proposed tissue engineered blood vessel fabricated using 3D bioprinting of a decellularized extracellular matrix powder blended with suitable natural polymer and vascular cells.

Vascularization is the major factor to be considered during blood vessel fabrication. Bioprinting of a vessel substitute is done with the help of a multi-nozzle bioprinter to have selective printing vascular cells and natural material depending upon the layer printed. For instance, the media layer may be printed employing hyaluronic acid, decellularized vessel powder and SMCs to most probably encourage vascularisation and impart elasticity to the tubular construct. This approach can lead to fabrication of a spatially defined heterogeneous blood vessel construct.

References

Atlan M, Simon-Yarza T and Ino J M *et al* 2018 Design, characterization and *in vivo* performance of synthetic 2 mm-diameter vessel grafts made of PVA-gelatin blends *Sci. Rep.* **8** 7417

Augustine R, Dan P and Sosnik A *et al* 2017 Electrospun poly(vinylidene fluoride-trifluoroethylene)/zinc oxide nanocomposite tissue engineering scaffolds with enhanced cell adhesion and blood vessel formation *Nano Res.* **10** 3358–76

Betsholtz C 2018 Cell–cell signaling in blood vessel development and function *EMBO Mol. Med.* **10**

Bourget J-M, Kérourédan O and Medina M *et al* 2016 Patterning of endothelial cells and mesenchymal stem cells by laser-assisted bioprinting to study cell migration *Biomed. Res. Int.* **2016** 3569843

Chlupác J, Filová E and Bacáková L 2009 Blood vessel replacement: 50 years of development and tissue engineering paradigms in vascular surgery *Physiol. Res./Acad. Sci. Bohemoslov.* **58** S119–39

Cui X and Boland T 2009 Human microvasculature fabrication using thermal inkjet printing technology *Biomaterials* **30** 6221–7

Dahl S L M, Koh J, Prabhakar V and Niklason L E 2003 Decellularized native and engineered arterial scaffolds for transplantation *Cell Transplant.* **12** 659–66

Drilling S, Gaumer J and Lannutti J 2009 Fabrication of burst pressure competent vascular grafts via electrospinning: effects of microstructure *J. Biomed. Mater. Res.* A **88A** 923–34

Funamoto S, Nam K and Kimura T *et al* 2010 The use of high-hydrostatic pressure treatment to decellularize blood vessels *Biomaterials* **31** 3590–95

Geudens I and Gerhardt H 2011 Coordinating cell behaviour during blood vessel formation *Development* **138** 4569–83

Gong W, Lei D and Li S *et al* 2016 Hybrid small-diameter vascular grafts: Anti-expansion effect of electrospun poly ε-caprolactone on heparin-coated decellularized matrices *Biomaterials* **76** 359–70

Gupta S, Bissoyi A and Bit A 2018 A review on 3D printable techniques for tissue engineering *Bionanoscience* 1–16

Gupta S and Bit A 2018 Rapid prototyping for polymeric gels *Polymeric Gels* (Amsterdam: Elsevier) pp 397–439

Haraguchi Y, Shimizu T, Yamato M and Okano T 2012 Scaffold-free tissue engineering using cell sheet technology *RSC Adv.* **2** 2184

Hasan A, Deeb G, Atwi K, Soliman S and Hasan A 2015 Electrospun PET-PU scaffolds for vascular tissue engineering *2015 Int. Conf. on Advances in Biomedical Engineering (ICABME)* ((Piscataway, NJ: IEEE) pp 217–21

Herbert S P and Stainier D Y R 2011 Molecular control of endothelial cell behaviour duringblood vessel morphogenesis *Nat. Rev. Mol. Cell Biol.* **12** 551–64

Jung Y, Ji H and Chen Z *et al* 2015 Scaffold-free, human mesenchymal stem cell-based tissue engineered blood vessels *Sci. Rep.* **5** 15116

Karaman O, Şen M and Demirci E A 2017 Electrospun scaffolds for vascular tissue engineering *Electrospun Materials for Tissue Engineering and Biomedical Applications* (Amsterdam: Elsevier) pp 261–87

Karamysheva A F 2008 Mechanisms of angiogenesis *Biochemistry (Moscow)* **73** 751–62

Kelm J M, Lorber V and Snedeker J G *et al* 2010 A novel concept for scaffold-free vessel tissue engineering: self-assembly of microtissue building blocks *J. Biotechnol.* **148** 46–55

Kolesky D B, Homan K A, Skylar-Scott M A and Lewis J A 2016 Three-dimensional bioprinting of thick vascularized tissues *Proc. Natl Acad. Sci. USA* **113** 3179–84

Konig G, McAllister T N and Dusserre N *et al* 2009 Mechanical properties of completely autologous human tissue engineered blood vessels compared to human saphenous vein and mammary artery *Biomaterials* **30** 1542–50

Koochaki A, Moghbeli M R and JavanNikkhah S 2018 Coil-to-globule transition of thermo-responsive γ-substituted poly (ε-caprolactone) in water: a molecular dynamics simulation study *Curr. Appl Phys.* **18** 1313–9

Koschwanez H E and Reichert W M 2013 Textured and porous materials *Biomaterials Science* (Amsterdam: Elsevier) pp 321–31

Labbé B, Marceau-Fortier G and Fradette J 2011 Cell sheet technology for tissue engineering: the self-assembly approach using adipose-derived stromal cells *Methods Mol. Biol.* **702** pp 429–41

Lee J S, Lee K and moon S-H *et al* 2014 Mussel-inspired cell-adhesion peptide modification for enhanced endothelialization of decellularized blood vessels *Macromol. Biosci.* **14** 1181–9

L'Heureux N, Dusserre N and Konig G *et al* 2006 Human tissue-engineered blood vessels for adult arterial revascularization *Nat. Med.* **12** 361–5

Liu J W, Dunoyer-Geindre S and Serre-Beinier V *et al* 2007 Characterization of endothelial-like cells derived from human mesenchymal stem cells *J. Thromb. Haemost.* **5** 826–34

Marcolin C, Draghi L, Tanzi M and Faré S 2017 Electrospun silk fibroin–gelatin composite tubular matrices as scaffolds for small diameter blood vessel regeneration *J. Mater. Sci. Mater. Med.* **28** 80

Marelli B, Alessandrino A, Farè S, Freddi G, Mantovani D and Tanzi M C 2010 Compliant electrospun silk fibroin tubes for small vessel bypass grafting *Acta Biomater.* **6** 4019–26

Melchiorri A J, Hibino N, Brandes Z R, Jonas R A and Fisher J P 2014 Development and assessment of a biodegradable solvent cast polyester fabric small-diameter vascular graft *J. Biomed. Mater. Res.* A **102** 1972–81

Metz R P, Patterson J L and Wilson E 2012 Vascular smooth muscle cells: isolation, culture, and characterization *Methods Mol. Biol.* **843** 169–76

Moschouris K, Firoozi N and Kang Y 2016 The application of cell sheet engineering in the vascularization of tissue regeneration *Regen. Med.* **11** 559–70

Oswald J, Boxberger S and Jørgensen B *et al* 2004 Mesenchymal stem cells can be differentiated into endothelial cells *in vitro Stem Cells* **22** 377–84

Papavasiliou G, Sokic S and Turturro M 2012 Synthetic PEG hydrogels as extracellular matrix mimics for tissue engineering applications ed R Sammour *Biotechnology—Molecular Studies and Novel Applications for Improved Quality of Human Life* (Rijeka: InTech)

Potente M, Gerhardt H and Carmeliet P 2011 Basic and therapeutic aspects of angiogenesis *Cell* **146** 873–87

Quint C, Arief M, Muto A, Dardik A and Niklason L E 2012 Allogeneic human tissue-engineered blood vessel *J. Vasc. Surg.* **55** 790–8

Quint C, Kondo Y, Manson R J, Lawson J H, Dardik A and Niklason L E 2011 Decellularized tissue-engineered blood vessel as an arterial conduit *Proc. Natl Acad. Sci. USA* **108** 9214–9

Rim N G, Yih A, Hsi P, Wang Y, Zhang Y and Wong J Y 2018 Micropatterned cellsheets as structural building blocks for biomimetic vascular patches *Biomaterials* **181** 126–39

Sekine H, Shimizu T and Sakaguchi K *et al* 2013 *In vitro* fabrication of functional three-dimensional tissues with perfusable blood vessels *Nat. Commun.* **4** 1399

Settembrini A M and Settembrini P G 2017 Medical, surgical therapy, and alternative treatment of infected vascular grafts *The Microbiology of Skin, Soft Tissue, Bone and Joint Infections* (Amsterdam: Elsevier) pp 287–301

Sheridan W S, Duffy G P and Murphy B P 2013 Optimum parameters for freeze-drying decellularized arterial scaffolds *Tissue Eng. Part C, Methods* **19** 981–90

Shimizu K, Ito A and Arinobe M *et al* 2007 Effective cell-seeding technique using magnetite nanoparticles and magnetic force onto decellularized blood vessels for vascular tissue engineering *J. Biosci. Bioeng.* **103** 472–8

Sohail M R, Wilson W R and Baddour L M 2010 Infections of nonvalvular cardiovascular devices *Mandell, Douglas, and Bennett's Principles and Practice of Infectious Diseases* (Amsterdam: Elsevier) pp 1127–42

Stocum D L 2012 Regenerative therapies for hematopoietic and cardiovascular tissues *Regenerative Biology and Medicine* (Amsterdam: Elsevier) pp 393–426

Suzuki Y, Nagai N and Umemura K 2016 A review of the mechanisms of blood-brain barrier permeability by tissue-type plasminogen activator treatment for cerebral ischemia *Front. Cellular Neurosci.* **10** 2

Taki A, Kermani A, Ranjbarnavazi S M and Pourmodheji A 2017 Overview of different medical imaging techniques for the identification of coronary atherosclerotic plaques *Computing and Visualization for Intravascular Imaging and Computer-Assisted Stenting* (Amsterdam: Elsevier) pp 79–106

Vaz C M, van Tuijl S, Bouten C V C and Baaijens F P T 2005 Design of scaffolds for blood vessel tissue engineering using a multi-layering electrospinning technique *Acta Biomater.* **1** 575–82

Wakuda Y, Nishimoto S, Suye S-I and Fujita S 2018 Native collagen hydrogel nanofibres with anisotropic structure using core–shell electrospinning *Sci. Rep.* **8** 6248

Wang K, Zhang Q and Zhao L *et al* 2017 Functional modification of electrospun poly(ε-caprolactone) vascular grafts with the fusion protein VEGF-HGFI enhanced vascular regeneration *ACS Appl. Mater. Interfaces.* **9** 11415–27

Wang P, Xu T Y, Guan Y F, Su D F, Fan G R and Miao C Y 2009 Perivascular adipose tissue-derived visfatin is a vascular smooth muscle cell growth factor: role of nicotinamide mononucleotide *Cardiovasc. Res.* **81** 370

Weinberg C B and Bell E 1986 A blood vessel model constructed from collagen and cultured vascular cells *Science* **231** 397–400

Wu P K and Ringeisen B R 2010 Development of human umbilical vein endothelial cell (HUVEC) and human umbilical vein smooth muscle cell (HUVSMC) branch/stem structures on hydrogel layers via biological laser printing (BioLP) *Biofabrication* **2** 014111

Wu P, Nakamura N and Morita H *et al* 2019 A hybrid small-diameter tube fabricated from decellularized aortic intima-media and electrospun fiber for artificial small-diameter blood vessel *J. Biomed. Mater. Res.* A **107** 1064–70

Yalcinkaya F, Yalcinkaya B, Pazourek A, Mullerova J, Stuchlik M and Maryska J 2016 Surface modification of electrospun PVDF/PAN nanofibrous layers by low vacuum plasma treatment *Int. J. Polym. Sci.* **2016** 1–9

Zhang H, van Olden C, Sweeney D and Martin-Rendon E 2014 Blood vessel repair and regeneration in the ischaemic heart *Open Heart* **1** e000016

Zhang W J, Liu W, Cui L and Cao Y 2007 Tissue engineering of blood vessel *J. Cell. Mol. Med.* **11** 945–57

Zhu C, Fan D and Duan Z *et al* 2009a Initial investigation of novel human-like collagen/chitosan scaffold for vascular tissue engineering *J. Biomed. Mater. Res.* A **89** 829–40

Zhu C, Fan D and Ma X *et al* 2009b Preparation and properties of novel human-like collagen-silk fibroin scaffold for blood vessel *Chin. J. Biotechnol.* **25** 1225–33

Zhu C, Fan D and Wang Y 2014 Human-like collagen/hyaluronic acid 3D scaffolds for vascular tissue engineering *Mater. Sci. Eng. C: Mater. Biol. Appl.* **34** 393–401

Zhu W, Qu X and Zhu J *et al* 2017 Direct 3D bioprinting of prevascularized tissue constructs with complex microarchitecture *Biomaterials* **124** 106–15

Chapter 11

Bioreactors for tissue engineered blood vessels

Arindam Bit, Jasjit S Suri and Khemraj Deskmukh

This chapter gives a detailed description of various bioreactors used for cell culture for blood vessel replacement. Diseases related to damage of vascular grafts lead to the majority of deaths in most countries. The basic treatment for such diseases is use of autografts that mimic the environment of native tissue. However, the major limitation is lack of sufficient donors and the mechanical and biochemical relation between the autograph which may result in the graft rejection. A mechanically developed vessel shows good mechanical properties, but has the risk of thromboembolism where a blood vessel is obstructed by a blood clot. In this case, these valves need anticoagulation throughout life. Lots of invention and research has been done in the development of tissues and vascular constructs. One of the major inventions in the field of the blood vessel is its application in various cardiovascular diseases. These kinds of blood vessels or heart valves developed *in vitro* have high mechanical and morphological properties which enable their functioning *in vivo*. Tissue engineered valves often have limitations like restricted growth in the lumen due to insufficient supply of the media throughout the graft. This can be controlled by developing the tissue construct in a bioreactor with controlled conditions and design of the bioreactor specific to the construct. Here we are discussing various types of bioreactors available for tissue engineered blood vessels.

11.1 Introduction

Tissue engineering (TE) is the most fundamental approach for the regeneration of damaged tissues or organs. The main reason for growth of TE in the field of developing artificial organs or tissues because of disorders is the shortage of donors. For an engineered tissue to develop, autologous cells will be seeded on a 3D biomaterial scaffold or matrix which is further grown *in vitro* (Naughton 2002, Sodian *et al* 2001). The major function of any blood vessel is the transfer of oxygenated blood to the heart. One of the challenging tasks that limits its applications is angiogenesis (Sodian *et al* 2002). The major drawback of the

doi:10.1088/978-0-7503-2088-7ch11

in vitro method is not having sufficient blood vessels (Novosel *et al* 2011). When *in vivo* signals are generated already existing blood vessels can be grown into new tissues. But the growth will be slow, which may affect the cells in the scaffold due to insufficient nutrients (Rouwkema *et al* 2008). Increasing cardiovascular diseases are also a major problem leading to death (Roger *et al* 2013). One of the present solutions for cardiovascular diseases is surgical implantation of vascular grafts. Coronary artery bypass and peripheral bypass are used in revascularization of myocardial and limb which involves small diameter vascular grafts (<6 mm) (U 2000). Now a days small diameter vascular graft is more needed. Though vascular grafts can be used as a solution for cardiovascular diseases, their availability is limited and very limited grafts can exhibit the natural properties of the native vessels (Opitz *et al* 2004). Although vascular grafts of small diameter can be made using synthetic biomaterials, the factors that limit their function is development of intimal hyperplasia, thrombogenesis and coagulation (Graham *et al* 1989). The most important factor to be considered while developing artificial blood vessel should full fill the all the properties like mechanical, physiological of the native vessels. This arises the need of developing tissue engineered blood vessels which are small in diameter and have all the functionalities (Borschel *et al* 2005, Schmedlen *et al* 2003). TE blood vessels can be apply in the bypass heart operation overcoming the limitations like the extraction of autologous tissues and finding suitable donor (Mitchell and Niklason 2003).

Tissue engineered blood vessel are one of great developments in the field of modern medicine. This process of developing blood vessels is still in the stage of developing and many ways are available in developing the tissue artificially which will mimic the blood vessels. One of the methods is development of tubes by culturing rat muscles on the surface of silicon tubes. The major problem developed in this method is that the cell layers that are present inner side will not receive do not have access to nutrients and other supplements when compared to outer layers. Due to this, the growth of outer layers is more compared to the inner ones. The key to minimize this problem is adaptation of a bioreactor where the media and supplements will be pumped throughout the inner surface so as to have constant growth of the cells or tissues (perfusion bioreactor). Tissue constructs grows better when supplied with laminar flow and uniform distribution of shear stress (Jockenhoevel *et al* 2002, Carrier *et al* 2002). Growth of the artificial tissues can be achieved by cell insertion in a synthetic support material called a scaffold and then continued growth *in vitro* (Jockenhoevel *et al* 2000, Naughton 2002, Sodian *et al* 2001). The scaffold on which the cells are seeded will communicate with the environment for better growth (Jockenhoevel *et al* 2000). The major part played in the growth of the cells is by designing systems that provide better growth of the vascular grafts *in vitro*.

For growth of the cells *in vitro*, suitable environmental conditions that mimic natural tissues have to be provided. As the proper *in vitro* conditions are clearly not known, several studies have suggested that providing mechanical preconditioning will help in the growth of the tissue engineered vascular constructs (Hoerstrup *et al* 2000, Niklason *et al* 1999). Some other studies proved that pulsatile fluid flow can also help in developing tissue engineered heart valves *in vivo* before implanting them for *in vivo* studies (Iaizzo *et al* 2013). Cardiovascular constructs show better growth when provided with laminar flow with uniform distribution and moderate shear stress (Jockenhoevel *et al* 2002).

11.2 Properties of a bioreactor

Bioreactor design is a complex task as one should keep in view both engineering and scientific aspects to develop a predefined environment for expansion of a tissue. By considering several factors we will design the appropriate bioreactor. The factors may change with the type of the tissue to be developed, but there are general factor requirements that a bioreactor has to meet. They are

- controlled physiochemical environment in the bioreactor;
- provides monitoring the quality of cell or tissue;
- ensured sterile conditions for the growth of the tissue;
- continuous supply of sufficient nutrients and waste exchange;
- basic cellular distribution and their attachment to the developing tissue;
- exposure of the dynamic tissue to mechanical and hydrodynamic forces;
- maintenance of a high degree of duplicability;
- controlled media flow;
- reduction of high turbulence in fluid flow;
- material used to fabricate the bioreactor should have compatibility with the cells or tissues;
- easy maintenance;
- compact design to place inside the globalize incubator;
- waste collection should be avoided.

The bioreactor that will be designed will depend on the tissue that has to be developed. During design of a bioreactor, major factors such as shear stress, pulsatile forces, flow of media pressure etc have to be simulated that will affect the growth of the tissue. Some other major factors that play a major role in the design of a bioreactor are removal of waste products and proper distribution of the nutrients for development of the tissue. The major factors that are to be considered in designing a bioreactor are biochemical controls such as nutrients, oxygen and shear stress, pressure, carbon dioxide levels, temperature, humidity and pH. At the same time, some biomechanical properties such as flow rate, volume etc also have to be considered in designing a bioreactor. The major role of blood vessels is transport of nutrients, CO_2, O_2 and waste material to various parts of the body. The major force that drives this function is diffusion (Van Luyn *et al* 2002, Lwigale *et al* 2000).

A bioreactor should be able to supply the nutrients to each and every part of the tissue thus preventing cell death (Galban and Locke 1999, Kofidis *et al* 2002, Ozawa *et al* 2002). Perfusion also plays a major role in thickness of the tissue engineered construct (Jockenhoevel *et al* 2002). Another factor that has to be maintained is pH. A pH in the range of 7–7.4 is suitable for developing tissue to maintain homeostasis. vascular endothelial cells play a crucial role in maintaining homeostasis (Guillaume *et al* 1997). Oxygen plays a major role in the cultivation of cells. Oxygen supply should be provided according to the rate it is consumed. Low oxygen levels affect cell growth and other activities and exposure of cells to low oxygen levels will lead to cell death. Oxygen should be provided at the same rate it is consumed. If the delivery of oxygen is slow compared to its consumption, it will affect the enzyme activities and also viability of cells (Catapano 1996). Media that is supplied to the cells has to be uniformly distributed (Carrier *et al* 2002, Van Luyn *et al* 2002, Lwigale *et al* 2000). Normal temperature of the human body is 37 °C. It may vary slightly but risk is associated if the temperature goes beyond 35 °C or above 40 °C. Therefore, there is a necessity to maintain optimum temperature conditions in the bioreactor, as there will be some heat loss generated by metabolically active cells. Although bioreactors are usually inserted in an incubator, a heat exchanger can additionally be placed to maintain the temperature of 37 °C.

When it comes to maintenance of biomechanical properties of a bioreactor, additional pumps are used to achieve the pulsatile flow rate. Mostly piston, centrifugal and peristaltic pumps are inserted for this purpose. A chamber or a diaphragm can be replaced by a piston to enable media flow through the bioreactor to avoid apoptosis caused when it comes directly in contact with the cells (Hoerstrup *et al* 2000, Sauer and Gerlach 2002). Peristaltic pumps are widely used in heart valve and blood vessels due to their low shear stress which would not affect the cells. Fluid dynamic factors such as shear rate, shear stress, turbulence and recirculation play a major role in hyperplasia, that is, increase in tissue with increase in reproduction rate of the cells.

Shear stress on endothelial cells is important in maintaining metabolic activities, orientation, and hemostasis of the vascular endothelium (Chien 2007, Davies 2009). In the case of arteries, the wall is covered with endothelial cells in the same direction of the blood stream (Yamamoto *et al* 2015). Many studies has proved that shear stress helps in stimulating nitric oxide (NO) which is a vascodilator released from endothelial cells protecting them from platelet aggregation (Vanhoutte 1989, Wootton and Ku 2002). Blood vessels are usually accustomed to shear stress fluctuations. There is an inverse relationship between flow rate and thickness of the wall of the lumen so that the shear stress can be normalized. Normal shear stress in blood vessels such as arteries is 15 dyn cm^{-2}. In the case of a decrease in shear stress up to <10 dyn cm^{-2}, thickening in the lumen of the vessel can be found (Weston *et al* 1999, Wootton and Ku 2002). Shear stress should be maintained in a normal range as high shear stress can damage the cells. Some numerical models were proposed by Begley and Kleis (2000), Weston *et al* (1999), Williams *et al* (2002) to observe the effect of shear stress on blood vessels. All the biochemical and biomechanical factors should be considered designing a tissue engineered product.

11.3 Blood vessel bioreactors

The major purpose of designing any bioreactor is to develop self-renewing tissue according to the requirements (Bader *et al* 2000, Dow *et al* 2005, Hoerstrup *et al* 2001, Ratcliffe 2000). In diseases like atherosclerosis where the blood vessel is blocked due to plaque, the major treatment is introducing autografts, xenografts and other synthetic grafts. In these cases, using natural vessels has a lower success rate of only up to 70% due to restrictions like mechanical properties, strength, dimensions and their connection with the adjacent blood vessels (Conklin *et al* 2002, Hoerstrup *et al* 2001). Another problem that arises in transplanted vessels is vasculopathy, that is, lesions occurring on vessels and capillaries (Hosenpud *et al* 1996). Synthetic grafts developed for large vessels are successful but with medium and smaller dimension grafts, loss of patency has been observed within a few days of the implantation. The major reason for failure of these vascular grafts is the presence of foreign material which has no ability for growth or repair.

In order to overcome all these limitations, studies have been done to develop vascular tissue with the ability to grow and repair under controlled conditions of mechanical properties and pulsatile flow conditions with similar properties of the native tissue.

11.3.1 Pulsatile perfusion bioreactor

A patented bioreactor was developed by Eberth (Shazly *et al* 2016) for culturing and development of engineered blood vessels. This bioreactor mostly consists of a chamber where the blood vessel can be placed, a monitoring system to measure and maintain properties such as tensile and shear stress on the vessel, and a pump for distribution of culture media throughout the vessel. The pump system is also available to maintain the steady flow and peristaltic pumps. It also has a facility of adjusting composite waveforms for maintaining stress levels. This bioreactor provides mechano-mimetic stimulation to the blood vessels.

11.3.2 Biaxial bioreactor

A biaxial bioreactor was developed where biaxial stretching of the engineered vessels was enabled during the culture. This bioreactor also provides the evaluation of the tissue engineered valves that are developed under static, uniaxial and biaxial conditions. The engineered valve grown under this bioreactor also showed the greatest thickness. Collagen fibers and mature elastin formation in the ECM show that biaxial stretching is one of the promising methods to develop tissue engineered valves. Using this bioreactor, biaxial stretching on tissue engineered vessels was tested. This bioreactor is made of glass. The tissue engineered valves that are incorporated in the bioreactor will be subjected to different mechanical stimuli. Biaxial stretching (circumferential and axial stretching) is provided by a linear motor and flow system to the bioreactor. The pulsatile pump is responsible for circumferential stretch maintained by the peristaltic pump and axial flow will be achieved once the motor stretches the connector of the bioreactor. This bioreactor plays a major role in optimizing the biomechanical aspects by improving the ECM properties of tissue engineered vessels. The schematical representation of the bioreactor is given in figure 11.1(a).

Figure 11.1(b) depicts the equivalent electrical circuit of a biaxial bioreactor. As a process diagram, it contains many components replaced by their equivalent electrical components, like phosphate buffer saline (PBS) reservoir and buffer media

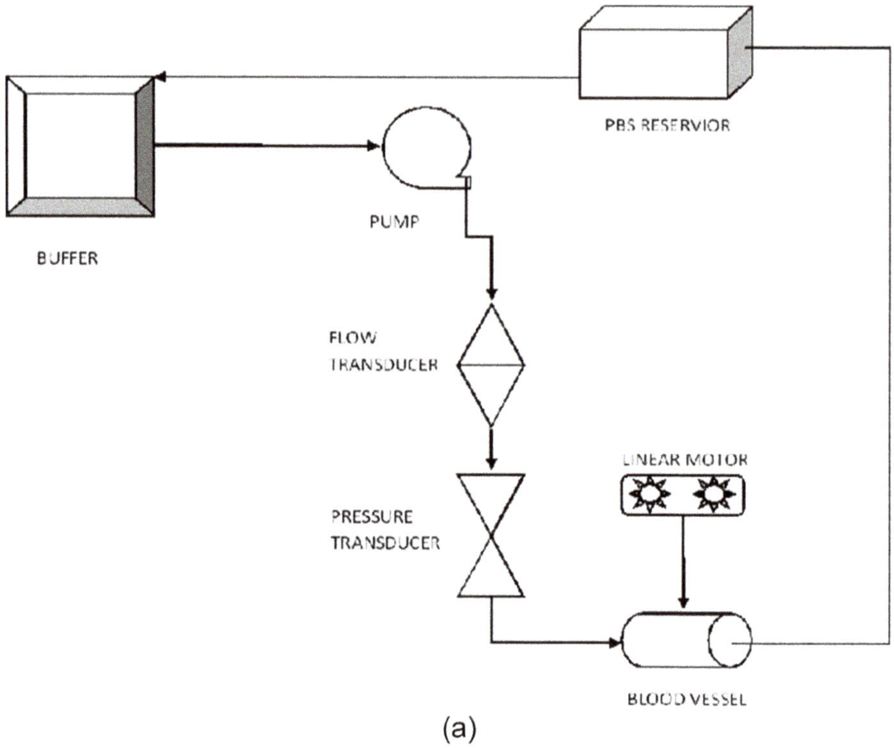

Figure 11.1(a). Top and side view of a biaxial bioreactor system. Arrows indicate the flow direction.

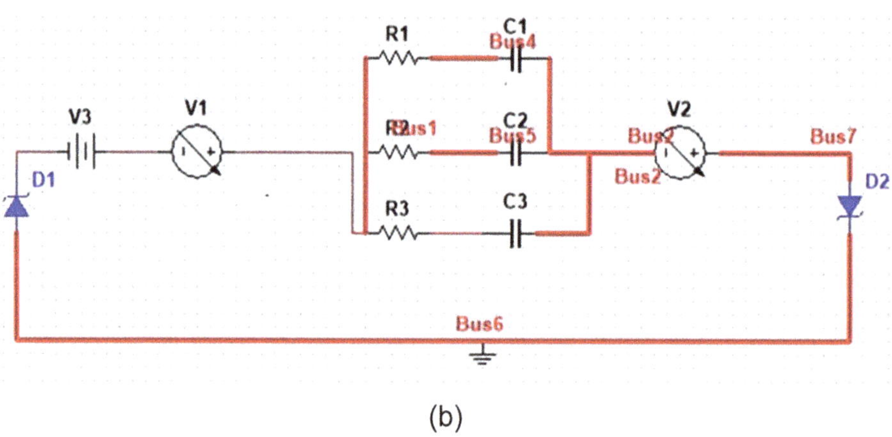

Figure 11.1(b). Electrical analogy of a biaxial bioreactor.

replaced by ground, peristaltic pump replaced by battery, pressure transducer replaced by voltage source, and flow is measured in terms of voltage drop. The flow of media is considered as the flow of current in equivalent circuits. The diode is used in the circuits for the valve. As in the process diagram, a blood vessel acts as an obstacle to the flow of media and hence it is represented by resistance, which opposes the flow of current. A linear motor is used for providing stretch in the blood vessel so in equivalent circuits the linear motor is replaced with a capacitor which is connected in series with a resistor.

After conversion, the circuit will be analyzed by principle of electrical circuits.

In the circuit, $R1$ is in series with $C1$, $R2$ is with $C2$ and $C3$ is in series. The resulting impedance is represented with $Z1$, $Z2$ and $Z3$.

$$Z1 = R1 + \frac{1}{sc1} \tag{11.1}$$

$$Z2 = R2 + \frac{1}{SC2} \tag{11.2}$$

$$Z3 = R3 + \frac{1}{SC3} \tag{11.3}$$

Now we observe that these $Z1$, $Z2$ and $Z3$ are connected in parallel. And the resulting impedance is represented with $Z(s)$.

$$Z = Z1\ ||Z2||Z3$$

When we solve that then we get the value of $Z(s)$ which is equal to

$$Z(s) = \frac{\frac{(1 + SC1R1)(1 + SC2R2)}{C2(1 + SC1R1) + C1(1 + SC2R2)}(1 + SC3R3)}{C3\left[\frac{(1 + SC1R1)(1 + SC2R2)}{C2(1 + SC1R1) + C1(1 + SC2R2)}\right] + C1(1 + SC2R2)} \tag{11.4}$$

Suppose I current is the flow from the circuit. Now Apply Kirchoff's current law in a closed loop and also take the Laplace transformation, then we get

$$-\frac{V3}{S} - \frac{V1}{S} + Z(S)I(S) \quad \frac{V2}{S} = 0 \tag{11.5}$$

Then solving the above equation (11.5) for $I(S)$ we get

$$I(S) = \frac{V1 + V2 + B3}{SZ(S)}$$

The bioreactor designed by Williams (Williams and Wick 2004) is a perfusion bioreactor, cylindrical in shape and made of glass. It is designed to develop small arterial constructs. Glass tubes are present where a scaffold is placed between them. There is a media reservoir for pumping of the media throughout the construct. Using peristaltic pumps, the media is pumped through the cavity of the arteries and also to its external surface with a flow rate of 40 mL min^{-1}. This bioreactor provides an environment to help cell culture. Smooth muscle and endothelial cells were placed on the scaffolds and are cultured for 25 days in the bioreactor using pulsatile flow. The results obtained after culturing the cells in the bioreactor show this bioreactor supporting the seeding of the desired constructs and forming an extracellular matrix in the tissue engineered arterial construct. Assembly of the bioreactor during cell seeding and development can be observed from figure 11.2(a).

Figure 11.2(b) represents the equivalent electrical circuit of a perfusion bioreactor. In the reactor process diagram many components involved in performing specific tasks are replaced by their equivalent electrical components, such as the media reservoir replaced by ground, peristaltic pump replaced by battery, flow is measured in terms of voltage drop between two resistors, R1 and R2, and R2 is considered as a variable resistor because it flows over the exterior surface of the

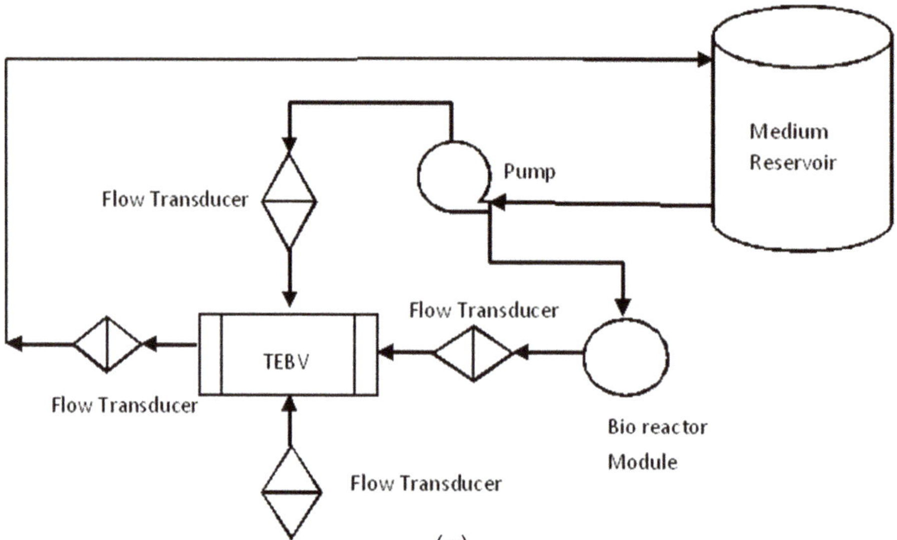

(a)

Figure 11.2(a). Assembly of bioreactor during cell seeding and growth of construct.

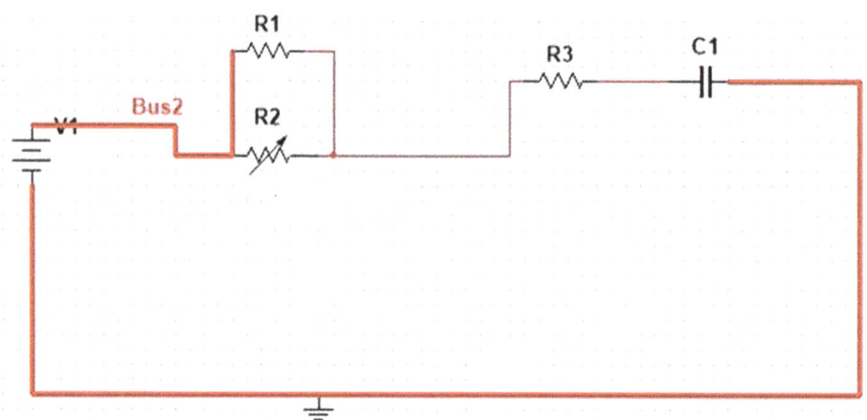

(b)

Figure 11.2(b). Electrical analogy of a perfusion bioreactor.

blood vessel in the bioreactor module. The flow of media is considered as current in equivalent circuits/electrical analogy. As in the process diagram, the blood vessel acts as an obstacle to the flow of media and hence it is represented by resistance, which opposes the flow of current. In electrical circuits we use a capacitor, C1, for the stretching phenomenon that occurs in the bioreactor module, and this capacitor is connected in series with resistor R3. Resistor R3 is used for the blood vessel.

After conversion, the circuit will be analyzed by the principle of electrical circuits.

In figure 11.2(b) we clearly see that R1 is parallel with R2. The resulting impedance is denoted by R.

So

$$R = R1\,||R2 \tag{11.6}$$

Suppose I current is flow from the circuit. Apply KCL in a closed loop and also take the Laplace transformation, then we get

$$-\frac{V1}{S} + RI(S) + R3I(S) + \frac{1}{SC1}I(S) = 0 \tag{11.7}$$

Then, solving the above equation (11.7) for $I(S)$ we get

$$I(S) = \frac{V1C1}{RSC1 + R3SC1 + 1}$$

11.3.3 VascuTrainer bioreactor

A mobile bioreactor called the VascuTrainer (figure 11.3(a)) was developed by Wolf *et al* (2018) for formation of a blood vessel outside the body. Major properties of this bioreactor are providing good flow of media, pressure and other mechanical properties. It is easy to assemble, transportable and disposable. The important parts of the bioreactor are the media reservoir which is made of polymethylpenthene, polypropylene and silicon, a centrifugal pump, pressure sensor, flow sensor and pump control unit. The potentiality of this developed bioreactor was that vascular grafts were introduced using a fibrin scaffold that is embedded with human umbilical artery smooth muscle cells (SMCs) and a macroporous mesh knitted from polyvinyledene fluoride. After 14 days the output of the bioreactor was observed and any change in the growth medium provided was measured. The media samples were tested every two days by measuring the pO_2, pCO_2, glucose and lactate levels. Cytocompatible testing was done for the SMCs and fibroblast cells seeded at various intervals by assessing the necrosis, apoptosis and proliferation rate of the cell. The media along with all the supplements are circulated at 37 °C for 72 h. The results no necrosis was observed and the apoptosis rate was lower. This shows that the bioreactor is cytocompatible. With the help of arterial pressure and shear stress, the fabricated blood vessel showed haemocompatibility. Histochemical studies and tissue analysis revealed the deposition of collagen I and III. Various parts and assembly of the bioreactor can be observed in figure 11.3(a).

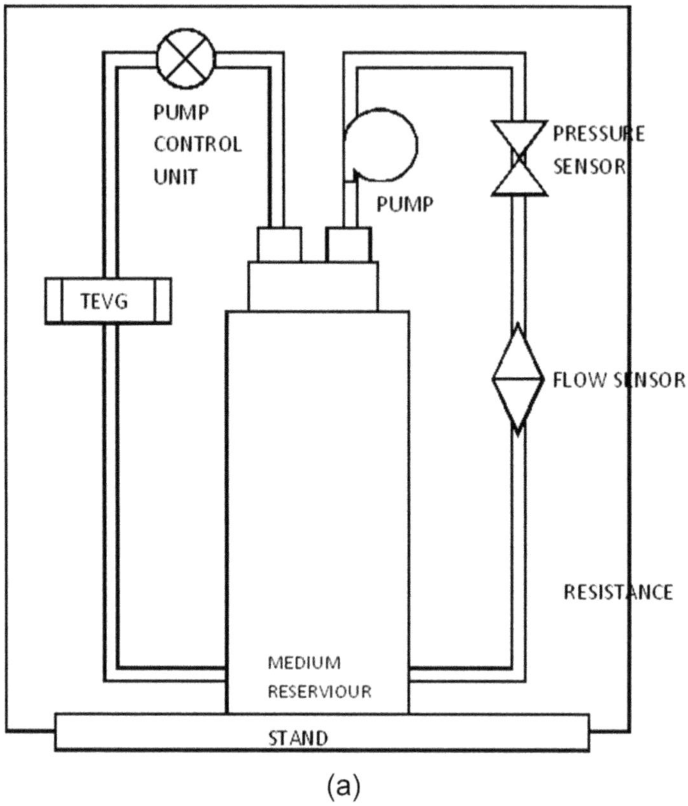

Figure 11.3(a). Diagrammatic representation of a VascuTrainer bioreactor.

Figure 11.3(b) shows the equivalent electrical representation of the VascuTrainer bioreactor. In the reactor process diagram, many components involved in performing specific tasks are replaced by their equivalent electrical components such as medium reservoir replaced by ground, centrifugal pump replaced by battery V2, pressure transducer replaced by voltage source V1, and flow is measured in terms of voltage drop across resistance R. The flow of media is considered as flow of current in equivalent circuits. As in the process diagram, the blood vessel acts as an obstacle to the flow of media and hence it is represented by resistance R, which opposes the flow of current. The electrical connection is drawn on the basis of the operating principle of the bioreactor.

The next step is to evaluate circuits on the basis of electrical principles. Suppose I current is flowing from the circuit. Apply KCL in a closed loop and also take the Laplace transformation, then we get

$$-\frac{V2}{S} - \frac{V1}{S} + RI(S) = 0 \tag{11.8}$$

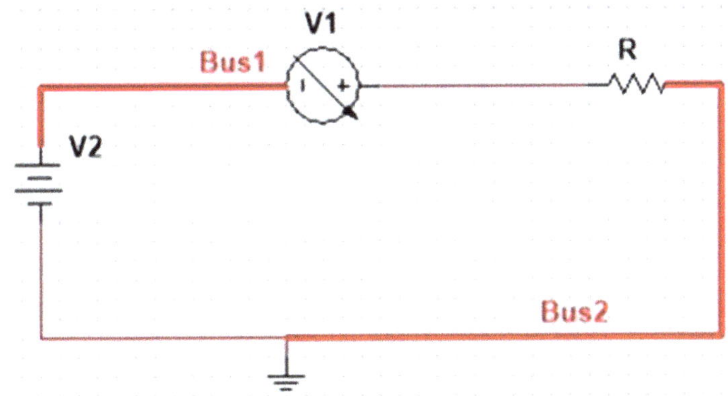

(b)

Figure 11.3(b). Electrical analogy of a VascuTrainer bioreactor.

Then, solving above equation (11.8) for $I(S)$ we get

$$I(S) = \frac{V1 + V2}{SR}$$

11.3.4 Perfusion bioreactor with longitudinal stretch

A perfusion bioreactor was designed by Mironov *et al* (2003) having the capacity of longitudinal strain. Longitudinal strain has an association with vascular development at the embryonic stage. The presented bioreactor has a closed area and two peristaltic pumps and piping. The two media perfusion structures are located in the presented bioreactor, one is used for inner perfusion and the other is for outer perfusion. Media will be continuously circulated through the closed system. The whole system is provided with a high sterility, gas supply, and can also be fitted easily in a humidified incubator to maintain the temperature and required CO_2 percentage. The control system includes transducers and digital cameras which control the biomechanical properties of the tissue construct. Longitudinal stretch supplied to silicon tubes will be in the range of 0 to 200% applying circumferential strain. This bioreactor provides properties to record data in both static and dynamic conditions. The performance output of the bioreactor was observed by using a bovine ceratoid artery as a tubular construct. Experimental setup and design of the bioreactor is shown in figure 11.4(a).

Figure 11.4(b) represents the equivalent electrical representation of the bioreactor. Figure 11.4(c) represents a simplified version of figure 11.4(b). In the reactor process diagram, many components involved to perform specific tasks are replaced by their equivalent electrical components such as reservoir replaced by ground, peristaltic pump replaced by battery V1 and V3, diode D1 and D2 is used for the valve. For the pressure vessel we consider resistance R1, the blood vessel is replaced by resistance R2 and resistance R3 is for some media flowing upon the surface of the blood vessel, R3 is considered as a variable resistor because it flows outer surface of

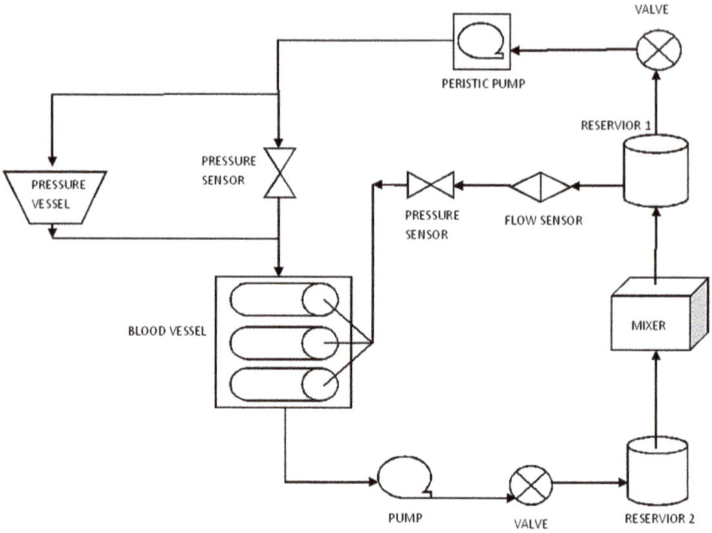

(a)

Figure 11.4(a). Diagram showing experimental setup of the bioreactor.

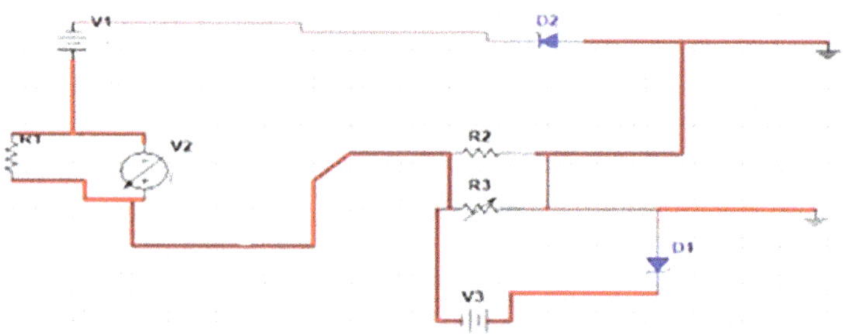

(b)

Figure 11.4(b). Diagram showing the equivalent circuit of the bioreactor.

blood vessel in the bioreactor. The flow of media is considered as a current in the equivalent circuits/electrical analogy. The pressure sensor is represented by V2. As in the process diagram, the blood vessel acts as an obstacle to the flow of media and hence it is represented by resistance, which opposes the flow of current.

Now we examine the electrical circuits.

Suppose I current is flowing from the circuit. Apply KCL in a closed loop and also take the Laplace transformation, then we get

$$-\frac{V1}{S} + R1I(S) - \frac{V2}{S} + R23I(S) - \frac{V3}{S} = 0 \tag{11.9}$$

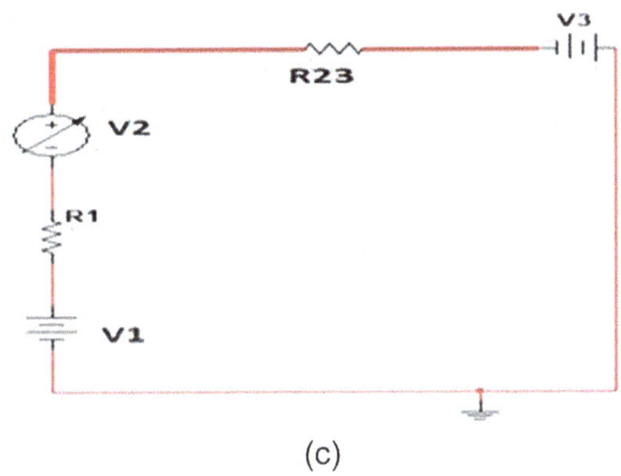

Figure 11.4(c). Simplified circuit of the bioreactor.

Then, solving equation (11.9) for $I(S)$ we get

$$I(S) = \frac{V1 + V2 + V3}{S(R1 + R23)}$$

11.3.5 Pulsatile flow bioreactors

Niklason and Langer (1997) have designed a bioreactor where continuous pulsatile flow is used. For the presented bioreactor SMCs and endothelial cells were incident on a PGA mesh and PLGA material. These cells were grown under static conditions. Under these conditions, vascular tissues formed with polymer and SMCs did not have the required physical strength for a vascular conduit. To improve the conditions, physical force such as pulsatile stress was introduced. The system is further developed by introducing pulsatile intra-luminal pressure and flow which helps in mimicking the *in vivo* conditions. In this design, vascular scaffolds are seeded parallel using pulsatile flow (figure 11.5(a)). Two pumps are attached to regulate continuous flow throughout the scaffolds. A mechanical conditioning called cyclic stretch is applied to the construct by pulsatile flow. The pulsatile applies cyclic stretch to the constructs by stretching the silicon tubing periodically. The constructs were cultured with nutrient media with continuous stirring and pressure was monitored. A highly sterile environment is maintained. SMCs are incident on a fabricated blood vessel and then cultured for certain time intervals. The vessels that are grown for three weeks under these conditions have shown the formation of fluid-tight tissue conduits which help in tubing of the flow system. It is also observed that there is improvement in physical strength of the cultured tubes due to the static culture.

Figure 11.5(b) represents the equivalent electrical circuit of a perfusion system supplying pulsatile flow to bioreactors. There are many components involved in the above process replaced by their equivalent electrical elements, such as reservoir gas exchange replaced by ground, continuous pump replaced by battery V1 used for

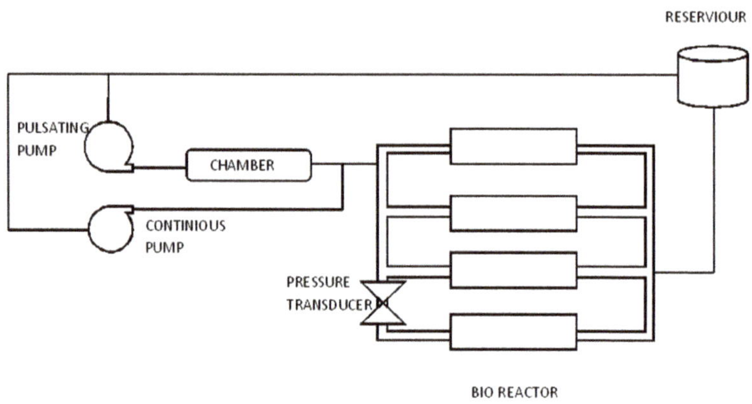

(a)

Figure 11.5(a). Arrangement of a perfusion system supplying pulsatile flow to bioreactors each containing a tissue engineered vessel. Gas exchange is provided by a reservoir.

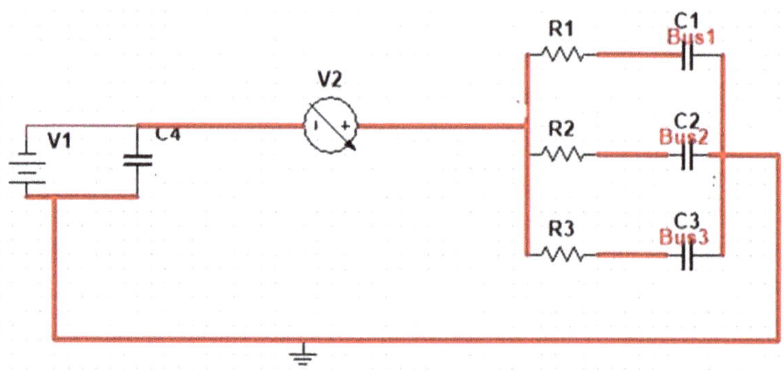

(b)

Figure 11.5(b). Electrical analogy of a perfusion system supplying pulsatile flow to bioreactors.

continuous flow of media, pressure transducer replaced by voltage source V2, and the pulsating pump used to provide stretching of the blood vessel replaced by capacitors C1, C2 and C3, and these capacitors are connected with resistance R1, R2 and R3 in series, respectively. The resistance R1, R2 and R3 is included due to the blood vessel. The flow of media is considered as flow of current in equivalent circuits. As in the process diagram, the blood vessel acts as an obstacle to the flow of media and hence it is represented by resistance, which opposes the flow of current. The resistance is connected in parallel with each other.

After conversion, the circuit is analyzed as follows.

In circuit, $R1$ is in series with $C1$, $R2$ is with $C2$ and $C3$ is in series. The resulting impedance is represented with $Z1$, $Z2$ and $Z3$.

$$Z1 = R1 + \frac{1}{SC1} \tag{11.10}$$

$$Z2 = R2 + \frac{1}{SC2} \tag{11.11}$$

$$Z3 = R3 + \frac{1}{SC3} \tag{11.12}$$

Now we observe that this $Z1$, $Z2$ and $Z3$ are connected in parallel. The resulting impedance is represented with $Z(s)$.

$$Z = Z1 \,||Z2||Z3$$

When we solve that, then we get the value of $Z(s)$ which is equal to

$$Z(s) = \frac{\frac{(1 + SC1R1)(1 + SC2R2)}{C2(1 + SC1R1) + C1(1 + SC2R2)}(1 + SC3R3)}{C3\left[\frac{(1 + SC1R1)(1 + SC2R2)}{C2(1 + SC1R1) + C1(1 + SC2R2)}\right] + C1(1 + SC2R2)} \tag{11.13}$$

A simplified version of the above diagram is figure 11.5(c).

Suppose I current is the flow from the circuit. Now apply KCL in a closed loop and also take the Laplace transformation, then we get

$$-\frac{V1}{S} + \frac{1}{SC4}I(S) - \frac{V2}{S} + Z(S)I(S) = 0 \tag{11.14}$$

Then, solving the above equation (11.14) for $I(S)$ we get

$$I(S) = \frac{(V1 + V2)C4}{1 + SC4Z(S)}$$

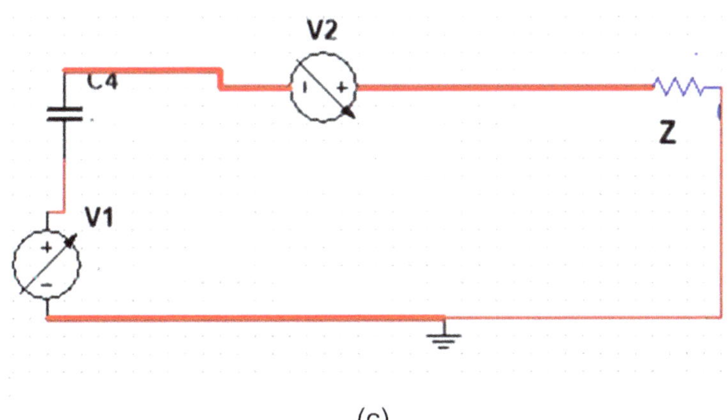

Figure 11.5(c). Simplified diagram of figure 11.5(b).

Hoerstrup *et al* 2000, Hoerstrup *et al* 2002) developed a bioreactor for heart valves conditioning which can also be used for small vascular grafts. In this study, a pulsatile flow system is developed which provides both biochemical and biomechanical signals which regulate the tissue development *in vitro*. This bioreactor made of plexiglas is transparent, which helps in observing the development of tissue during *in vitro* culturing. The bioreactor has three chambers, one air chamber (1) and two culture medium chambers (2 and 3). Chamber 1 is separated from chamber 2 by a diaphragm used to provide isolation for chamber 1 and chamber 2 and joined to a respiratory pump. Chambers 2 and 3 are divided by a patch construct surrounded by continuously circulated culture media. The media reservoir is attached with chamber 2. The media reservoir is connected to and supplied with an inlet and outlet. When air is pumped into air chamber 1, the diaphragm is expanded upward to direct the cell culture media against the tissue construct in chamber 3. Simultaneously, culture media in chamber 2 is pumped into the reservoir through the outlet. In this condition, the shear stress in the air chamber 1 ranges between 0.2–10 dyne cm^{-2}, which is related to different flow rates. In this, vascular grafts are seeded in parallel. The whole bioreactor is placed in an incubator. The temperature of the incubator is settled at 310 °F and 5% CO_2, and the incubator is connected to the reservoir containing the culture media. Pulsatile flows and biomechanical stress were developed by regulating stroke volume and stroke rate. For creating shear stress inside the blood vessel or lumen surface we pass pulsed flow media at a flow rate of 115 mL min^{-1} at 30 mmHg to 750 mL min^{-1} at 55 mmHg.

With some advancement in the bioreactor of Hoerstup *et al*, Sodian *et al* (2002) modified the design where cell seeding and perfusion systems were combined providing sterile conditions as well as biomechanical stimuli for the dafting of the blood vessel. The bioreactor consists of three chambers, air chamber (1), cell culture media chamber (2) and another chamber (3) for cell seeding and perfusion (figure 11.6(a)). Chamber 3 is separated from chamber 2 and a vascular scaffold is fixed in chamber 3. Chamber 3 can be connected with chamber 2 after cell seeding for the pulsatile flow system. The inlet of the cell culture media is connected to chamber 2 and the outlet that is present on top of cell seeding and perfusion chamber is connected to the reservoir. The whole bioreactor is placed in an incubator. The temperature of the incubator is settled at 310 °F and 5% CO_2, and the incubator is connected to a reservoir containing culture media. The vascular construct that has to be developed is continuously supplied by the media. The following parameter of the media is set to below the given value.

Flow rate = 100 mL min^{-1} to 3 L min^{-1};
pressure = 3 to 150 mmHg; and
shear stress = 1.11 to 32.45 dyn cm^{-2}.

An equivalent electrical analogy of a pulsatile bioreactor is shown in figure 11.6(b). There are many components involved in the above process replaced by their equivalent electrical elements. In figure 11.6(b) we have replaced the air chamber with ground because it acts as a reservoir. There is one valve between chamber 1 and chamber 2, which is why we use a diode in figure 11.6(b). For the blood vessel we use resistance R1. Because the system is a pulsatile bioreactor we use

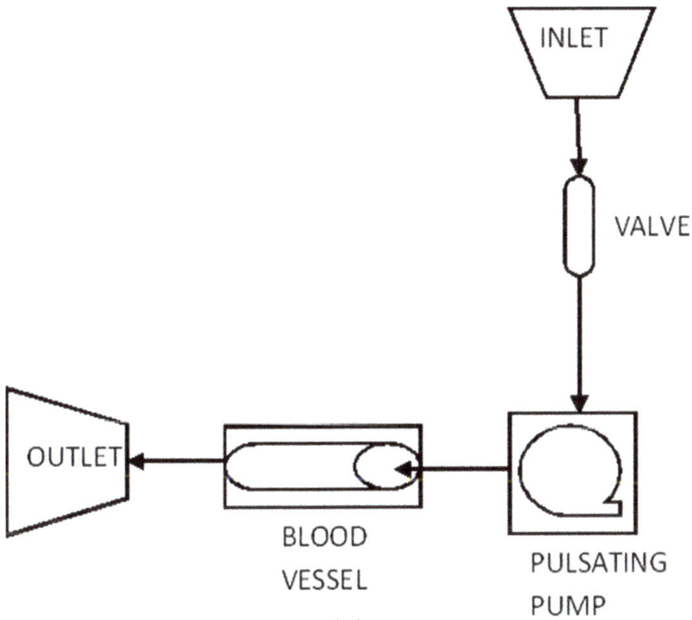

(a)

Figure 11.6(a). Schematic diagram of a pulsatile bioreactor.

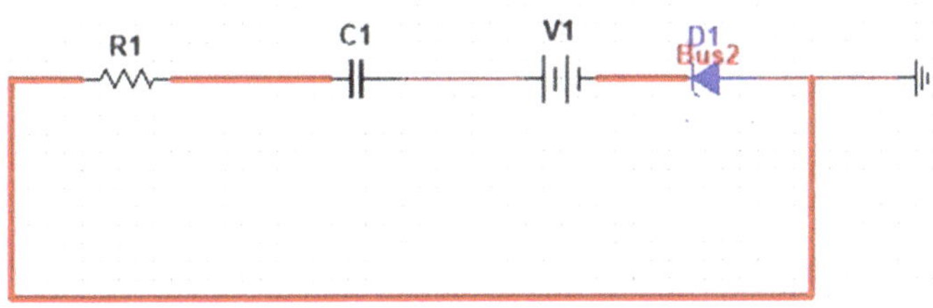

(b)

Figure 11.6(b). Electrical circuit diagram of a pulsatile bioreactor.

capacitor C1. All the elements are connected in series in figure 11.6(b) because in figure 11.6(a) the whole operation is in series.

Suppose I current is flowing from the circuit. Now apply KCL in a closed loop and also take the Laplace transformation, then we get

$$-\frac{V1}{S} + R1I(S) + \frac{1}{SC1}I(S) = 0 \tag{11.15}$$

Now determine the value of $I(S)$ from equation (11.15)

$$I(S) = \frac{V1C1}{1 + SR1C1}$$

11.3.6 Bioreactor with cyclic strain

Eliktar *et al* (2000) (Zajonc 2019) developed a bioreactor to impose cyclic strain on the vascular graft that is developing in the bioreactor. Thin walled silicon tubes are present over which the vascular constructs are cultured. Cell culture was prepared by isolating adult rat aorotic SMCs. Cells cultured over silicon tubes are supplied with culture medium under pneumatic control. The tissue engineered vessel is continuously cultured in collagen-based cell suspension and cast into gels. Tubular constructs are made by the use of a silicon sleeve. The concentration of collagen in collagen-based suspension is maintained at 2 mg ml^{-1}. Once the collagen is formed into a gel, constructs are removed and placed in growth media for 48 h. These constructs are then subjected to dynamic mechanical conditioning and are exposed to the frequency of 1 Hz inflation and deflation. The constructs are subjected to mechanical testing and histological analysis. This bioreactor provides improved mechanical strength of TE. Diagrammatic representation of the bioreactor is shown in figure 11.7(a).

An equivalent electrical analogy of the bioreactor design to develop cyclic strain on the tissue construct is shown in figure 11.7(b). There are many components involved in the above process replaced by their equivalent electrical elements. In figure 11.7(a) we have replaced the gas chamber valve with diodes D1 and D2. For the pressure transducer we use voltage sources V1 and v2. For cell seeding construct we use resistance R1 in figure 11.7(b), which is just like the blood vessel.

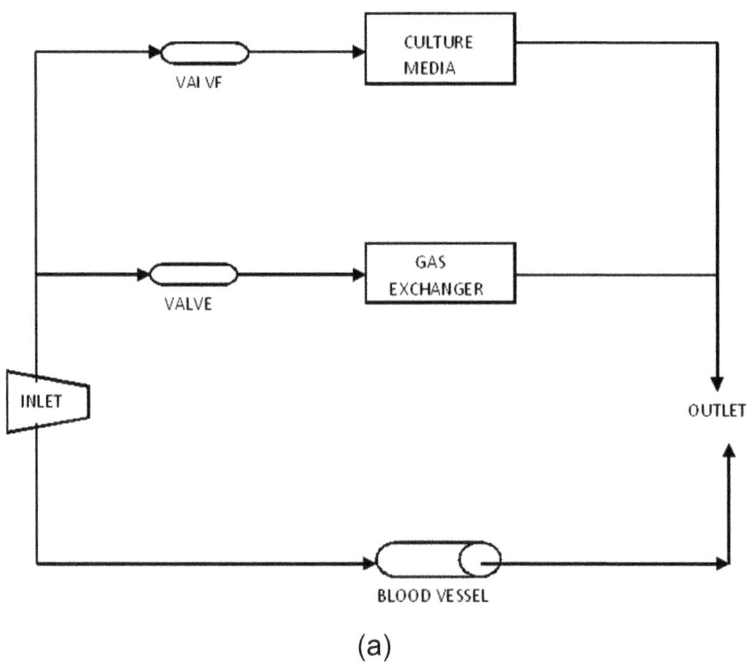

Figure 11.7(a). Bioreactor design to develop cyclic strain on the tissue construct.

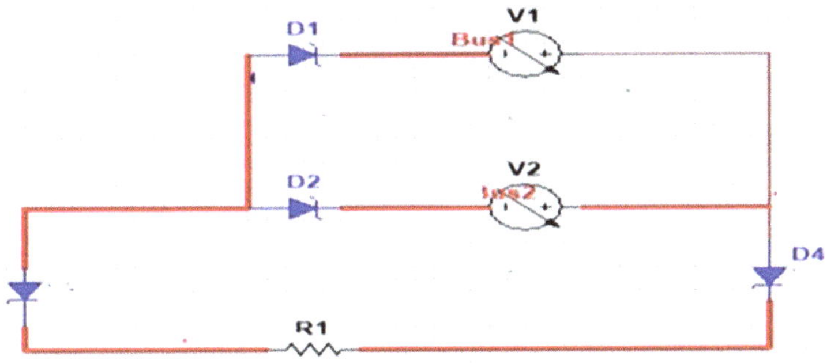

(b)

Figure 11.7(b). Electrical analogy of the bioreactor design to develop cyclic strain on the tissue construct.

Here $V1 = V2$. Suppose I current is flowing from the circuit. Apply KCL in a closed loop and also take the Laplace transformation, then we get

$$\frac{V}{S} - I(S)R = 0 \tag{11.16}$$

Then, solving equation (11.16) for $I(S)$ we get

$$I(S) = \frac{V}{SR}$$

11.3.7 Multi-cue bioreactor

McCulloch *et al* (2004), designed a bioreactor called a multi-cue bioreactor where physiological conditions to the vessels are supplied by providing an environment of *in vivo* pressure and strain. The cell seeded construct that has to be developed is placed in a chamber called a tissue cassette which has one inlet for entry of the culture media and an outlet. This tissue cassette is connected to main bioreactor (figure 11.8). Strain is provided by a pulsatile pump through the pump in the lumen. The pressures are regulated in the range between 70 mmHg to 110 mmHg and flow rate of 282 mL min^{-1}. The force developed of strain induction mechanism oscillates with an amplitude of 9, 25, 40 and 60 grams, respectively, at 1 Hz over series of time. Then the tissue product is passed for the casual testing and some other analysis. Further, the data obtained shows the ability of supporting long term cell seeded substrates under the influence of mechanical properties. Cell alignment was also observed after seven days of continuous pulsation.

Peterson *et al* (2000) has patented a bioreactor system where the major parts are reservoir, pump, processing chamber and pressure source. An expandable tube is situated on the processing chamber. The artificial blood vessel scaffold is located inside the expandable tube. The tube is connected to the pressure source where it expands and contracts. This leads to the stress on the scaffold simulating physiological conditions. The major purpose of this bioreactor is to analyze the effect of

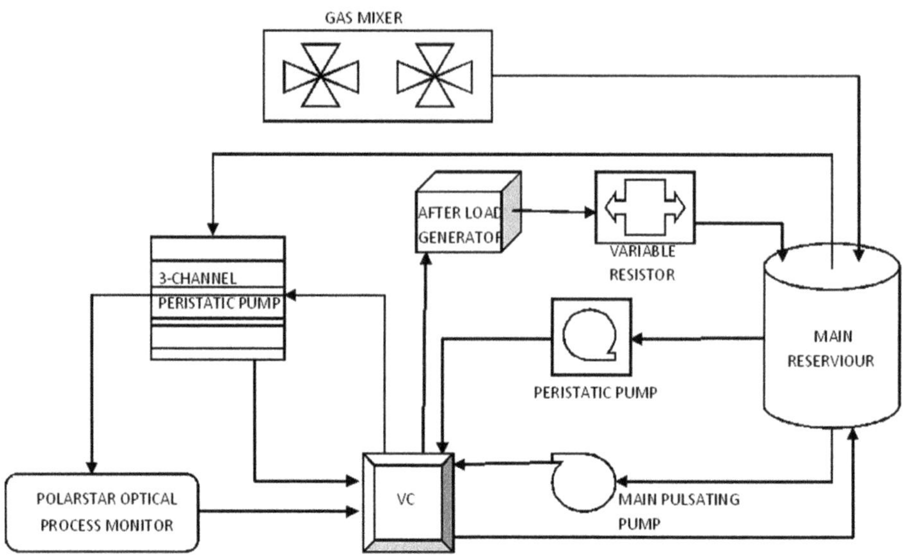

Figure 11.8(a). Schematic representation of a bioreactor showing various components.

axial stress to the biological construct at the time of its seeding and culturing. This axial stress supplied results in tissue engineered constructs that possess long term dimensional stability and physiological properties of native vessels.

11.4 Bioreactors specifically designed for tissue engineered heart valves

Heart valves related to TE have more advantages over commercially available ones such as heterografts and homografts (Stock *et al* 2002, Sutherland *et al* 2002). These help in dealing with side effects of anticoagulation, immune rejection, and remodeling. Mechanical heart valves will have good durability and strength but require continuous anticoagulation due to the risk of thromboembolism (Bader *et al* 2000, James *et al* 2006). Due to these reasons, TE is considered a better option for development of heart valves over commercial methods. Tissue grafts also have the risk of degradability due to tissue stiffening, which ultimately leads to stenosis (Jamieson 1993). To overcome these problems and to generate potential tissue engineered grafts mimicking the mechanical properties of native tissues, bioreactors are specifically designed by subjecting the tissue to shear stress along with the blood flow.

11.4.1 Cardiac valve bioreactor

A cardiac valve bioreactor was developed to minimize the limitations of tissue engineered valves located inside the heart. With the help of this bioreactor, hydrodynamic performance of the tissue constructs can be assessed. The major components of a bioreactor are pulsatile pump, peristaltic pump, medium reservoir, afterload generator with compliance chamber and relief valve, variable resistors, flow probe and a controller. This whole system is operated by software. A flow rate of 18 l min^{-1} at a temperature of 37 °C, valve chamber of pressure not exceeding 20 mmHg and aortic pressure in the range of 60–140 mmHg are maintained throughout the bioreactor by the software. Hydrodynamic conditions were tested using a bioprosthesis and cell culture was maintained for about two weeks. This bioreactor was designed to see the maturation of the valve located inside the heart. Here we consider biological and mechanical maturation (Tefft *et al* 2019). The design of the bioreactor was made depending on the Windkessel approach. Parts of the bioreactor are explained in figure 11.8(a).

In figure 11.8(b), we replace the main reservoir with ground and the peristaltic pump is replaced by battery V2. In figure 11.8(a) the after load generator is used, which is replaced by amplifier and here we use a non-inverting amplifier whose gain is represented by R3. VC is a valve chamber which acts as resistance and is replaced with R1. The pulsating pump is used to provide stretching on the VC so it can be replaced with capacitor C1. The equivalent circuit is shown in figure 11.8(c). Now let us examine the circuit.

$R3$ is an amplifier, here we consider it as a non-inverting amplifier whose gain is $\frac{-RF}{Rin}$

From the circuit it is clear that $R1$ and $C1$ are in series and resultant impedance is parallel with $R3$. So when we solve it then we get total impedance as

$$\text{Total impedence is} = \frac{RF(R1C1S + 1)}{[RFC1S - Rin(R1C1S + 1)]} \tag{11.17}$$

Suppose I current is flowing from the circuit (figure 11.8(d)). Apply KCL in a closed loop and also take the Laplace transformation, then we get

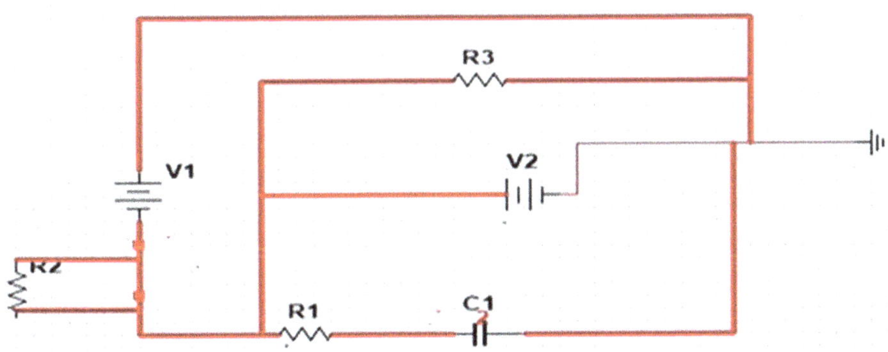

(b)

Figure 11.8(b). Electrical analogy of a bioreactor.

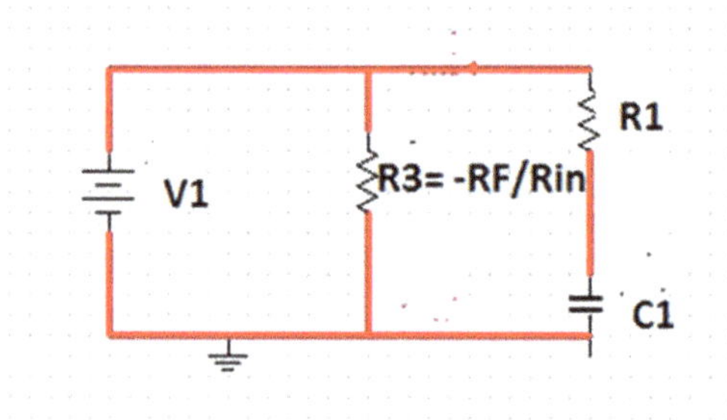

(c)

Figure 11.8(c). Simplified structure of figure 11.8(b).

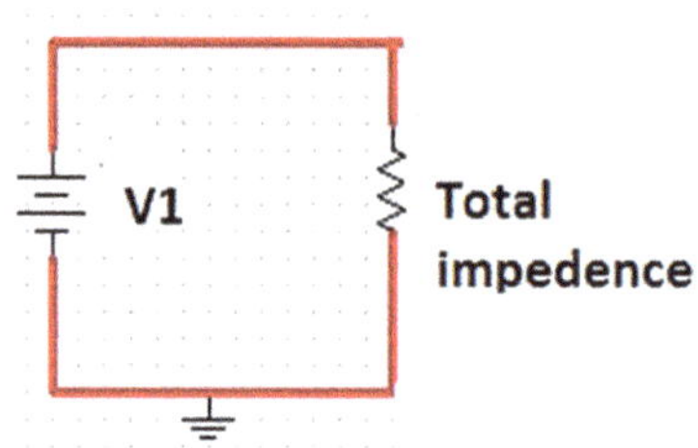

(d)

Figure 11.8(d). Simplified structure of figure 11.8(c).

$$\frac{V1}{S} - \frac{RF(R1C1S + 1)I(S)}{[RFC1S - Rin(R1C1S + 1)]} = 0 \quad (11.18)$$

Now determine the value of $I(S)$

$$I(S) = \frac{V1[RFC1S - Rin(R1C1S + 1)]}{SRF(R1C1S + 1)}$$

11.4.2 Pulsatile bioreactors for cardiac valves

Hoerstrup *et al* have developed a pulsatile bioreactor which provides physiological pressure along with nutrient media flow to the tissue construct (Hoerstrup *et al* 2000). This bioreactor has two chambers which are divided by a silicon diaphragm. The air chamber is located on the upper level which is connected to a respiratory pump while the lower chamber is for nutrient media for the cells deposited (figure 11.9(a)). The two chambers and a reservoir of culture media are kept in an incubator. The temperature of the incubator is maintained at 310 °F and 5% CO_2. A simple respirator outside the incubator will connect the air driven system, functioning as an air pump. The flow rate range of the pulsatile pump is settled in between 50

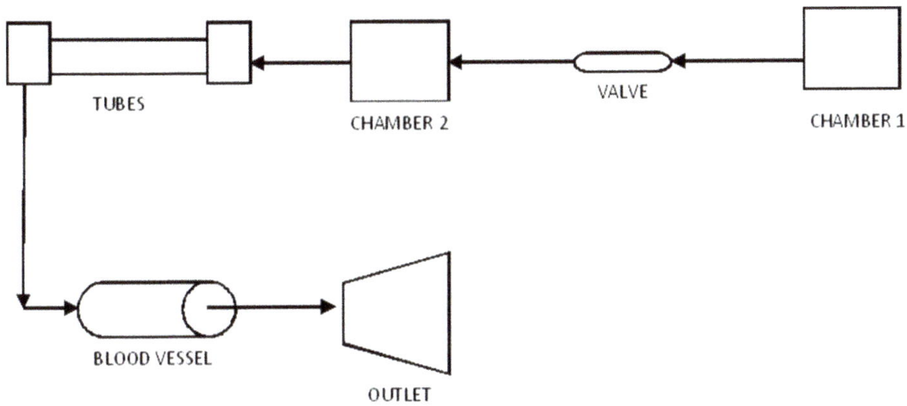

(a)

Figure 11.9(a). Diagrammatic representation of a bioreactor (Hoerstrup *et al* 2000)

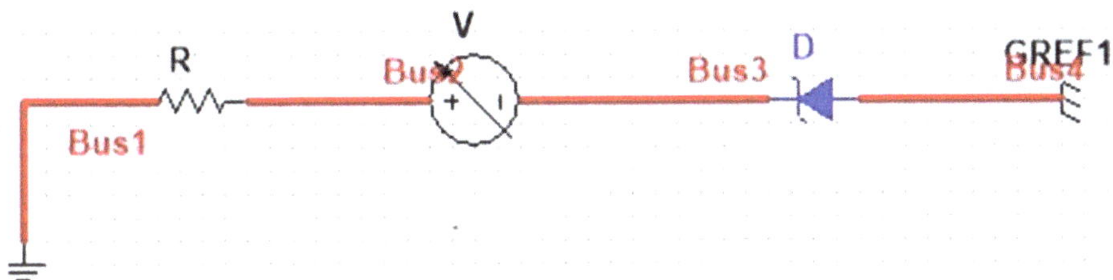

(b)

Figure 11.9(b). Electrical analogy of a representation bioreactor.

to 2 L min^{-1} and the range of pressure is maintained from 10 to 240 mmHg. Human SMCs were isolated and generated by TE *in vivo*. These developed valves show mechanical properties of a native heart valve (Hoerstrup *et al* 2000).

Figure 11.9(b) represents the equivalent electrical circuit of the representation of the bioreactor. In the reactor process diagram, many components involved to perform specific tasks are replaced by their equivalent electrical components, such as chamber 1 and chamber 2 act as a reservoir and are replaced by ground. The flow of media is considered as current in equivalent circuits/electrical analogy. As in the process diagram, the blood vessel acts as an obstacle to the flow of media and hence it is represented by resistance R, which opposes the flow of current. In electrical circuits we use voltage source V for the pressure transducer. For the valve we use diode D. All elements are connected in series.

Simplifying the above circuit, we get figure 11.9(c).

Suppose I current is flow from the circuit. Apply KCL in a closed loop and also take the Laplace transformation, then we get

$$RI(S) - \frac{V}{S} = 0 \tag{11.19}$$

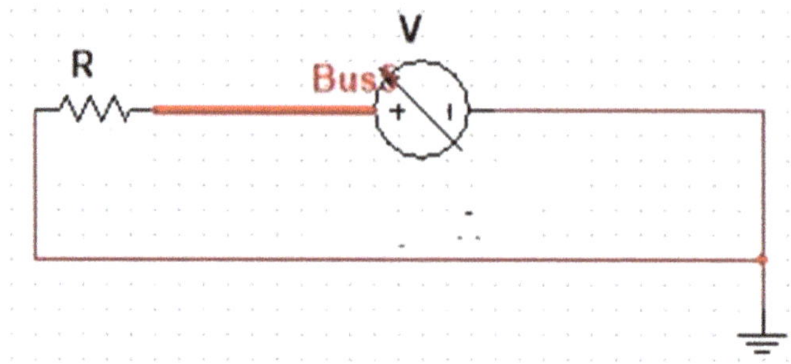

(c)

Figure 11.9(c). Simplified version of figure 11.9(b).

Solving equation (11.19) for $I(S)$ gives

$$I(S) = \frac{V}{RS}$$

Dumont (Sauer and Gerlach 2002) have designed a bioreactor which was based on the pulsatile flow loop of the left ventricle. This bioreactor mainly has two parts. The first part is called the left ventricle (LV), after load consisting of compliance, the second one is called resistance. The LV is made of silicon rubber having 80 ml of stroke volume. A piston is operated with the help of an external circuit to regulate the media flow through the tissue graft. A chamber is filled with air which provides the compliance representing the function of large arteries. This whole system will provide hydrodynamic conditions by controlled conditions like resistance, stroke volume, frequency and compliance. Biocompatibility studies were conducted using endothelial cells supplemented with cell culture media and cultured for a period of two weeks. The results obtained show that biomechanical factors generated by the bioreactor are very useful for the cell culture and proliferation.

A bioreactor for heart valves designed by Jockenhoevel *et al* (2002) has a peristaltic pump which produces continuous laminar flow rate of 250–500 ml min^{-1} throughout the chamber and provides growth factor for scaffolds. This bioreactor was designed to add cross the main flow. This flow is induced by a dynamic roller pump with flow rate of 500 ml min^{-1}. A nutrition flow is pumped using a perfusion pump with flow rate of 3 ml h^{-1}. The purpose of this bioreactor is to see the effect of laminar flow on the formation of the ECM. The total bioreactor is divided into two chambers (figure 11.10(a)), where the lower chamber provides nutrition to the tissue with lower flow rate and the upper chamber provides shear stress through a high flow rate. PGA scaffolds were coated poly-L-lysine seeded with myofibroblasts and then cultured for two weeks in static condition. Simultaneously, a control group is also incubated. For testing the collagen production, hydroxyproline assay was conducted. The increase in the content of hydroxyproline in the incubated constructs shows that ECM production can be enhanced with applied mechanical stress.

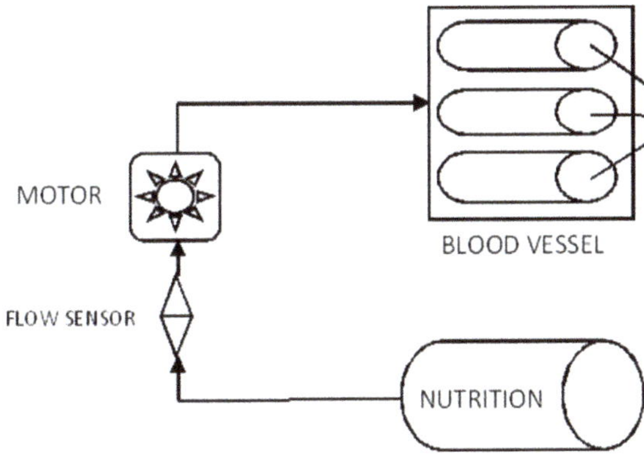

(a)

Figure 11.10(a). Diagram representing laminar flow in the bioreactor.

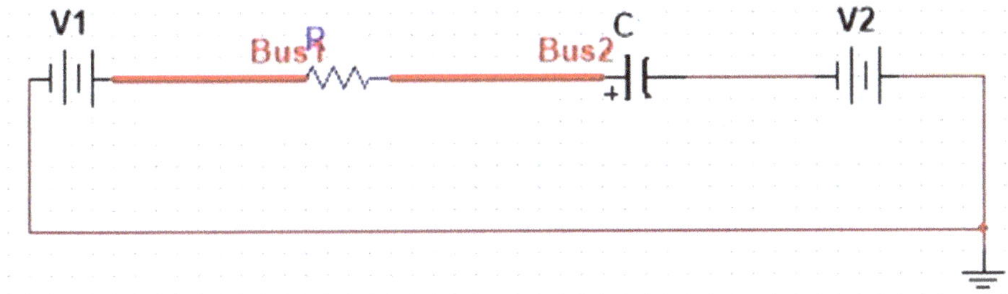

(b)

Figure 11.10(b). Electrical analogy of the diagram representing laminar flow in the bioreactor.

Figure 11.10(b) represents the equivalent electrical circuit of representation of a laminar flow bioreactor. In the reactor process diagram, many components involved in performing specific tasks are replaced by their equivalent electrical components. The lower part of fig 11.10(a) represents a nutrition chamber which acts as reservoir and is represented as ground. Then this nutrition passes through a valve which is represented by a diode in the electrical analogy (figure 11.10(b)). The pump is replaced by batteries V1 and V2. The flow of media is considered as the flow of current in equivalent circuits. As in the process diagram, the blood vessel acts as an obstacle to the flow of media and hence it is represented by resistance, which opposes the flow of current. All the elements are connected in series as shown in figure 11.10(b).

Suppose I current is flowing from the circuit. Apply KCL in a closed loop and also take the Laplace transformation, then we get

$$\frac{V1}{S} - RI(S) - \frac{1}{CS} \times I(S) + \frac{V2}{S} = 0 \qquad (11.20)$$

Solving equation (11.20) for $I(S)$

$$I(S) = \frac{V1}{R\left(S + \frac{I}{RC}\right)} + \frac{V2}{R\left(S + \frac{1}{RC}\right)}$$

A patented bioreactor was developed by Goldstein (Kirby *et al* 1999), who used a pulsatile flow loop for the developing tissue valves. The major parts of this bioreactor are mechanical mitral valve, a section in which heart valve samples are placed and oxygenated, a heat exchanger, a resistance element and a bladder which simulates the left ventricle. A bladder which is made of rubber with flexibility is placed inside the water-filled chamber that is similar to the left ventricle. A piston is available which squeezes the bladder and allows the circulation of the media throughout the engineered heart valve. The flow is maintained that mimics the flow of the cardiac cycle with a frequency of 60–110 bpm and 2–7.5 L min^{-1} cardiac output. This bioreactor is useful in studying cell differentiation in tissue constructs when subjected to mechanical loading, and also creates an environment of dynamic flow to the constructs mimicking the aortic valve.

The patented bioreactor developed by Elizondo (Campbell and Totten 1999), has three major parts that are: an environmental control unit which provides controlled temperature, a nutrient supply circuit which helps in the distribution of the nutrients throughout the system and a processing chamber in which the tissue engineered construct is placed. A pulse evaluation circuit is also available which helps in providing pulsatile flow in the system. This bioreactor provides a system for the processing of biological materials. Here the biological materials are usually collagen-containing material which may be derived from mammalians. This bioreactor provides all the required conditions for developing heart valves and vascular grafts.

One more bioreactor available for cardiovascular grafts was patented by Wolfinbarger (Weber 2002). This bioreactor is mainly developed for recellularizing and re-endothelializing of vascular grafts with the major purpose of replacing the defective tissues. In this, an inflatable piston is used to control the media flow through the lumen of the tissue graft. A pulsatile pump is also available for providing pulsatile flow through the graft. The vascular graft is treated with growth factors for binding of the tissue matrix. Unbound or loosely bound growth factors are removed by washing. The vascular graft is then seeded with endothelial cells. Luminal volume is pressurized so as to compress the viscous cell mixture in the luminal area. Slow pulsatile flow of nutrient rich media is supplied to achieve a viable vascular graft. The application of pulsatile flow of nutrient rich media will result in the stress field on the tissue graft. This stress will help in physiological stimulation of the cell population in the vascular graft. These vascular grafts are further preserved using cryopreservation or nutrient preservation methods.

11.5 Conclusion

Bioreactors for the development of an artificial blood vessel equivalent to a native blood vessel *in vivo* has to be developed. Bioreactors provide all the biochemical and biomechanical properties of native vessels that are necessary in engineering of a tissue construct. Designing of bioreactors varies according to the mode of scaffold fabrication, its material properties, orientation of scaffold printing, and density of cell seeding for generating a particular type of tissue construct. The use of bioreactors results in development of both tissue engineered vessels and a cardiovascular construct with essential biomechanical and physiological characteristics of native blood vessels. These tissue engineered constructs can be used in treating various cardiovascular diseases. In such circumstances, growth of such constructs require micro-environmental support for the growth and sustainability of the seeded cells. Therefore, rigorous design of the most appropriate bioreactor makes it feasible for the implantation of the construct with greater likelihood of success.

References

Bader A, Steinhoff G and Haverich A 2000 Tissue engineering of vascular grafts: human cell seeding of decellularised porcine matrix *Eur. J. Vasc. Endovasc. Surg.* **19** 381–6

Begley C M and Kleis S J 2000 The fluid dynamic and shear environment in the NASA/JSC rotating-wall perfused-vessel bioreactor *Biotechnol. Bioeng.* **70** 32–40

Borschel G H, Huang Y-C, Calve S, Arruda E M, Lynch J B, Dow D E, Kuzon W M, Dennis R G and Brown D L 2005 Tissue engineering of recellularized small-diameter vascular grafts *Tissue Eng.* **11** 778–87

Campbell T D and Totten R P 1999 *United States Patent (19)* (19)

Carrier R L, Rupnick M, Langer R, Schoen F J, Freed L E and Vunjak-Novakovic G 2002 Perfusion improves tissue architecture of engineered cardiac muscle *Tissue Eng.* **8** 175–88

Catapano G 1996 Mass transfer limitations to the performance of membrane bioartificial liver support devices *Int. J. Artif. Organs* **19** 18–35

Chien S 2007 Mechanotransduction and endothelial cell homeostasis: the wisdom of the cell *Am. J. Physiol. – Heart Circul. Physiol.* **292** H1209–24

Conklin B S, Richter E R, Kreutzinger K L, Zhong D S and Chen C 2002 Development and evaluation of a novel decellularized vascular xenograft *Med. Eng. Phys.* **24** 173–83

Davies P F 2009 Hemodynamic shear stress and the endothelium in cardiovascular pathophysiology *Nature Clin. Pract. Cardiovasc. Med.* **6** 16–26

Eliktar D R O R S, Ichard R, Lack A B, Aymond R, Ito P V, Obert R and Erem M N 2000 Dynamic mechanical conditioning of collagen-gel blood vessel constructs induces remodeling *in vitro***28** 351–62

Galban C J and Locke B R 1999 Analysis of cell growth kinetics and substrate diffusion in a polymer scaffold *Biotechnol. Bioeng.* **65** 121–32

Graham L M, Brothers T E, Darvishian D, Vincent C K, Burkel W E and Stanley J C 1989 Effects of thromboxane synthetase inhibition on patency and anastomotic hyperplasia of vascular grafts *J. Surg. Res.* **46** 611–5

Guillaume T, Rubinstein D B and Symann M 1998 Immune reconstitution and immunotherapy after autologous hematopoietic stem cell transplantation *Blood J. Am. Soc. Hematol.* **92** 1471–90

Hoerstup S P, Zünd G, Cheng S, Melnitchouk S, Kadner A, Sodian R, Kolb S A and Turina M 2002 A new approach to completely autologous cardiovascular tissue in humans *ASAIO J.* **48** 234–8

Hoerstrup S P, Sodian R, Sperling J S, Vacanti J P and Mayer Jr J E 2000 New pulsatile bioreactor for *in vitro* formation of tissue engineered heart valves *Tissue Eng.* **6** 75–9

Hoerstrup S P *et al* 2002 Tissue engineering of functional trileaflet heart valves from human marrow stromal cells *Circulation* **106** I143–50

Hoerstrup S P, Zünd G, Sodian R, Schnell A M, Grünfelder J and Turina M I 2001 Tissue engineering of small caliber vascular grafts *Eur. J. Cardio-Thorac. Surg.* **20** 164–9

Hosenpud J D, Morris T E, Shipley G D, Mauck K A and Wagner C R 1996 Cardiac allograft vasculopathy. preferential regulation of endothelial cell-derived mesenchymal growth factors in response to a donor-specific cell-mediated allogeneic response *Transplantation* **61** 939–48

Iaizzo P A, Bianco R W, Hill A J and Louis J D S 2013 Heart valves: from design to clinical implantation *Surgery* 1–429

James A H, Brancazio L R, Gehrig T R, Wang A and Ortel T L 2006 Low-molecular-weight heparin for thromboprophylaxis in pregnant women with mechanical heart valves *J. Mater.-Fetal Neonat Med.* **19** 543–9

Jamieson W R E 1993 Modern cardiac valve devices—bioprostheses and mechanical prostheses: state of the art *J. Card. Surg.* **8** 89–98

Jockenhoevel S, Zund G, Hoerstrup S, Schnell A and Turina M 2002 Cardiovascular tissue engineering: a new laminar flow chamber for *in vitro* improvement of mechanical *ASAIO J.* **48** 8–11

Jockenhoevel S *et al* 2000 *Ye et al 2000; Swirzeland.Pdf* **17** 449–54

Kirby S, Thoracic J and Isabella D J 1999 Primary Examiner *United States Patent (19)* (19)

Kofidis T, Akhyari P, Boublik J, Kraft T, Leyh R and Haverich A 2002 *In vitro* engineering of heart muscle: Artificial myocardial tissue *J. Thoracic Card. Surg.* **124** 63–9

Lwigale P Y, Thurmond J E, Norton W N, Spooner B S and Wiens D J 2000 Simulate microgravity and hypergravity attenuate heart tissue development in explant culture *Cells Tissues Organs* **167** 171–83

McCulloch A D, Harris A B, Sarraf C E and Eastwood M 2004 New multi-cue bioreactor for tissue engineering of tubular cardiovascular samples under physiological conditions *Tissue Eng.* **10** 565–73

Mironov V, Kasyanov V, McAlister K, Sherrell O, Sistino J and Markwald R 2003 Perfusion bioreactor for vascular tissue engineering with capacities for longitudinal stretch *J. Craniof. Surg.* **14** 340–7

Mitchell S L and Niklason L E 2003 Requirements for growing tissue-engineered vascular grafts *Cardiovasc. Pathol.* **12** 59–64

Naughton G K 2002 From lab bench to market: Critical issues in tissue engineering *Ann. N.Y. Acad. Sci.* **961** 372–85

Niklason L E, Gao J, Hirschi K K, Houser S, Marini R and Langer R 1999 Functional arteries grown *in vitro Science* **284** 489–93

Niklason L E and Langer R S 1997 Advances in tissue engineering of blood vessels and other tissues *Transpl. Immunol.* **5** 303–6

Novosel E C, Kleinhans C and Kluger P J 2011 Vascularization is the key challenge in tissue engineering *Adv. Drug Delivery Rev.* **63** 300–11

Opitz F, Schenke-Layland K, Richter W, Martin D P, Gegenkolbe I, Wahlers Th and Stock U A 2004 Tissue engineering of ovine aortic blood vessel substitutes using applied shear stress and enzymatically derived vascular smooth muscle cells *Ann. Biomed. Eng.* **32** 212–22

Ozawa T, Mickle D A G, Weisel R D, Koyama N, Ozawa S, and Li and R-K 2002 Optimal biomaterial for creation of autologous cardiac grafts *Circulation* **106** 176–82

Peterson A *et al* 2000 *US Patent* 6,121,042 (Washington, DC)

Ratcliffe A 2000 Tissue engineering of vascular grafts *Matrix Biol.* **19** 353–7

Rouwkema J, Rivron N C and van Blitterswijk C A 2008 Vascularization in tissue engineering *Trends Biotechnol.* **26** 434–41

Roger V L *et al* 2013 Heart Disease and Stroke Statistics — 2011 Update *A Report From the American Heart Association* PMC4418670

Sauer I M and Gerlach J C 2002 Modular extracorporeal liver support *Artific. Organs* **26** 703–33

Schmedlen, Rachael H, Wafa M Elbjeirami, Andrea S Gobin, and Jennifer L West 2003 Tissue engineered small-diameter vascular grafts *Clin. Plastic Surg.* **30** 507–17

Shazly T *et al* 2016 (12) Patent Application Publication (10) Pub. No.: US 2016/0298073 A1. 1(19).

Sodian R, Lemke T, Loebe M, Hoerstrup S P, Potapov E V, Hausmann H, Meyer R and Hetzer R 2001 New pulsatile bioreactor for fabrication of tissue-engineered patches *J. Biomed. Mater. Res.* **58** 401–5

Sodian R, Lemke T and Fritsche C 2002 Tissue-engineering bioreactors: a new combined cell *Tissue Eng.* **8** 863–70

Sodian R, Hoerstrup S P, Sperling J S, Schoen F J, Vacanti J P and Mayer J E Jr 2002 Tissue engineering of heart valves: *in vitro* experiences *Ann. Thor. Surg.* **70** 140–44

Stock U A, J. P. Vacanti J E M and Wahlers T 2002 Tissue engineering of heart valves – current aspects *Thorac. Cardiovas. Surg.* **50** 184–93

Sutherland F W H, Perry T E, Nasseri B A, Wang J, Kaushal S, Guleserian K J, Martin D P, Vacant J P and Mayer J E Jr 2002 Advances in the mechanisms of cell delivery to cardiovascular scaffolds: Comparison of two rotating cell culture systems *ASAIO J.* **48** 346–9

Tefft B J *et al* 2019 Cardiac valve bioreactor for physiological conditioning and hydrodynamic performance assessment *Cardiovasc. Eng. Technol.* **10** 80–94

van Luyn M J A, Tio R A, Gallego y van Seijen X J, Plantinga J A, de Leij L F M H, Dejongste M J L and van Wachem P B 2002 Cardiac tissue engineering: characteristics of in unison contracting two- and three-dimensional neonatal rat ventricle cell (co)-cultures *Biomaterials* **23** 4793–801

Vanhoutte P M 1989 Endothelium and control of vascular function: state of the art lecture *Hypertension* **13**(6 Pt 2) 658–67

Weber J P 2002 Data, Related U S Application, and Primary Examiner *(12) United States Patent* 11018

Weston M W, LaBorde D V and Yoganathan A P 1999 Estimation of the shear stress on the surface of an aortic valve leaflet *Ann. Biomed. Eng.* **27** 572–9

Williams C and Wick T M 2004 Perfusion bioreactor for small diameter tissue-engineered arteries *Tissue Eng.* **10** 930–41

Williams K A, Saini S and Wick T M 2002 Computational fluid dynamics modeling of steady-state momentum and mass transport in a bioreactor for cartilage tissue engineering *Biotechnol. Progr.* **18** 951–63

Wolf F, Rojas González D M, Ulrich S, Obdenbusch M, Herfs W, Brecher C, Jockenhoevel S and Schmitz-Rode T 2018 VascuTrainer: a mobile and disposable bioreactor system for the conditioning of tissue-engineered vascular grafts *Ann. Biomed. Eng.* **46** 616–26

Wootton D M and Ku D N 2002 Fluid mechanics of vascular systems, diseases, and thrombosis *Annu. Rev. Biomed. Eng.* **1** 299–329

Yamamoto K, Takahashi T, Asahara T, Sokabe T, Kamiya A and Ando J 2015 Proliferation, differentiation, and tube formation by endothelial progenitor cells in response to shear stress *J. Appl. Physiol.* **95** 2081–8

Zajonc R B 2019 Social facilitation *Science* **149** 269–74

IOP Publishing

Flow Dynamics and Tissue Engineering of Blood Vessels

Arindam Bit and Jasjit S Suri

Chapter 12

An artificial blood vessel and its controlling aspects—I

Jasjit S Suri, Karabi Ganguly and Swati Sikdar

The capillary system, also called the blood vessels, is responsible for carrying blood within the entire living anatomy. The blood vessels are three types: artery, capillary and vein which form the major divisions of the vascular tree. The vascular tree begins from the aorta. The aorta branches repeatedly and gives rise to narrower arteries which again branch to give rise to arterioles. Arterioles are free within capillaries and capillaries are open within venules, which are very fragile veins. The combination of venules unite to form veins, and the combination of two small size veins form other large size veins called superior venacava and inferior venacava. The two venacava are free inside the right atrium. The systemic circuit and greater or systemic circulation are the areas of vascular tree and circulation, respectively.

12.1 Introduction

The pulmonary artery is broken into two parts called right and left lung. These two lungs force the blood into a left atrium with the help of four pulmonary veins. The pulmonary circuits and lesser or pulmonary circulation are the areas of vascular tree and circulation, respectively.

The vascular tree is usually divided into four major groups:

(a) Windkessel vessel/elastic artery/aorta/big arterioles;
(b) Pre capillary resistance vessel or muscular arteries/arterioles;
(c) Pre capillary sphineters;
(d) Exchange vessel/capillaries;
(e) Post capillary resistance vessels/venules;
(f) Large veins/capacitance vessels.

doi:10.1088/978-0-7503-2088-7ch12

General structure of the blood vessel:

The big artery, e.g., the aorta or venacava has three coats: tunica intima, tunica media and tunica adventitia (externa).

Tunica intima is the innermost coat. Internally it has a lining of a single layer of the epithelium called the vascular endothelium. This layer is very smooth and silky. This layer is in contact with the blood. External to the endothelial layer, lie mainly some elastic fibres and this layer of elastic fibres is called the lamina propria. The lamina propria is externally bounded by a layer of elastic fibres called fenestrated membrane.

External to the fenestrated membrane lies the tunica media. In the tunica media, there are smooth muscles as well as elastic fibres.

External to the tunica media lies the tunica externa which is made up of collagenous fibres mainly.

- In smaller arteries the amount of elastic tissue both in tunica intima and tunica media are much less, and the proportion of smooth muscular tissue in the tunica media increases.
- In arterioles, the elastic fibres have disappeared from the tunica media and this layer consists only of smooth muscles.
- In capillaries, there is only a layer of endothelium standing on a basement membrane. There is no tunica adventitia or media.
- The wall of the vein is much thinner than the artery but thicker than the venule. The elastic property of the wall of the vein is very low compared to the arteries. Their smooth muscle content is also very low.

Functions of the smooth muscle in the wall of the vessels:

- Smooth muscles are placed circularly round the vessels and when they contract the diameter of the vessel is reduced.
- Smooth muscles are supplied by sympathetic fibres. These muscles have α-receptors which combine with noradrenalin, adrenaline or epinephrin and in turn α-receptors are stimulated and vasoconstriction occurs. Vasoconstriction leads to an increase of blood pressure (BP).
- Phenoxybenzamine acts as a α-receptor inhibitor and thereby causes vasodilatation and decrease of BP.
- Some arterioles contain β_2 receptors, in addition to the α receptor. Adrenaline injection causes vasodilatation in those arterioles.
- There are some blood vessels which contain receptors which combine with acetylcholine (Ach) (figure 12.1). These cholinergic receptors, strangely enough are not attached with any nerve fibre. When Ach is given by intravenous injection, these cholinergic receptors' are stimulated and cause powerful vasodilatation of the underlying arterioles and produce a fall of arterial BP.
- Vascular smooth muscles obey the Frank–Starling law. Sympathetic stimulation → chronic stretching force → hypertrophy of smooth muscle → powerful contraction of hypertrophied smooth muscle → increase of peripheral resistance → increase of BP.

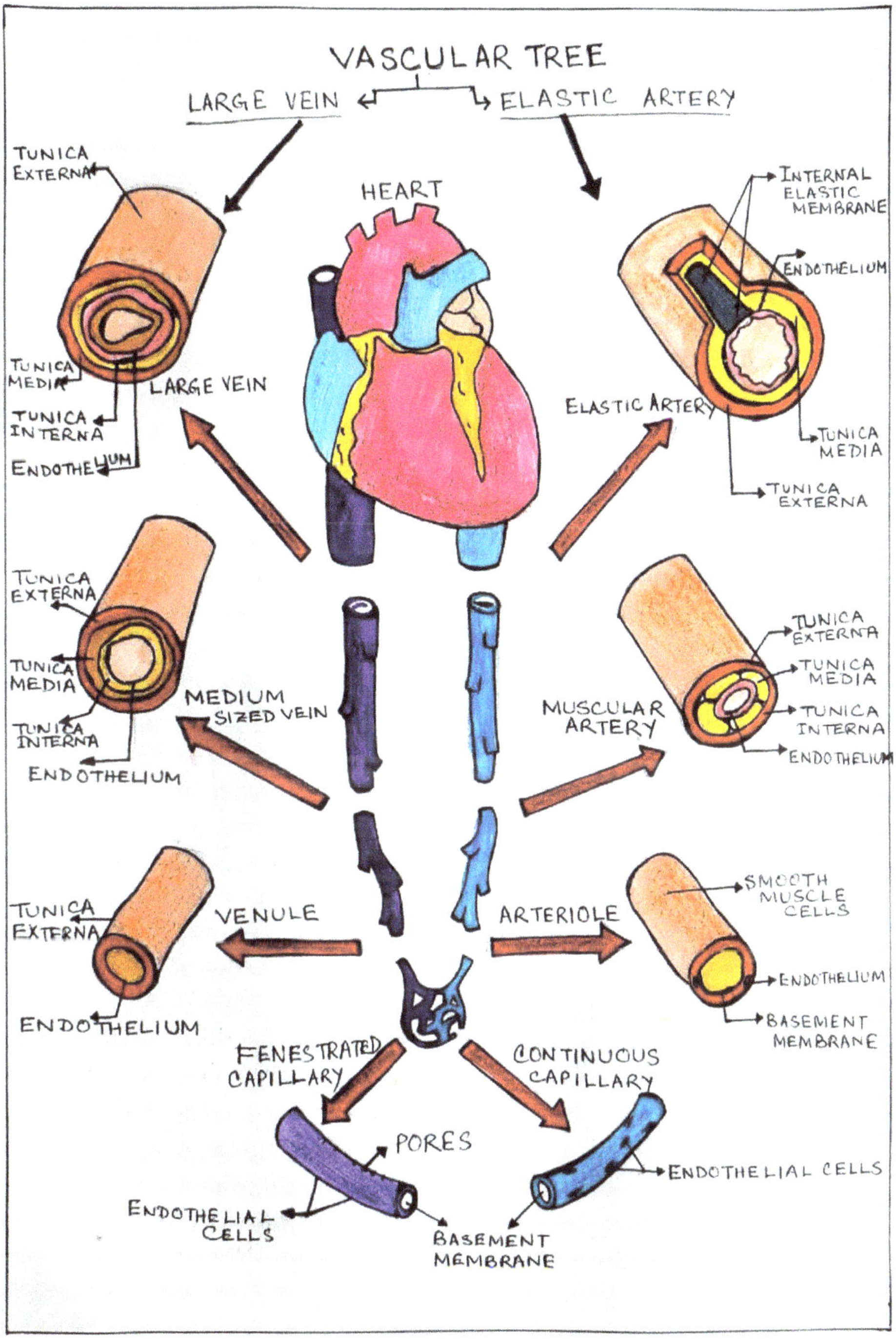

Figure 12.1. Structure of blood vessels.

- Sympathetic stimulation may cause various effects on smooth muscles of the blood vessels.

	Sympathetic stimulation ↓	
Constriction of	No effect on	Dilatation of
↓	↓	↓
Splanchnic vessels, Cutaneous vessels	Vessels supply to Brain	Coronary vessels, Skeletal vessels.

- The smooth muscles can be stimulated or inhibited by number of agents like the local O_2 tension, lactic acid, adenosine, histamine, bradykinin, prostaglandins, ADH, etc.

12.2 Formation of blood vessels in the human body

Blood vessels are formed from the middle embryonic layer of cells called the mesoderm which lies between the ectoderm and the endoderm. endothilial cells (ECs) are helpful for forming the lining of the blood vessel. These ECs have a salient feature to manipulate their quantity and rearrange according to the needed structure–function of blood vessels. ECs can migrate into almost every part of the body to create an adaptable life support system. Tissue growth and repaire are done through extending and remolding the network of blood vessels by ECs.

ECs are responsible for the formation of a single cell layer. That single cell layer lines all blood vessels. These single cell layers alter the transaction between blood stream and surrounding tissue. Signals generated from the ECs alter the expansion of connecting tissue. And these surrounding tissues are responsible for formation of the blood vessel. The new blood vessel is generated by the angiogenesis process. The protein structure of arteries and veins is different due to ECs and the protein structure is controlled or alterered accordingly to create a capillary bed.

After growth of the vessel, signals generated from ECs play a very important role for manipulating the vessel's activity and its structure. For example, the shear stress generated by the blood vessel is sensed by ECs called mechanoreceptors and converted into a signal. This signal is transferred to the surrounding cells to enable the blood vessel to be made compatible for blood flow. ECs are also responsible for adjust the neural signal for blood vessel dilation as well as making the smooth muscle relax inside the blood vessel by draining nitric oxide gas.

The formation of new ECs is governed by pre-existing ECs. The ECs have a property for cell breaking and migration. For example if the wall of the aorta is damaged and its existing ECs are damaged, then ECs located alongside drift in order to overlap the area where ECs are damaged.

Most of the time we can see in the adult that ECs are grown very slowly from a few months to years in the liver and brain, respectively. The ECs are not only responsible for generating new cells but also for repairing blood vessels. From the literature we find that there is a call for rapid blood vessel growth; the strength of

ECs is stepped up by blood flow because blood flow carries a small number of EC particles obtained from the bone marrow.

12.3 New capillaries formed from sprouting

Specific cell signals simulate new vessel formation in the form of capillaries from existing small vessels (figure 12.2). This process of angiogenesis is controlled by cell signaling and is conveniently observed in naturally transparent structures.

During the process of angiogenesis, the endothelial cells extend long pseudopodia, in order to form a new vessel sprout from the existing vessel. ECs by creating a hollow sprout form a tubular structure from the side of an existing small vein or

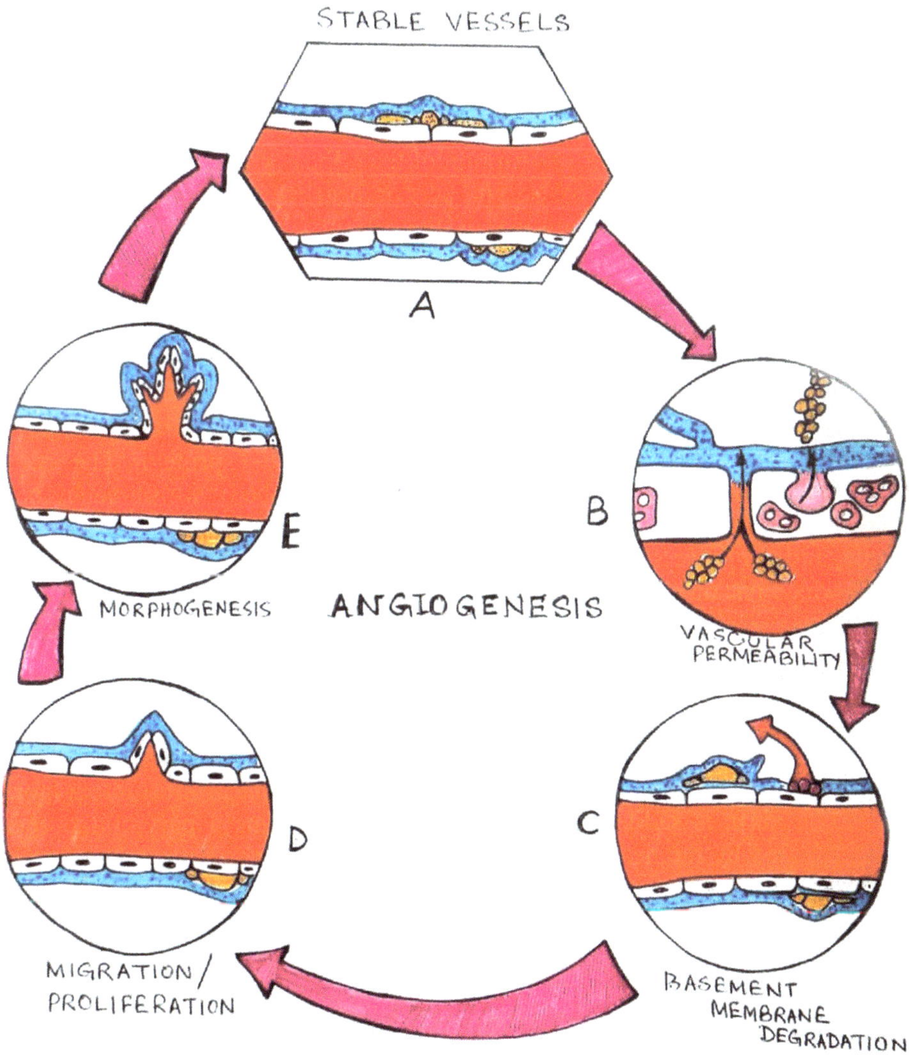

Figure 12.2. Formation of blood vessels.

capillary. The surface property of the EC on the arterial and venous side of the newly formed vascular network is different and also causes a difference in surface properties of ECs. The trans-membrane protein ephrin-B2 is present in the plasma membranes of the arterial cells while Eph-B4, which is a receptor tyrosine kinase, remains in the membranes of the venous cells. The signal mediated by ephrin-B2 and Eph-B4 at the site of cell–cell contact plays an important role in the formation process of the network of vessels.

12.4 Angiogenesis controlling factors

ECs invade tissue which in turn provides a signal of this entry and **vascular endothelial growth factor (VEGF)** which is a special platelet-derived growth factor (PDGF) that starts playing a role in angiogenesis. VEGF controls growth of a blood vessel depending on the needs of the tissue by changing the stability of its mRNA and in turn its rate of transcription. A gene regulatory protein called **hypoxia-inducible factor 1 (HIF-1)** gets stimulated when there is a shortage of oxygen supply in tissues and HIF stimulates transcription of the VEGF gene, the VEGF protein is secreted, diffuses through the tissue, and acts on nearby endothelial cells.

ECs first digest the basal lamina of the parent blood vessels by producing proteases that make their path to enter the vessels, next these cells migrate to the site and source of the signal, ECs then proliferate and form tubes and differentiate. VEGF triggers the process by acting selectively on ECs. Members of the fibroblast growth factor family and other growth factors stimulate angiogenesis by influencing other cell types besides ECs.

When cells have their required oxygen supply, the concentration of HIF-1 remains low due to continual degradation of the HIF-1 protein. The oxygen concentration rises in the tissues when the tissue is supplied with blood by newly formed vessels and HIF-1 activity declines. As a consequence, VEGF production is stopped, and the process of angiogenesis terminates. Angiogenesis is a cell signaling process which has fine-tuned control in the body.

12.5 Functions of blood vessels

Windkessel vessel/big arteries/aorta:

(1) The heart pumps out its stroke volume into the aorta (which already contains some blood), an additional 70 ml or the stroke volume is thrust into it (aorta). The aorta is suddenly more distended. This means a part of the energy released by the heart during the ventricular systole is stored as pulmonary embolism (PE) in the walls of the aorta. In the next moment during diastole the aortic wall recoils due to its elasticity and the PE is released into the blood, causing it to surge forward with a renewed vigor. This is called Windkessel effect. Due to this effect, velocity of blood is reduced to some extent during systole and velocity is increased during diastole.

The Windkessel effect reduces the requirement of energy expenditure by the heart during the systole.

(2) The pumping action of the heart and the elastic recoil of the aorta, taken together are the driving force which drives the blood forward and is called 'vis-a-tergo'.
(3) Windkessel vessels do not offer any serious resistance to the flow of blood.

Precapillary resistance vessels/arterioles:
(1) The wall of the arterioles does not contain elastic fibre and is mostly formed by smooth muscles. They are richly supplied by the sympathetic system. Due to the action of the sympathetic system and naturally occurring vasoactive agents, the diameter of the arterioles alter from time to time, which in turn causes a change of total peripheral resistance from time to time.
(2) Critical closing pressure (of Burton): there is a critical diameter for every blood vessel and if the diameter is further reduced below this critical diameter due to the action of sympathetic nervous system and strong vasoconstrictors, the blood vessel will totally collapse. Arterioles therefore face the possibility of closing when subjected to the excessive action of vasoconstrictors.
(3) Auto regulation : in some organs (brain, heart, kidney) the blood flow does not alter unless the perfusion pressure is severely changed. Even minor or moderate rise or fall of perfusion pressure does not affect the perfusion of the organ. Thus ability of the individual organ to keep its blood flow reasonably constant despite a fair degree of changes in the perfusion pressure is called autoregulation.

Some popular theories for autoregulation are as follows:

(a) A tissue produces its own metabolites some are vasodilators

---

If now perfusion pressure increases

↓

more blood flows in the tissue

↓

Washing out of the metabolites

↓

Less vasodilatation

↓

Reduction of the diameter of the blood vessels

↓

Reduction of blood flow

(b) If the perfusion pressure in the feeding artery of the organ rises

---

↓

This causes more stretch of the arterial wall. In the arterioles this leads to stretch of the smooth muscle wall.

↓
Vigorous contraction of the smooth muscle wall occurs because smooth muscle responds to stretch by contracting.
↓
Diameter of the arterioles reduced below the critical diameter
↓
Arterioles collapse and this causes reduction of blood flow.

(c) More Perfusion pressure

More pressure in the capillary
↓
More formation of the tissue fluid
↓
Increased tissue fluid causes increased pressure on the capillary wall
↓
Increased pressure on the capillary wall trying to collapse them
↓
Reduction of blood flow.

(4) Arterioles and the fall of arterial pressure

The mean arterial pressure in the aorta is about 100 mmHg and this pressure remains at the beginning of the arterioles about 85 mmHg. Fall of the pressure is not great. The pressure at the beginning of the capillaries is about 30 mmHg which shows a sharp fall of BP in the arterioles. Arterioles therefore are the major site of peripheral resistance.

The diameter of the arterioles can and does change during sympathetic stimulation
↓
The peripheral resistance offered by the arterioles also changes
↓
Inflow of the capillaries thus, changes.

- Precapillary sphincter: this sphincter when it is constricted, closes the capillaries, the reverse occurs when they dilate. The sphincter muscles are influenced by metabolites, local O_2 tension. They have sympathetic innervations too. Normally when tissues are not working hard, a large number of capillaries remain closed. But when the tissue is working (O_2 requirement is increased) the capillaries in the tissue open up. The opening and closing of the capillaries are regulated by the precapillary sphincters. The opening of the capillaries is called recruitment.

- Exchange vessels, means the capillaries. In this region various exchanges, those of O_2 and CO_2, nutrients, waste products etc, occur between the blood and the tissue.
- Capacitance vessels/large veins. These veins have thin walls and low elasticity. When the volume of blood in them increases, up to a point the venous blood pressure rises only slightly. So veins can hold a large amount of blood without developing a unique rise of venous BP. 70% of the total blood volume remains in the veins. The veins act as a reservoir of blood.

Sympathetic stimulation causes vasoconstriction leading in turn to release of blood from the venous reservoir.

Segments of the blood vessels

↓

Windkessel vessels	Resistance vessels	Exchange vessels	Venules	Capacitance vessels
(Aorta and bigger arteries)	(Arterioles)	(Capillaries)		(Big veins)
↓	↓	↓	↓	↓
Percentage of elastic fibres high; elastic recoil is maximum.	Elastic fibres almost nil; smooth muscle content is more.	Single layer of endothelium	wall is only slightly thick	Wall is thicker than venules but much thinner than aorta
↓	↓	↓		↓
Function: Windkessel effect; Vis-a-tergo	Function: Seat of Peripheral resistance	Function: Exchange of nutrients, gases and waste products		Function:Holds extra fluid (acts as a reservoir) without BP elevation.
↓	↓			
App. Physiology: Arteriosclerosis	App. Physiology: Essential hypertension			

12.6 Biological control of blood vessel structure and its effects on various physiological parameters

Regulation of BP:

BP = cardiac output × peripheral resistance. It has been shown that $Q = \frac{\Delta P}{R}$, the Q may be regarded as cardiac output and ΔP is the difference of BP between the beginning of the aorta and the end of the vena cava. Now, as the BP at the end of the vena cava is nearly zero, the value of ΔP becomes identical with the blood pressure at the beginning of the aorta.

Further, the resistance R is encountered by the flowing blood as it moves down the vessels. When all these issues are considered together, then it is observed that

$$\text{Cardiac output} = \text{Blood pressure/Peripheral resistance}$$

or,

$$\text{blood pressure} = \text{cardiac output} \times \text{peripheral resistance}$$

Again, peripheral resistance

$$(R) = 8\eta l / r^4$$

where η = viscosity of the blood (the greater the viscosity, the less the flow);
r = radius of the tube (the flow varies as the fourth power of the radius. A slight vasodilatation causes great increase in flow);
l = the length of the tube (as every segment of the tube is offering resistance to the flow; the longer the length, the greater will be the total resistance offered).

This means that the factors controlling BP are: (1) cardiac output (2) peripheral resistance. Peripheral resistance, in turn, depends on: (a) viscosity (b) length of vascular tree (c) radius of the blood vessels. Cardiac output depends on: (a) myocardial contractility (b) end diastolic volume (EDV). Therefore, these factors affect the BP. To have stable BP during the process of homeostasis, the mechanism for regulation of BP is usually divided into two groups viz. (A) short term (B) long-term controls.

(A) Short-term control is a neural mechanism. It comprises of baroreceptor reflex or sino-aortic reflex along with reflexes arising from atrial receptors, left ventricular stretch reflexes etc, baroreceptors and chemoreceptors of the heart and arterial tree. Short term control plays an important role in maintenance of BP homeostasis.

Baroreceptor reflex or sino-aortic reflex is so termed because it has two major wings: (a) carotid sinus reflex; (b) aortic arch reflex and baroreceptors as the specialized cells sensitive to pressure change (stimulus).

Sino aortic mechanism/reflex: the carotid sinus and aortic arch have numerous baroreceptors. When the BP fluctuates, these baroreceptors are stimulated and sense the change. The impulse is sent to the afferent traffic set up. The afferent impulses ultimately terminate in three regulatory parts of the nervous system: (i) VMC (vasomotor centre, present in medulla); (ii) nucleus ambiguous (i.e., the origin of cardiac parasympathetic of vagus); (iii) respiratory centre in the cerebral cortex. During hypertension (rise in BP) these baroreceptors are stimulated, (i) the VMC is inhibited, (ii) the nucleus ambiguus is excited, and (iii) the respiratory centre is depressed. As a result: (i) the sympathetic system is inhibited; (ii) the vagus is stimulated and (iii) the respiration is depressed. These in turn cause bradycardia and vasodilatation, and the elevated BP is corrected (BP homeostasis). Respiratory depression leads to reduction of pumping action of respiration that causes reduction of cardiac inflow and consequently reduction of cardiac outflow (via the Frank–Starling mechanism) and ultimately corrects hypertension.

Anatomical details:
In the beginning of the internal carotid artery, a dilatation called carotid sinus is present. Afferent impulses from carotid sinus form the sinus nerve → the sinus nerve joins the IXth nerve → the IXth nerve → the nucleus tractus solitaries (NTS) → to nucleus ambiguus and VMC. From the nucleus ambiguus, the cardiac vagal fibres arise and from VMC fibres arise to terminate on the spinal centres (lateral horns of the spinal cord).

From the barorceptors of the aortic arch, afferent impulses arise, travel through the aortic nerve, are next transmitted through the vagus nerve and then follow the same neural path as in the carotid sinus.

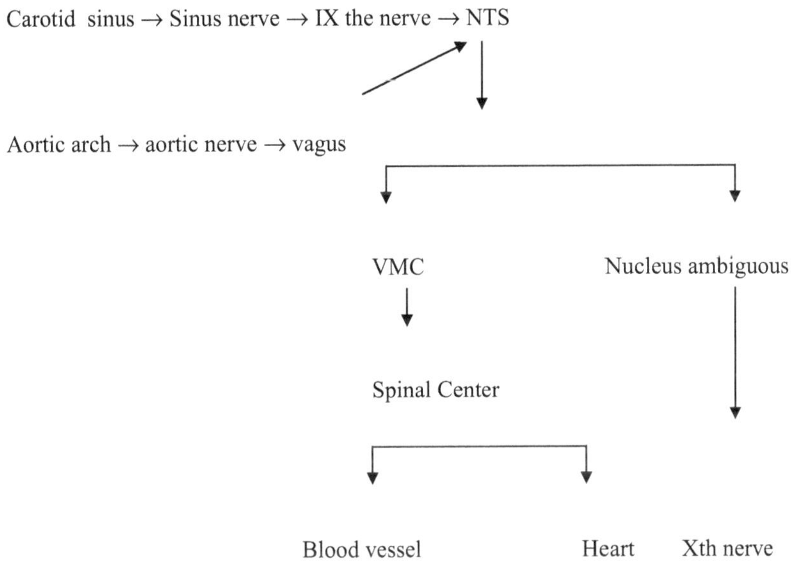

Chemoreceptor reflex: carotid bodies (located near the root of external carotid arteries) and aortic bodies (lie in the aortic arch) constitute the chemoreceptors of the arterial tree. They come in direct contact with arterial blood allowing them to calculate the O_2, CO_2 concentration or tension, as well as the hydrogen ion concentration of the arterial blood. The chemoreceptors are simulated with decrease in O_2 level or increase in CO_2 or with decrease in pH value of blood. With stimulation of chemoreceptors there is a possibility of: (i) hyperpnea; (ii) vasoconstriction and elevation of systemic BP; and (iii) changes in the heart rate.

(B) Long-term control
Suppleness of blood vessel and BP: the Windkessel vessels turn out to be supple. If their suppleness drops, systolic hypertension builds and systolic blood pressure increases, whereas the diastolic BP remains same. A prevailing increase in diastolic blood pressure shows an increase in peripheral opposition, which is a feature of essential hypertension where systolic BP shows atherosclerotic degradation of windkessel vessels.

12.7 Need for artificial blood vessels

Blood vessels play a major role in every physiological system and event. There are various diseases and abnormal conditions where the structure and functions of blood vessels get disrupted and this in turn evokes the need for artificial blood vessels. The following are a few common disease conditions where one may need artificial blood vessels or vascular prostheses.

The most fatal cardiovascular abnormality in the Western world is development of lipid lumps (atheromas) within the blood vessel wall, which is known as **atherosclerosis**. It is a common cause of mortality in industrialized countries. The deteriorating changes in the vessel wall cause arteriosclerosis, which further progresses towards condensing, degeneration of plasticity and ultimately to stenosis or blockage of the blood vessels. The constriction or stenosis of coronary arteries is a commonly occurring medical issue which is dealt with either by the use of natural vessels such as veins from the leg to form a circumvent, or by using vascular prostheses.

Atherosclerosis and peripheral heart disease: the main function of coronary arteries is to provide blood to heart muscles. Blood is carried to other organs and tissues with the help of peripheral arteries. Coronary arteries and peripheral arteries may both accumulate plaques which are fat deposits, cholesterol and other matter within their walls. Over time, this plaque formation makes the vessel narrow and obstructs blood flow. Over time, plaque can build up, so the vessel becomes narrow and it is harder for blood to flow, and tissues and peripheral organs do not get sufficient blood. This can cause different symptoms such as:

- obstruction in coronary arteries may be a reason for chest pain (angina) or a heart attack;
- the condition of a transient ischemic attack (TIA) may occur if there is any blockage in the carotid arteries that provide blood to brain, which is a form of stroke;
- obstruction in the kidneys can hamper its function and can lead to issues like uncontrolled high blood pressure and heart failure;
- obstruction in the legs can cause pain and cramps while one is active; this situation is known as claudication and can cause change in skin color, sores or ulcers, and tiredness in legs;
- the tissues in our body may die due to lack of proper blood flow. In extreme conditions this may lead even to loss of an organ or limb.

Blood clot in vein (VTE): A blood clump in a vein within a muscle—normally in the thigh, pelvis or lower leg—is known as deep vein thrombosis (DVT). It is very dangerous because these clots can break loose and travel to the lungs, which becomes a pulmonary embolism (PE). These types of clumps within veins are known as venous thromboembolisms, or VTE.

- Vasculitis or inflammation within vessel walls may happen because of an autoimmune disease for which any infection can be fatal.

In a blood vessel of any type a bulge can be seen which is known as an **aneurysm**, it is mostly visible in the primary blood vessel leaving the heart which is called the aorta. Hence an aortic aneurysm is formed in one's thoracic or abdominal area. There is not much danger if aneurysms are small but still some risk is associated with them such as:

- there can be plaque formation near an aneurysm.
- a clump can be developing near an aneurysm which may also break and get caught at other places, which is very dangerous.
- if the size of the aneurysm is larger then it can be a reason for pain and put pressure on other organs.
- at the place of the aneurysm the artery wall is elongated, thin and also fragile. Under pressure and stress it can burst, which can be fatal.

12.7.1 Buerger's disease

This disease happens in our arms and legs within the small arteries and veins and medium arteries and veins. These arteries may swell and form a clump which then prevents the blood supply from reading the arms and legs. These parts even hurt when at rest and in extreme cases one can lose one's fingers and thumbs also. However, it is a rare condition.

12.7.2 Peripheral venous disease and varicose veins

Veins contain flaps within them which are known as valves and these valves open at muscle contraction and let the blood flow via the tubes. The valves then close when the muscle is at rest or relaxed, which keeps the flow of blood in only one direction. These valves do not shut properly even when muscles are relaxed if they are damaged, and this causes blood flow in both ways which causes it to pool, and this happens with varicose veins. These veins may also look like purple threads when bulged under our skin or small red bursts on thighs, knees and calves. The main cause of these veins is swollen capillaries which are small blood vessels. This problem occurs more in women than men and causes legs to ache, sting, twitch and swell. These are hereditary as well and trouble more during pregnancy. Taking exertion for longer duration may also be a cause as blood moves slowly. This can remain in veins and form clumps as well.

12.7.3 Raynaud's disease or Raynaud's syndrome

In the cold, the small arteries in fingers may convulse and also obstruct the blood flow, these symptoms are related to diseases including Raynaud's disease or Raynaud's syndrome, lupus, rheumatoid arthritis and scleroderma. Raynaud's disease can happen in people with Buerger's disease. It becomes essential to replace a blood vessel in these situations. This can be done with the help of artificial vascular prostheses or by taking another vessel from the body.

12.8 Development of artificial blood vessels through tissue engineering

Coronary occlusion due to atherosclerosis, stroke and aortic aneurysm are the cardiovascular major killer diseases across the world. For small-caliber (<6 mm) vascular grafts the saphenous vein is most often used for vascular prosthesis in these cardiovascular pathological conditions. Size mismatch and venous disease become a hindrance to accepting a saphenous vein suitable for prosthetic replacement in 10%–40% of patients.

Tissue engineering approaches are being used to develop artificial blood vessels using biosynthesized bacterial cellulose (BC) as a material for scaffolding. Structural and functional features of the vessels, as well as the coating of the scaffolding that is cellulolytic tubular exhibit applicable mechanical properties, and also their morphology shows connected phases such as porous and nonporous as a support to mimic, respectively, the blood vessel layers such as tunica interna and tunica media. Vascular smooth muscle cells (SMCs) in the presence of tubular scaffolds supports colonization by vascular smooth muscle cells when cultured in *in vitro* conditions.

BC vessels can be developed using selected bacteria grown in a medium consisting of mannitol, peptone and yeast extract in a container, having a template tube (tube made of silicon with inner diameter of 4 mm) attached to a circulation air pump. Cells are cultured under static conditions for 12 days and BC biosynthesis is allowed all over the template tube. These artificial vessels are washed and sterilized.

Evaluation of morphology of the vessels is done using a scanning electron microscopy (SEM) while the mechanical properties are assessed by tensile testing. According to [51], vascular tissue engineering aims to apply the principles of engineering and life sciences towards the development of a vascular construction that demonstrates biological and mechanical properties as close as possible to those of a native vessel.

For a perfect tissue engineered vascular graft (TEVG), having large and small diameter, the requirements are related to both reconstruction of the functional endothelium and also the resemblance between the mechanical properties of natural blood vessels and TEVG. The major strategy that lies in vascular tissue engineering is the designing and production of suitable scaffolds for proliferation, cell adhesion, differentiation and cell type preference. The perfect cells for human application should possess properties that make them functional, non-immunogenic and simple for isolation and expansion (figure 12.3).

Two methods are mainly developed. In the first method, a bioreactor is used for generating physiological-like stimuli on the scaffolds with cell seeding for maturation of TEVG *in vitro* before it is implanted *in vivo.* The second method involves the implantation of cell seeded scaffolds directly into the body, which works as the bioreactor for maturation of TEGV. Recent studies have mainly concentrated on the necessity for off-the-shelf grafts for regenerating blood vessels with small diameter, evaluating the probability of directly implanting acellular scaffolds into the body. The main objective of this approach is to develop grafts that will be readily available for vascular surgeries in the case of emergency (figure 12.4).

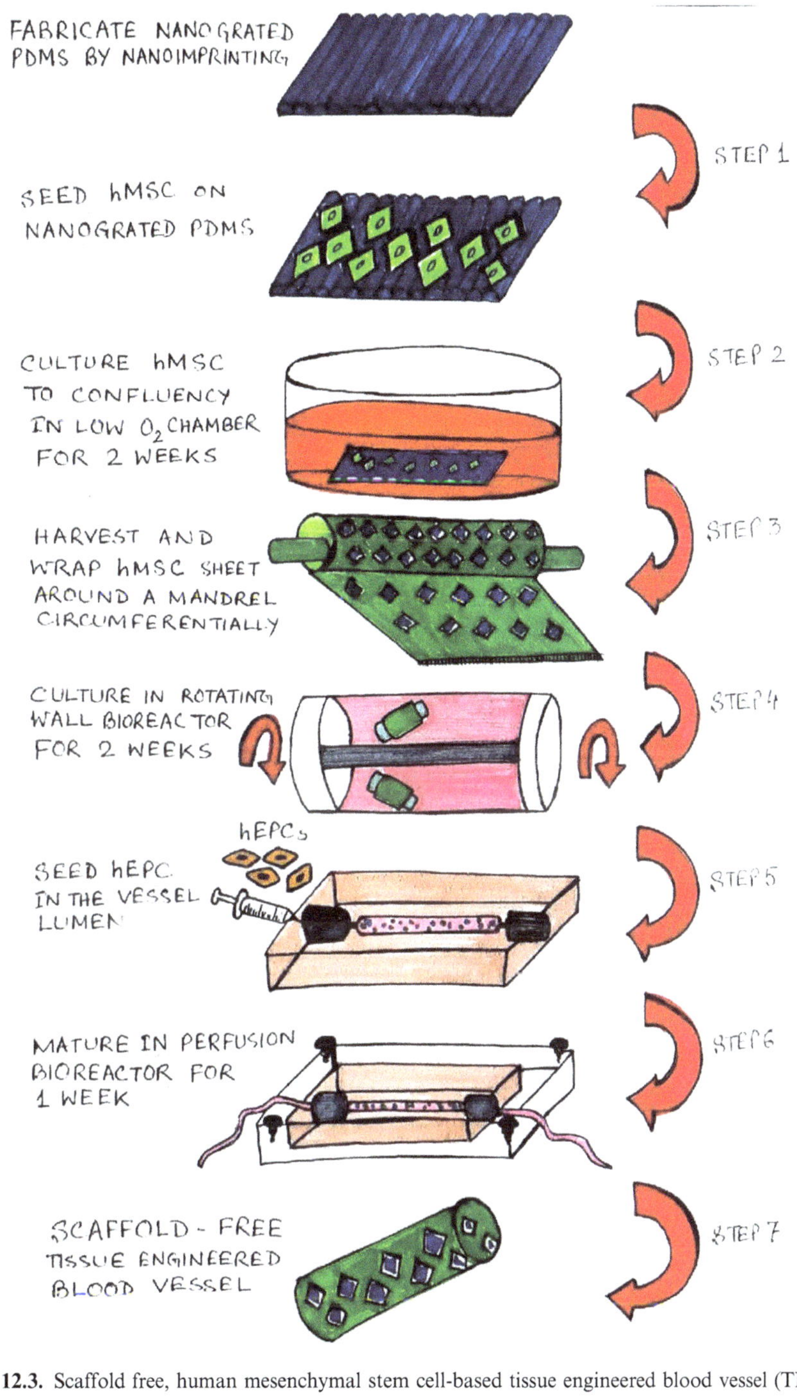

Figure 12.3. Scaffold free, human mesenchymal stem cell-based tissue engineered blood vessel (TEBV).

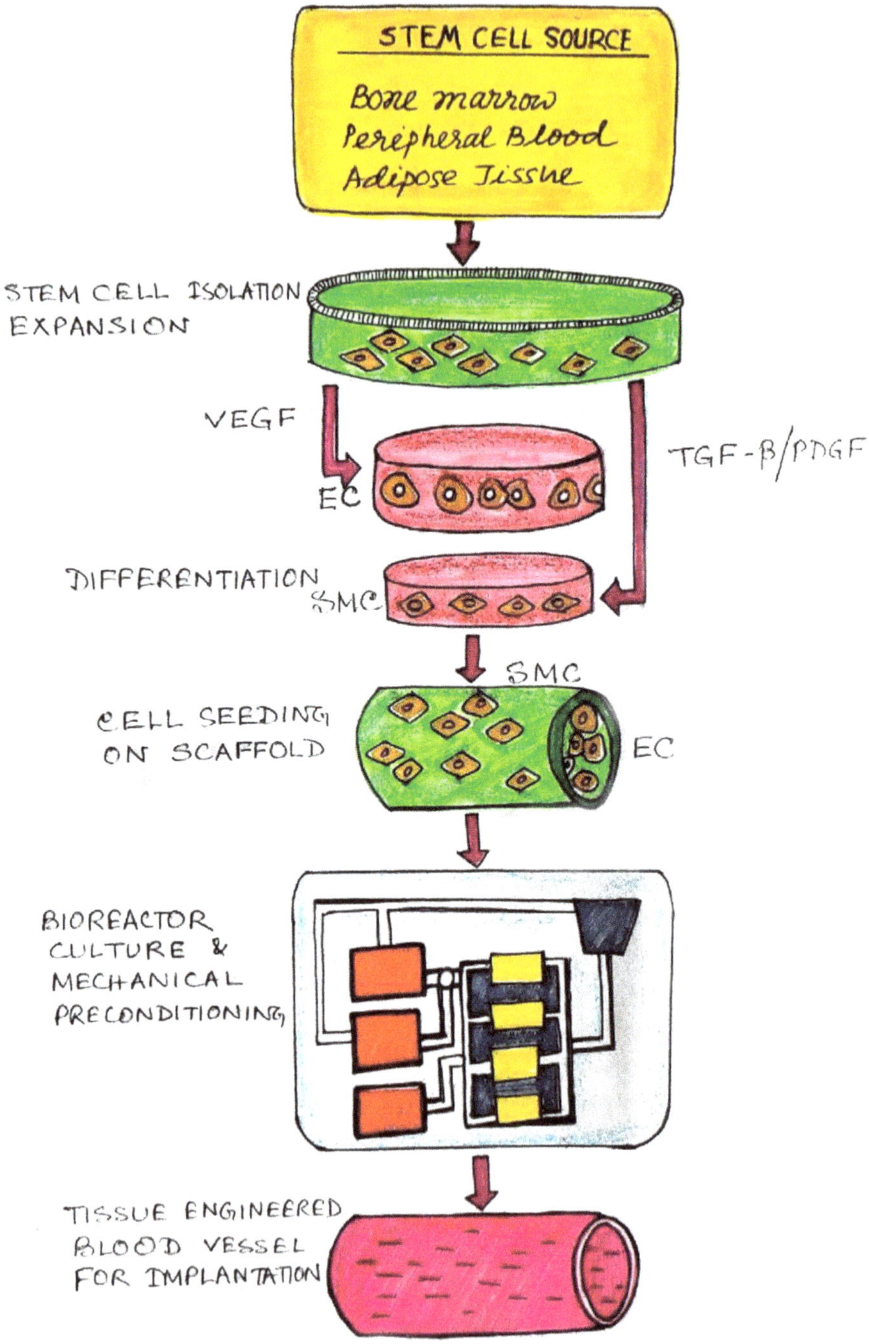

Figure 12.4. Scaffold-based TEBV.

12.9 Vascular tissue regeneration

12.9.1 Decellularized matrices based scaffolds

Cellular and nuclear matter is removed with the help of a process called decellularization. This minimizes negative effects on biological activity, composition and mechanical activity on the rest of the extracellular matrix (ECM) and helps in developing new tissue. Various organs and tissues (like skin, liver and ureter) from humans and animals (like sheep, monkey, bovine, rabbits) which act as a good source of ECM are decullalarized for regeneration of blood vessels. Advantages of decellularized matrix can be related to its properties such as its original three-dimensional ultra structure and the structural and functional proteins that are necessary for migration, proliferation, differentiation and cell adhesion. The particular concentration of ECM correlates to source of the tissue. Matrices of ECM are majorly effected by factors such as health status and age of the harvested animal, process of manufacturing which influences the quality, biocompatibility, biochemical and mechanical properties and clinical performance. The potential antigenic compound traces such as DNA, lipids and glycosylation products present are responsible for inflammation. Additional to this, procedures used for decellularization may remove vital ECM components like collagen resulting in the decrease of mechanical properties.

12.9.2 Scaffolds from natural polymers

Polymers obtained from natural sources exhibit exceptional biological activities. These polymers will not be involved in the activation of toxicity or chronic inflammation. Some of the natural polymers which are used in vascular tissue engineering are collagen, silk fibroin, elastin, hyaluronan and fibrin. They exhibit attractive characteristics used for these functions and are also the most commonly used natural polymers. Fibrin in an insoluble protein used for tissue repair and wound healing purposes. Fibrin clot is a fibrillar network gel formed by polymerization of fibrinogen by thrombin. This network facilitates structural support for proliferation, migration and adhesion of the cells that are associated in the healing. The fibrin clot is further resorbed by a process called fibrinolysis, which is a fibrinolytic process that cuts down the fibrils of fibrin. Fibrinogen can also be purified from autologous blood. This purified fibrinogen is used for fabrication of a scaffold to avoid immunological problems.

Elastin is another extensive ECM protein in the arterial wall. It confers elastic recoil, resilience, and durability. Elastin, when used as a coating for vascular devices which are made of made of ePTFE, PET (copolymer of ePTFE and polyethylene, and a polycarbonate polyurethane) exhibits less thrombogenicity along with less activation and platelet adhesion.

Non-sulfated anionic glycosaminoglycan (GAG) (hyaluronan) can be produced in large amounts by microbial fermentation. It consists of glucuronic acid and N-acetyl glucosamine. The significance of hyaluronic acid is that it is not only recognized as hydrophilic, non-adhesive and biocornpatible, but is also known for its biodegradable properties.

Silk fibroin is one of the familiar proteins made up of glycine (43%), alanine (30%), and serine (12%). It is produced by silkworms and spiders. This biocompatible protein shows excellent mechanical properties. The contents of silk degrade gradually and proteolytically *in vivo*, while preserving more than half of its original mechanical properties after a couple of months.

Collagen is the abundant fibrous protein in the ECM in the body which provides mechanical sustenance to numerous tissues. It exhibits major characteristics such as low antigenicity, low inflammatory response, biocompatibility, biodegradability, and exceptional biological properties. Amongst the major components of the vascular wall collagen type I is considered as an important component. The foremost area of utilization of collagen is in vascular tissue engineering as a scaffold.

Fabrication of TEVGs with natural polymers, structural proteins like fibrin, elastin, and collagen after improving their mechanical properties is getting paramount importance. Concurrently, silk fibroin is extensively studied for its vascular applications because of its greater mechanical properties in comparison to other natural polymers, such as fibrin.

12.9.3 Scaffolds from biodegradable synthetic polymers

Biodegradable artificial polymers have a general tendency to exhibit customizable mechanical properties, such as high reproducibility in comparison with natural polymers. They can also be manufactured in bulk with minimal effort.

Some of the best examples for biodegradable synthetic polymers are polyglycolic acid (PGA), polylactic acid (PLA), poly-ε-caprolactone (PCL), and polyglycerol-sebacate (PGS). They are used for application in vascular tissue engineering.

The ring-opening polymerization of glycolide synthesizes as semi crystalline, thermoplastic aliphatic polyester, PGA which degrades quickly (in 6 months) *in vivo* by hydrolysis to glycolic acid, metabolized and removed as CO_2 and water. The Food and Drug Administration (FDA) has accredited it for clinical utilization in humans.

Appreciable biocompatibility and mechanical properties are exhibited by PLA, a thermoplastic aliphatic polyester. Its characteristics enable it to dissolve in common solvents for processing and it is produced by ring-opening polymerization of lactic acid. PLA has slower degradation rate and is more hydrophobic in nature than PGA. It is an FDA approved polymer for human clinical use.

It takes months or sometimes even years for PLLA to lose its mechanical integrity. The ring-opening polymerization of ε-caprolactone produces PCL as an emicrystalline, aliphatic polyester which demonstrates good mechanical properties, high elongation, strength, and good biocompatibility. PCL is one of the FDA approved polymers.

An elastomer (PGS) is amalgamated by polycondensation of glycerol and sebacic acid. PGS exhibits decent biocompatibility and mechanical properties, explicitly during high elongation and low modulus, signifying an elastomeric and hard behavior. PGS takes two months to degrade *in vivo* by hydrolysis. The FDA approved glycerol and polymers contain sebacic acid for medicinal applications.

PCL due to its deformability and mechanical strength is selected for fabrication of TEVGs using biodegradable synthetic polymers. Using PCL as PGS tube coating increases the mechanical characteristics of the substrate and ultimate strength stress.

12.9.4 Synthetic and natural scaffolds

TEVGs have suitable biological performance and mechanical properties for clinical use. TEVG biological execution provides wide-ranging integration of the grafts in the human body, avoiding the orientation of a lasting inflammatory response during the material deprivation process. TEVG mechanical properties must align with those of the native blood vessels and they should cope precisely in terms of deformability, acquiescence, and strength.

For vascular tissue engineering, it is mandatory to preserve the structural and mechanical integrity of the neovascular tissue overall in the regeneration process. Generally, in comparison to the counterpart synthetic polymer, natural polymers exhibit excellent biological behavior in spite of their less competent mechanical performance. Tropoelastin electrospun scaffolds confirm low mechanical strength compared to native blood vessels, in terms of decisive tensile strength (MPa) and burst pressure (mmHg).

ECM deposition causes partial remodeling of the structure when prior to *in vivo* implantation, natural scaffolds are *in vitro* seeded and cultured with cells. This *in vitro* remodeling creates a path to increase the graft mechanical properties. The combination of fibringels along with the entrapped SMCs can be dynamically cultured in a bioreactor for a recommended period (usually a few weeks) to achieve a circumferential alignment similar to native arteries. This alignment improves the mechanical properties of the grafts, reaching values similar to native blood vessels. Present day studies related to natural materials for vascular applications are focused on the improvement of graft mechanical properties. Some approaches consist of tissue *in vitro* maturation to reach the goal of improvement. A combination of natural materials with synthetic polymers (such as tropoelastin with PCL) are incorporated in other methods, accomplishing noticeable results like mechanical properties and cell/material interaction at times.

12.9.5 Hybrid scaffolds from synthetic and natural polymers

The preference of the scientific community has shifted recently towards hybrid scaffolds, made of synthetic and natural polymers, to combine the adequate mechanical properties of synthetic materials with the excellent biological behavior of natural polymers.

Fabrication of TEVGs using hybrid scaffolds, particularly using PCL with natural polymers such as elastin and collagen produces satisfactory results. For illustration purposes, the addition of PCL to tropoelastin electrospun scaffolds advances the mechanical properties, making them more parallel to native blood vessels. Specifically, compliance and burst pressure of PCL/tropoelastin scaffolds decreases and increases, respectively in comparison with tropoelastin-made scaffolds.

12.9.6 TEVGs without scaffolds

Completely biological TEVGs without the use of scaffolds have paramount importance in evading difficulties that may be related to synthetic materials, such as inflammation, stenosis, and infection, permitting complete graft integration and increase of the patencyrate. There exist three major approaches such as the *in vivo* bioreactor, the cell sheet-based, and the cell ring-based methods.

The *in vivo* bioreactor handles an approach of utilizing the body as a bioreactor, as an outcome of the reaction made by a foreign body due to the implantation of a 'non-self' material. The important restrictions of this method are the prerequisite of double surgery and the possible TEVG adhesion during maturation and, hence, the damage to the external layer. In order to produce an abundant quantity of ECM *in vitro*, cell growth is monitored in the presence of ascorbic acid for the sheet-based approach. One of the noticeable constraints of this method is the time factor. This method takes prolongated time for the *in vitro* TEVG growth before the commencement of the implantation process. The manufacturing of TEVGs from the amassed cells and cell-derived ECM are a part of the cell ring-based approach due to the use of annular agarose wells.

Vascular tissue engineering for small diameter blood vessel applications has taken a large leap in recent years, since the demand for TEVGs manufactured with the help of decellularized matrices and natural and/or biodegradable synthetic polymers and awareness about the limited use of scaffolds has grown exponentially. Due to the unique and better mechanical and biological performance, scaffolds made of natural and/or biodegradable synthetic materials draw special attention citing the advantages over synthetic materials.

The main deficiency of an ePTFE artificial limb is related to neointimal hyperplasia, remarkable calcification, thrombosis, and external body reaction. The contest for vascular tissue engineering still remains in the area of developing TEVGs, which play an important role in reviving functional endothelium and exhibiting mechanical properties similar to native blood vessels. Economic feasibility is still a major hurdle in spite of the encouraging results attained by TEVGs in clinical trials. The prolongated time and expensive production unit makes manufacturing of absolute biological TEVGs without the use of scaffolds difficult.

There are different scaffold based methods available to us but one of the methods called the direct method has many advantages over the others, as it provides a cell seeded scaffold without use of a bioreactor. Consequently, it minimizes the overall cost and cuts down the steps of execution. The working principle behind this method is that cells are directly seeded inside the scaffold and these seeded cells play a crucial role for tissue reproduction inside the living body. The tissue reproduction with the help of a cellular scaffold by the host cell, escapes limits correlated to the *in vitro* cellularization of scaffolds, namely the selection of cells and their proliferation and the operation of the bioreactor because of its handling and high-tech complication.

There are various types of biomaterials available for the production of very fine radius types of blood vessel. Among them, hybrid scaffolds (synthetic polymer) are most widely used because of their higher mechanical characteristics as well as higher biological results compared to available natural polymers. If we have good knowledge about the molecular pathogenesis of artery disease then we can directly insert the anti-inflammatory and anti-proliferative drugs into the TEVGs.

For better cell colonization we use a BC vessel. For formation of fabricated blood vessels we use BC as a biomaterial because of its potential to fulfil the requirements of fabricated blood vessels.

12.10 Selection of biomaterials and design parameters

It creates a fatigue-resistant tissue with long-term durability. The endothelial lining in the native vasculature acts as a protective, thromboresistant barrier between blood and the surrounding tissue. It also controls vessel tone, platelet activation and leukocyte adhesion.

Cardiovascular disease is one of the major causes of mortality in the modern world. As healthy autogenous vascular substitutes are not always readily available, fabrication of coronary and peripheral vascular grafts has gained attention in the field of reconstructive arterial surgery to provide medical support. Polytetrafluoroethylene (ePTFE), polyethylene terephthalate (dacron) and polyurethane are used to counterfeit synthetic vascular grafts, although these materials that make artificial vessels are prone to get thrombus formation, compliance mismatch and fail to make grafts less than 6 mm in diameter to be used to replace the saphenous vein, internal mammary or radial artery as a vascular substitute. The tensile strength and viscoelasticity of the blood vessel as the important normal physiologic responses of the vascular wall to prevent thrombosis and inflammation have provided the guideline in designing advanced vascular prosthesis using alike structural components, collagen and elastin. Again, the endothelial cell lining of the native vessel not only controls the vessel tone, platelet activation and leukocyte adhesion but also resists thrombosis. The pertinent factors that help to determine the ideal biomaterials for designing of a vascular graft are biocompatibility, resistance to infection, suturability and ready availability.

The first tissue engineered blood vessel alternative was fabricated by Weinberg and Bell in 1986 [19]. Using collagen gel as a matrix, with a dacron mesh for enhancing strength they created cultures of bovine endothelial cells, fibroblast and SMCs. This initiated the strategy to create convenient material for vascular grafts by concentrating on three areas of research: biodegradable scaffolds, coatings and surface chemical modifications of synthetic materials and selection of suitable biopolymers. In each group, tissue engineering strategies for *in situ* regeneration of vascular tissue are adopted. Under these strategies, using material design and fabrication, one modulates the body's natural response to the process of healing. The next major strategy is *ex vivo* formation of artificial blood vessels as a

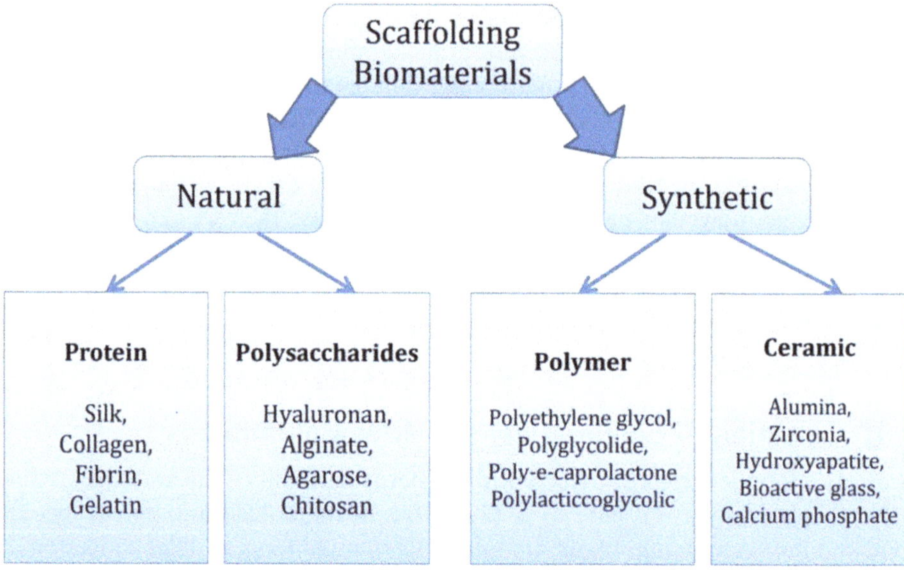

Figure 12.5. Biomaterials used in vascular grafts.

part of which polymer substrates along with human cells are layered *in vitro* culture. The implantation of this vessel prosthesis stands as a mark of their mechanical and biological properties to mimic the native blood vessel.

12.11 Biomaterials preferred for preparing vascular grafts

12.11.1 Synthetic nondegradable polymers (ePTFE, dacron and polyurethanes)

Numerous reasons and properties of synthetic materials make them potential candidates for design of vascular grafts, such as the ease and flexibility of tailoring their mechanical properties (figure 12.5). ePTFE is a porous polymer having a non-degradable electronegative luminal surface. In standard grafts of ePTFE, the internodal distance or fibril length is about 30 μm, and does not occur significantly either in transanastomotic or transmural endothelialization. In recent trends, dacron has been used for aortic replacement and as a conduit for femoropopliteal bypass surgery. Peculiarly, a knitted graft is incorporated in a velour finish. This knitted graft comprises loops of yarn upward, perpendicular to the surface, and thus the surface area is enhanced and the anchoring of fibrin and cells to regenerate tissue is also improved. However, these properties are lost after implantation. ePTFE and dacron share similar patency rates. Polyurethane is a copolymer of three different polymers: a soft domain (diol), a chain extender and a hard domain (diisocyanate). the soft domain provides flexibility, whereas the hard domain imparts strength in physiological temperatures. The most general medical grade polyurethanes are based on the soft domain made up of polyester, polycarbonate or polyether.

12.11.2 Polymer functionalization

Synthetic polymers have low patency rates, which leads to more strategies for direct tissue regeneration and functionalization of the luminal surface grafts. To improve endothelialization and also to inhibit inflammation, various strategies like coating, chemical and protein modifications, seeding of ECs in inert materials are used. Thus, technologies like carbon deposition, plasma discharge and photo-discharge are utilized for deposition of reactive groups onto polymer surfaces so that they can interact with cell specific peptides and enhance protein adsorption to the surface.

12.11.3 Degradable scaffolds

Biodegradable polymers as scaffold is an alternative tissue engineering approach for the development of functional vascular grafts. These biodegradable polymer scaffolds degrade and are replaced by an ECM secreted by the seeded cells. PGA is the most widely used biodegradable polymer used for tissue engineering application. The mechanism of its degradation is based on hydrolysis of ester bonds and glycolic acid, eliminating water and carbon dioxide molecules. The rate of biodegradation is controlled by copolymerization with polymers such as polyethylene glycol, oly-l-lactic acid (PLLA), polycaprolactone-co-polylactic acid and polyhydroxyalkanoate.

Polyhydroxyalkanoate is a linear polyester that is produced by bacterial fermentation of lipids or sugars and used for graft designing. This polyester can be modified to a wide range of degradation rates and mechanical strengths.

Polycaprolacton (PCL) is another versatile polymer used for tissue engineering application. It degrades on hydrolysis of ester linkages. The biodegradable polymer system provides an advantage on spatial and temporal release of the various growth factors which are essential for the vascular wall regeneration. For example, VEGF is released from PGLA scaffold which promotes angiogenesis *in situ*. Poly(ester urethane) urea scaffolds release FGF-2 amalgamates the mechanical properties of polyurethane with bio-active angiogenic proteins. Hydrolytic degradation is a powerful tool in tissue engineering for release of bio-active molecules from the polymeric matrix, incorporation of proteolytic sites can be further modified for presentation of these molecules to the surrounding environment via the cell degradation modification mechanism.

Biodegradable polymeric scaffolds used for making tissue engineered vascular tissue grafts are yielding promising results but still have some challenges. Cell sourcing is developed from time consuming culturing methods of almost 2–6 months that also have low proliferation capacity of cells.

12.11.4 Biopolymers

Manipulation of proteins constituting with the structure of a native ECM is an attractive approach of tissue engineering for the development of synthetic and degradable scaffold based vascular grafts. The approach for the generation of protein polymers which are capable of mimicking the native structural protein as well as adopting the characteristics and properties of arterial walls is found to be very useful for the development of vascular grafts. In the final analysis, the success of the approach relies on adequate migration, adhesion, proliferation of cells, and ECM production on bio-mimetic scaffolds.

Collagen is the most commonly used biopolymer in tissue engineering application, whose function is to limit high strain deformation and also to eliminate the probability of rupture of the vascular walls. Due to low inflammatory and antigenic response, the collagen fibers and gels are considered to be ideal for artificial blood vessel development. Adaptability such as cross linking conditions, fibre orientation and cell seeding techniques has improved the mechanical integrity of collagen based constructs. The cyto-toxicity of this chemical led to the alternative cross linking mechanisms.

The new upcoming stiff collagen based constructs promote exploration of the more potential elastic fibrin gels for vascular tissue engineering. Fibrin is formed when in addition to thrombin, fibrinogen polymerizes with the fibrillar mesh. The main advantage of this biopolymer is that it can be produced with the patient's own blood and thus does not show any inflammatory response upon implantation. Fibrin also binds to some critical proteins that control the cell's direct fate, such as fibronectin and VEGF.

12.11.5 Nanocomposites

The latest evolution in the field of nanotechnology has make it possible for vascular tissue engineering to mimic the nanostructure of native vascular architecture, consequently directing biological and mechanical performance of the bulk material. One such application is constructing nanofibers using a synthetic polymer via electro-spinning. The main advantage of this approach is the ability to develop a highly porous scaffold with high surface-to-volume ratio, thus stimulating the dimensions and structure of native collagen and fibrin fibers.

Nanocomposites are the result of enchantment of base material properties by adding filler materials. In the main, these materials have revealed a reduction of thrombogenecity and improvement of the mechanical properties. Despite the fact that the composite reduces the thrombogencity on the surface, toxicity of carbon nanotubes remains a concern.

12.11.6 Alternative tissue sources

Human umbilical vein and bovine and porcine carotid arteries comprise decellularized allogenic or xenogenic tubular tissue that depicts an intact and structurally organized ECM as vascular conduit. Decellularization refers to the removal of cellular antigenic components in xenogenic and allogenic tissue. A combined procedure including physical agitation, chemical surfactant removal and enzymatic digestion leads to disruption of the cell and removes protein, lipids and nucleotide remnants. To reduce the immunogenecity and to enhance the mechanical strength, chemical cross-linking is used in addition with decellularization. Dacron mesh is used essentially as external support to increase the mechanical strength and also to prevent late dilation. Enhancing the healing properties and durability of the decellularized scaffold includes coating with heparin and FGF, as well as seeding with bone marrow derived cells, ECs, and adipose derived stem cells.

12.12 Controlling factors and functional requirements in blood vessel tissue engineering

For TEBV design, mechanical and biological performance and biological–mechanical interconnectivity are considered as the most desired requirements. Any maladaptive biological responses lead to premature biodegradation, whereas degradable-scaffold grafts are the result of mechanical failure.

12.12.1 Mechanical requirements

Burst pressure, fatigue-resistance, suture retention, kinking radius, and compliance are important mechanical requirements which set the international standards for biomaterials like dacron and ePTFE used in TEBV. A vascular prosthesis or graft must have that ability to withstand physiologic variations in pressure and it is verified by testing the burst pressure and mechanical resistance to catastrophic rupture/tearing of the grafts.

Internal pressure loads are increased and simultaneously outer diameter and internal pressure are recorded using a system to test burst pressures of TEBV. Compliance and fatigue may be evaluated with similar test fixtures. Either the pores of the tubular construct are sealed with a material of negligible mechanical strength or porous constructs are inflated with a flexible elastomer which has negligible stiffness compared to that of the material being tested in the porous tubular construct. The mechanical response of constructs is dependent on the rate at which scaffold inflation is performed due to time-dependent phenomena such as stress relaxation.

12.12.2 Biological requirements

Blood clot or thrombus formation on the graft luminal surface due to the interaction of blood and material, in turn non-specific protein absorption results in acute response. The chronic response to vascular graft implantation causes growth of pannus tissue from the anastamotic regions or transmurally along with biodegradation and scar tissue formation in long term material interactions with the host.

To prevent thrombus formation, protein adsorption is resisted through surface modifications or using a luminal layer of ECs.

An inactive EC layer is important for promoting an anti-platelet, anti-coagulant and pro-fibrinolytic surface. Conversely, occlusion due to activation of the present or neo-ECs is of major concern.

SMCs are another cellular component of the vascular wall and are responsible for vascular tone and medial compliance of the vessel in normal pulsatile blood flow.

12.13 Vascular grafts and their main applications

Advancements in medical science and changes in lifestyle in modern society are the main cause for the increasing age of the population. It is a natural phenomenon that with increased age, the number of people with arteriosclerosis and cardiovascular disease is also increasing. When the blood vessel of a person loses its function due to disease or injury, vascular transplantation is to be carried out. Therefore, artificial blood vessels or vascular grafts play a vital role in successful treatment of the people who are in need of vascular prosthesis. Studies carried out by the National Institute of Health, USA, state that peripheral vascular disease is common in people who are 50 years of age and above in a pattern of 1:20. These conditions always demand vascular grafts that are structurally sound, versatile in nature and that do not clog easily, can prevent blood stream infections, avoid triggering of immune responses and can maintain a patient's life for years. Prosthetic grafts are used for patients who do not have a suitable blood vessel that can be used for grafting. Commonly available types of vascular prosthesis are biological blood vessels and synthetic/artificial blood vessel. Some advantages and disadvantages of the biological grafts are mentioned in table 12.1.

Due to the disadvantages of biological grafts, a major clinical application is found with synthetic or artificial ones because of their vascular convergence closer to the human body and long-term patency and more stable performance to meet the requirements of anticoagulation and anti-thrombosis. These are possible when the artificial vascular materials have three basic requirements, namely lasting strength, proper pore and good compliance. Failures of artificial blood vessels are commonly associated with thrombosis, intimal hyperplasia, atherosclerosis, or infection. Synthetic or tissue engineered vascular grafts have the potential to provide a commendable solution for the future of vascular surgery. Any tissue engineered blood vessels with the ability to grow, remodel, and repair *in vivo*, have noticeable advantages and are of great benefit. Vascular grafts can be of many shapes such as straight, Y-shaped, multi-branch bifurcation, and have 324 mm caliber. Each shape serves some specific treatment purpose. The straight shapes are used in curing arterial diseases, venous diseases, arterio-venous fistula, coronary artery stenosis,

Table 12.1. Pros and cons of biological graft.

Advantages and disadvantages of the biological grafts	
Advantages	Disadvantages
• Closest approximation	• Off-the-shelf availability
• Less diameter mismatch	• Donor site morbidity
• Excellent function	• Harvest injury
• Durable and versatile	• Infection risk
• Better resistance to infection	• Graft deterioration
• Good result	• Chronic rejection

Table 12.2. Applications of various biomaterial-based vascular graft.

Sl. no.	Types	Materials used	Application area
1	Natural material-based vascular grafts	Decellularized human iliac vein seeded with autologous cells	Graft as an extra-hepatic portal vein bypass.
		Decellularized bovine carotid artery graft	Lower extremity bypass surgery
2	Synthetic material-based vascular grafts	Decellularization of PGA scaffolds seeded with cadaver SMCs.	Arteriovenous (AV) shunt for heamodyalisis
3	Self-assembled vascular grafts	Cell-sheet of human fibroblast in a shape of conduit. ECs were seeded in the graft after devitalization of the luminal side	AV shunt for heamodyalisis access
		Cell-sheet of human fibroblast in a shape of conduit, without further endothelialization. Dehydrated and stored (−80 °C) before clinical application.	AV shunt for heamodyalisis access

bypass surgery etc. Y-shaped grafts are used for abdominal aortic aneurysm, or endovascular exclusion. The multi-branch bifurcation type is mainly used for reconstruction in heart aortic stenosis. A general overview of the applications of the vascular grafts gives a clear idea on the pathological conditions leading to the use of specific grafts. The most common application is coronary artery bypass graft. It is used when one of the arteries of the patient's heart is blocked. The surgeon has to harvest arteries from other areas of the patient's body. The next area of application is peripheral bypass grafts. This finds application with occurrence of atherosclerosis that narrows down the arteries. Another branch for application of the vascular prosthesis is kidney dialysis shunts. The natural shunt wears out due to repeated needle pricking by physicians. In this case, the artificial dialysis shunts serve an easily accessible vessel with fast moving blood for people on dialysis. The major applications of these prosthetic or tissue engineered blood vessels are summarized in table 12.2.

References and further reading

[1] Pappenheimer J R and Bohr D F 1980 *Handbook of Physiology: A Critical, Comprehensive Presentation of Physiological Knowledge and Concepts. Section 2, The Cardiovascular System, vol 2, Vascular Smooth Muscle* (Bethesda, MD: American Physiological Society)

[2] Drake C J 2003 Embryonic and adult vasculogenesis *Birth Defects Res. C Embryo Today* **69** 73–82

[3] Zygmunt M, Herr F, Munstedt K, Lang U and Liang O D 2003 Angiogenesis and vasculogenesis in pregnancy *Eur. J. Obstet. Gynecol. Reprod. Biol.* **110** S10–8

[4] Carmeliet P 2003 Angiogenesis in health and disease *Nat. Med.* **9** 653–60
[5] Charnock-Jones D S, Kaufmann P and Mayhew T M 2004 Aspects of human fetoplacental vasculogenesis and angiogenesis. I. Molecular regulation *Placenta* **25** 103–13
[6] Fraser H M and Duncan W C 2005 Vascular morphogenesis in the primate ovary *Angiogenesis* **8** 101–1016
[7] Fraser H M and Lunn S F 2000 Angiogenesis and its control in the female reproductive system *Br. Med. Bull.* **56** 787–97
[8] Girling J E and Rogers P A 2005 Recent advances in endometrial angiogenesis research *Angiogenesis* **8** 89–99
[9] Brown M D and Hudlicka O 2003 Modulation of physiological angiogenesis in skeletal muscle by mechanical forces: involvement of VEGF and metalloproteinases *Angiogenesis* **6** 1–14
[10] Williams J L, Weichert A, Zakrzewicz A, Da Silva-Azevedo L, Pries A R, Baum O and Eggington S 2006 Differential gene and protein expression in abluminal sprouting and intraluminal splitting forms of angiogenesis *Clin. Sci. (Lond)* **110** 587–95
[11] Gould J, Aramburo C, Capdevielle M and Scanes C G 1995 Angiogenic activity of anterior pituitary tissue and growth hormone on the chick embryo chorio-allantoic membrane: a novel action of GH *Life Sci.* **56** 587–94
[12] Gerhardt H and Betsholtz C 2003 Endothelial-pericyte interactions in angiogenesis *Cell Tissue Res.* **314** 15–23
[13] Hershey J C, Baskin E P, Glass J D, Hartman H A, Gilberto D B, Rogers I T and Cook J J 2001 Revascularization in the rabbit hindlimb: dissociation between capillary sprouting and arteriogenesis *Cardiovasc. Res.* **49** 618–25
[14] Kakisis J D, Breuer C and Sumpio B E 2005 Artificial blood vessel: The holy grail of peripheral vascular surgery *J. Vasc. Surg.* **41** 349–54
[15] Lloyd-Jones D *et al* 2010 Heart disease and stroke statistics—2010 update: a report from the American Heart Association *Circulation* **121** e46–215
[16] Mitchell S L and Niklason L E 2003 Requirements for growing tissue-engineered vascular grafts *Cardiovasc. Pathol.* **12** 59–64
[17] Nerem R M 2000 Tissue engineering a blood vessel substitute: The role of biomechanics. National Center for Biotechnology Information. National Center for Biotechnology Information *Yonsei Med. J.* **41** 735–9
[18] Weinberg C B and Bell E 1986 A blood vessel model constructed from collagen and cultured vascular cells *Science* **231** 397–400
[19] Hoerstrup S P, Kadner A, Breymann C, Maurus C F, Guenter C I, Sodian R, Visjager J F, Zund G and Turina M I 2002 Living, autologous pulmonary artery conduits tissue engineered from human umbilical cord cells *Ann. Thor. Surg.* **74** 46–52
[20] Hashi C K, Zhu Y, Yang G-Y, Young W L, Hsiao B S, Wang K, Chu B and Li S 2007 Antithrombogenic property of bone marrow mesenchymal stem cells in nanofibrous vascular grafts *Proc. Natl. Acad. Sci. USA* **104** 11915–20
[21] Hoerstrup S P, Zünd G and Sodian R 2001 Tissue engineering of small caliber vascular grafts *Eur. J. Cardiothorac. Surg.* **20** 164–9
[22] Campbell G R and Campbell J H 2007 Development of tissue engineered vascular grafts *Curr. Pharm. Biotechnol.* **8** 43–50
[23] Zhang W J, Liu W and Cao Y 2007 Tissue engineering of blood vessel *J. Cell. Mol. Med.* **11** 945–57

[24] Ziegler T and Nerem R M 1994 Tissue engineering a blood vessel: regulation of vascular biology by mechanical stresses *J. Cell. Biochem.* **56** 204–9
[25] Syedain Z H, Meier L A, Bjork J W, Lee A and Tranquillo R T 2010 Implantable arterial grafts from human fibroblasts and fibrin using a multi-graft pulsed flow-stretch bioreactor with noninvasive strength monitoring *Biomaterials* **32** 714–22
[26] Wagenseil J E and Mecham R P 2009 Vascular extracellular matrix and arterial mechanics *Physiol. Rev.* **89** 957–89
[27] Hong Y, Ye S-H, Nieponice A, Soletti L, Vorp D A and Wagner W R 2009 A small diameter, fibrous vascular conduit generated from a poly(ester urethane)urea and phospholipid polymer blend *Biomaterials* **30** 2457–67
[28] Webb A R, Macrie B D, Ray A S, Russo J E, Siegel A M, Glucksberg M R and Ameer G A 2007 *In vitro* characterization of a compliant biodegradable scaffold with a novel bioreactor system *Ann. Biomed. Eng.* **35** 1357–67
[29] L'Heureux N, Dusserre N, Marini A, Garrido S, de la Fuente L and McAllister T 2007 Technology insight: the evolution of tissue-engineered vascular grafts—from research to clinical practice *Nat. Clin. Pract. Cardiovasc. Med.* **4** 389–95
[30] Soletti L, Hong Y, Guan J, Stankus J J, El-Kurdi M S, Wagner W R and Vorp D A 2010 A bilayered elastomeric scaffold for tissue engineering of small diameter vascular grafts *Acta Biomater.* **6** 110–22
[31] Drilling S, Gaumer J and Lannutti J 2009 Fabrication of burst pressure competent vascular grafts via electrospinning: effects of microstructure *J. Biomed. Mater. Res.* A **88** 923–34
[32] Wang S, Zhang Y, Wang H, Yin G and Dong Z 2009 Fabrication and properties of the electrospun polylactide/silk fibroin-gelatin composite tubular scaffold *Biomacromolecules* **10** 2240–4
[33] Nieponice A, Soletti L, Guan J, Deasy B M, Huard J, Wagner W R and Vorp D A 2008 Development of a tissue-engineered vascular graft combining a biodegradable scaffold, muscle-derived stem cells and a rotational vacuum seeding technique *Biomaterials* **29** 825–33
[34] Shaw J A, Kingwell B A, Walton A S, Cameron J D, Pillay P, Gatzka C D and Dart A M 2002 Determinants of coronary artery compliance in subjects with and without angiographic coronary artery disease *J. Am. Coll. Cardiol.* **39** 1637–43
[35] van Andel C J, Pistecky P V and Borst C 2003 Mechanical properties of porcine and human arteries: Implications for coronary anastomotic connectors *Ann. Thorac. Surg.* **76** 58–64 discussion 64–5
[36] Roeder R *et al* 1999 Compliance, elastic modulus, and burst pressure of small-intestine submucosa (SIS), small-diameter vascular grafts *J. Biomed. Mater. Res.* **47** 65–70
[37] Laurent S, Caviezel B, Beck L, Girerd X, Billaud E, Boutouyrie P, Hoeks A and Safar M 1994 Carotid artery distensibility and distending pressure in hypertensive humans *Hypertension* **23** 878–83
[38] Tai N R, Giudiceandrea A, Salacinski H J, Seifalian A M and Hamilton G 1999 *In vivo* femoropopliteal arterial wall compliance in subjects with and without lower limb vascular disease *J. Vasc. Surg.* **30** 936–45
[39] Greenwald S E and Berry C L 2000 Improving vascular grafts: the importance of mechanical and haemodynamic properties *J. Pathol.* **190** 292–9
[40] Wise S G, Byrom M J, Waterhouse A, Bannon P G, Ng M K C and Weiss A S 2011 A multilayered synthetic human elastin/polycaprolactone hybrid vascular graft with tailored mechanical properties *Acta Biomater.* **7** 295–303

[41] Dahl S L *et al* 2011 Readily available tissue-engineered vascular grafts *Sci. Transl. Med.* **3** 68ra9
[42] Herring M, Baughman B S, Glover J, Dilley R, Evan A and Gardner A 1984 Endothelial seeding of Dacron and polytetrafluoroethylene grafts: the cellular events of healing *Surgery* **96** 745–55
[43] Alobaid N, Salacinski H J, Sales K M, Kannan R Y, Hamilton G and Seifalian A M 2006 Nanocomposite containing bioactive peptides promote endothelialisation by circulating progenitor cells: An *In vitro* evaluation *Eur. J. Vasc. Endovasc. Surg.* **32** 76–83
[44] Anderson J M 2001 Biological responses to materials *Ann. Rev. Mater. Res.* **31** 81–110
[45] Konig G *et al* 2009 Mechanical properties of completely autologous human tissue engineered blood vessels compared to human saphenous vein and mammary artery *Biomaterials* **30** 1542–50
[46] Solan A, Dahl S L and Niklason L E 2009 Effects of mechanical stretch on collagen and cross-linking in engineered blood vessels *Cell Transplant* **18** 915–21
[47] Sarkar S, Hillery C, Seifalian A and Hamilton G 2006 Critical parameter of burst pressure measurement in development of bypass grafts is highly dependent on methodology used *J. Vasc. Surg.* **44** 846–52
[48] Isenberg B C and Tranquillo R T 2003 Long-term cyclic distention enhances the mechanical properties of collagen-based media-equivalents *Ann. Biomed. Eng.* **31** 937–49
[49] Zong L and Ji X 2004 The application of nonwoven materials for surgical implants *Nonwovens* **9** 12
[50] Jia L, Wang L and Lin K 2003 The developing course and direction of artificial blood vessel *Shanghai Textile Sci. Technol.* **6** 31
[51] Guesnet P *et al* Blood lipid concentrations of docosahexaenoic and arachidonic acids at birth determine their relative postnatal changes in term infants fed breast milk or formula *Am. J. Clin. Nutrition* **70** 292–8

IOP Publishing

Flow Dynamics and Tissue Engineering of Blood Vessels

Arindam Bit and Jasjit S Suri

Chapter 13

Control aspects of the circulatory system

Arindam Bit, Jasjit S Suri and Khemraj Deshmukh

Primarily, the circulatory system is designed to perform functions such as to transport O_2 and metabolites, CO_2 and waste etc, according to the different demands of the body; the blood flow is adjusted by itself. This demand may vary with a step-up in metabolic rate during work out, changes in ambient temperature, and many other factors.

On the other hand, distribution of pressure throughout the system must guarantee the capillary pressure stays close to the optimal level for diffusion of O_2, CO_2, and metabolites. Proper distribution of both pressure and flow can be achieved by regulating the heart and the vascular system with nerve control and humoral agents.

13.1 Introduction

Many regulatory mechanisms in the circulatory system have not been investigated from a system engineer point of view so far. Details of auto-regulation of flow in the atrial bed, control of local flow according O_2 to concentration in blood, and the central mechanism involved in a work out are poorly understood. These are the areas of investigation where strong cooperative relations between physiologists, medical scientists, and engineers are required.

For the better understanding of the behavior of a biological system, the automatic control approach may help significantly such as in the circulatory system. The introduction of a controlling system in the cardiovascular model is only possible if a quantitative explanation is available. This means that at least the transfer relation between variables sensed by the measuring device and the final parameter values set by this system must be known quantitatively. Quantitative information on the influence of variation in individuals and combined parameters on the variables of the system should form the basis for studies on control of the circulatory system.

This work deals with a mathematical explanation of the circulatory system. The relationship between pressure, volume and flow in various parts of the system will be derived. Mathematical formulation of different control systems whose main task is

doi:10.1088/978-0-7503-2088-7ch13

the control of blood pressure and flow will be given. Circulatory behavior during a left ventricular (LV) bypass will be discussed based on a mathematical model. LV bypass aims at allowing the failing heart to recover by reducing the work of the heart while maintain the circulation. This can be achieved by implanting an artificial pump in parallel with the left heart. Circulatory behavior is discussed, because this is an outstanding example of multi-level integrative control of cardiovascular performance by the nervous system under a drastic change of environment.

13.2 Passive control system of circulatory system

With the help of the cardiovascular system we can transport CO_2, O_2, nutritive and waste material, heat and hormones to the various part of body. Figure 13.1 shows the rough circulatory system.

The circulation system consists of two pumps equivalent to the left heart and right heart and two circuits as a systemic and pulmonary organized series. The discussion of the passive circulatory system deals with the hemodynamic aspects of the systemic and pulmonary vascular tree. At the end, the system is divided into numerous segments. For every segment, the pressure–volume relationship, among inflow–outflow volume and pressure, outflow and pressure in the next segments are derived.

13.2.1 The ventricles

A description of hemodynamic behavior of the ventricles basically relates ventricular volumes, pressure and outflow. The relationship between volumes and pressure is time-dependent, because contraction and relaxation of heart muscles are time dependent process. If we plot the graph than it is concluded that pressure versus volume for different preloads and afterloads under constant inotropic action [7–9], points on the pressure–volume locus representing a constant time from the start of the systolic period are distributed on or near a line.

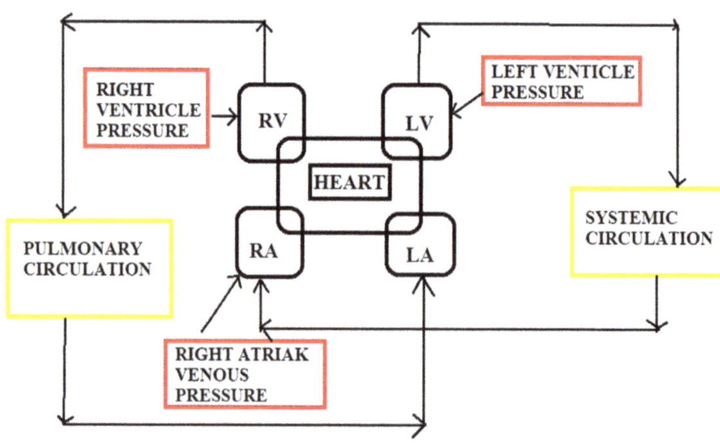

Figure 13.1. A representation of the circulation system.

This means that the contractile nature of each cardiac chamber is characterized by an instantaneous elastance. Equation (13.1) represents the rough relationship between pressure and volume of ventricles.

$$p = \alpha(t)(v - v_v) \tag{13.1}$$

where $\alpha(t)$ represents the time dependent elastance, which is small in diastole and large in systole, and v_v represents the unstressed ventricular volume found at the point of intersection of the tangent. The following relationship between duration of ventricular systole (T_{SYS}) and total cycle length (T_C) is found for the normal operating range, as shown by equation (13.2).

$$T_{SYS=0.28.}\bar{T}_c \quad \text{(Sec)} \tag{13.2}$$

Figure 13.2 represents the pressure volume relationship of ventricles and the time varying ventricular elastance [7].

13.2.2 The atria

An approximation description of the atria is acceptable if it includes end diastolic ventricular volumes and pressures of normal amplitudes and in the appropriate cardiac cycle. The equation for left and right atria is given as described by equations (13.3) and (13.4), respectively:

$$P_{LA}(t) = \alpha_{LA}(t)(V - V_V)_{LA} \tag{13.3}$$

$$P_{RA}(t) = \alpha_{RA}(t)(V - V_V)_{RA} \tag{13.4}$$

where $\alpha(t)$ represents the time dependent elastance and V_V is unstressed atrial volume. The systolic and diastolic value of the two atrial elastances has been determined to yield proper end diastolic ventricular volume and pressure values:

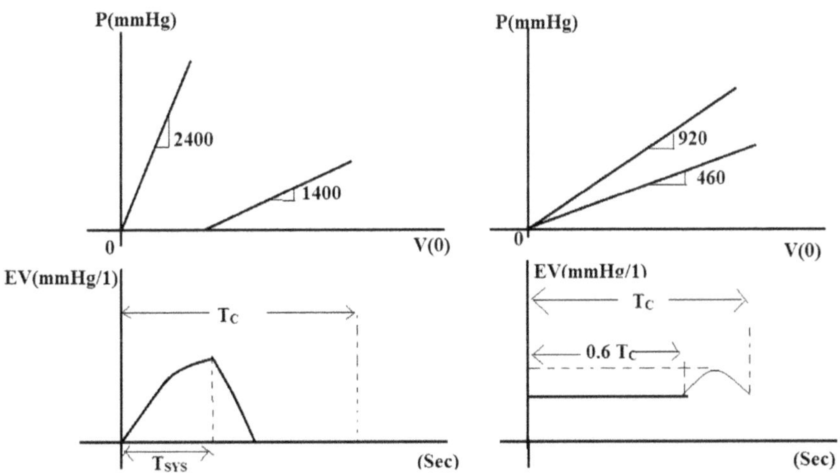

Figure 13.2. Pressure–volume ratio of LV and left atrium (LA).

$\alpha_{LA} = 0.92$ mmHg ml^{-1}
$\alpha_{LA} = 0.46$ mmHg ml^{-1}
$\alpha_{LA} = 0.49$ mmHg ml^{-1}
$\alpha_{LA} = 0.16$ mmHg ml^{-1}

The duration of the atrial systole is considered to be matched to the time between the onset of the P-wave and a point halfway between the R-wave and S-wave. A linear approximation of this duration, T_{AS} is evaluated by equation (13.5).

$$T_{AS} = 0.10 + 0.09T_c \tag{13.5}$$

Figure 13.2 shows the pressure and volume relationship between LA and LV.

13.2.3 The systemic and pulmonary vessels

An equivalent network of a subdivision of a circulatory system is shown in figure 13.4. In a segment of cylindrical elastic, pressure flow dynamics for an incompressible fluid tube may be roughly describe by a pair of differential equations (13.6a) and (13.6b).

$$P_n = R_n(f_n - f_{n+1}) + \frac{1}{C_n}\int (f_n - f_{n+1})dt \tag{13.6a}$$

$$P_n - P_{n+1} = L_{n+1}\frac{df_{n+1}}{dt} + R_{n+1}f_{n+1} \tag{13.6b}$$

which interrelate pressure P_n *and* P_{n+1} and flows f_n *and* f_{n+1} at the end of segments. Here C_n is the compliance, as determined by the dimension and elasticity of the vessel wall. R_n is viscosity coefficient, which is property of the wall material [6]. For all symmetric segments, the product of RC is evaluated to be 0.04 s. L_{n+1} is the inertance, determine by the mass of blood in segment, and R_{n+1} is the viscous resistance term. The capillaries are modeled as ideal purely resistive elements and the superior and inferior vena cava as a series resistance with shunt capacitance. An electric analog of the systemic arteries based on this description consist of 125 segments similar to the one shown in figure 13.3.

Figure 13.4 shows an equivalent circuit of the pressure flow simulation of the complete circulatory system. The effect of blood inertia is not included. This is a closed loop model, i.e., blood volume is converted and the sum of the product of pressure and compliance in each element represents only the portion of total blood volume f_n elastic storage.

Parameters for the systemic and pulmonary loads and valvular resistance have been selected so as to produce a result comparable to dog data. For a dog weight 10 kg, cardiac output is about 1.5 l min^{-1} and mean atrial pressure is around 100 mmHg. Then the total peripheral resistance can be evaluated by equation (13.7).

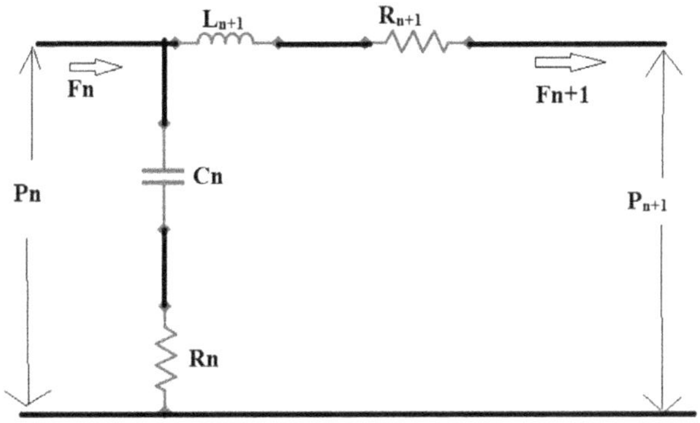

Figure 13.3. Electrical representation of an arterial uncontrolled segment.

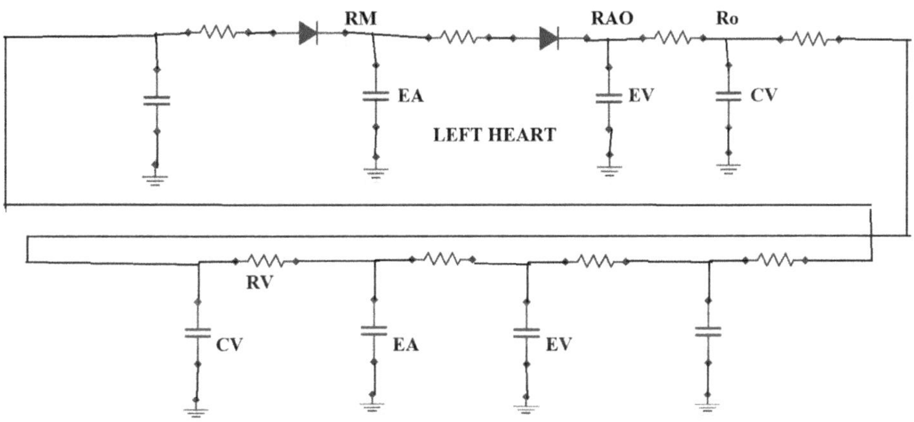

Figure 13.4. Electric analog of the cardiovascular system.

$$\begin{aligned}\text{Total peripheral resistance} &= \frac{\text{Mean arterial pressure}}{\text{Cardiac output}} \\ &= \frac{100\ \text{mmHg}}{1.5\ \text{L min}^{-1}} \\ &= 4000\ \text{mmHg s L}^{-1}\end{aligned} \tag{13.7}$$

The difference between venous pressure and mean RA pressure is about 5 mmHg. Hence venous resistance (R_v) is

$R_v = 5$ mmHg/1.5 L min^{-1}, hence $R_v = 200$ mmHg s L^{-1}

Therefore, atrial resistance (R_A) is 3800 mmHg s L^{-1}. Atrial compliance is calculated from the time constant of change in atrial pressure during the diastolic period. If heart rate is 100 beats min^{-1}, diastolic period is about 400 ms, atrial pressure decreases according to equation (13.8).

$$80 = 120\exp(-0.4/\tau) \tag{13.8}$$

where τ is time constant, thus $\tau = 0.74$ s, and $\tau = R_A C_A$.

$C_A = \tau/R_A = 0.74$ s/3800 mmHg-s $L^{-1} = 0.195 \times 10^{-1} \times 18$ L mm^{-1}-Hg^{-1} $= 3.51 \times 10^{-3}$ L $mmHg^{-1}$

The vascular resistance can be calculated based on pressure drop through the valve. We maintained ventricular pressure near about 4 mmHg at the time of LV ejection period, and this pressure is higher than aortic pressure. The mean flow for a 10 kg dog is 150 mL s^{-1} during valve opening.

$R = 4$ mmHg/150 mL $s^{-1} = 27$ mmHg s L^{-1}

Mitral valve resistance was assumed to be equal to aortic valve resistance.

13.3 Neural and humoral control system

Distribution of pressure throughout the system must guarantee that capillary pressure stays close to the optimal level for the diffusion of O_2, CO_2 and metabolic. The hydraulic pressure required in most capillary beds is between 15 and 30 mmHg and should be maintained even when flow changes dramatically. Capillary pressure depends upon the central aortic pressure, the venous pressure and the resistance to flow at the arteriolar level; also, because of the high resistance of the arterioles small iluctance in central pressure does not appear at the capillary level. For the same reason, the central pressure cannot change rapidly in the response of peripheral dilation. The particular distribution of pressure throughout the vascular system is determined by many different factors.

13.3.1 Baroreceptor reflexes

Factors affecting central arterial pressure can be understood from the simple uncontrolled cardiovascular system, the systematic circulation is represented by arterial storage capacity C_{as}, peripheral resistance R_s, and venous distensibility C_{vs}. An expression for the static flow–pressure relationship can be obtained by equation (13.9) as follows. If cardiac output $I = 0$

$$P_{as}C_{is} + P_{vs}C_{vs} = P_0(C_{as} + C_{vs}) \tag{13.9}$$

where P_0 is pressure assumed to exist in all compartments at $t = 0$ and $I = 0$, the so-called mean circulatory pressure. If $i > 0$

$$P_{as} = R_{s*}I + P_{vs} \tag{13.10}$$

Combination of equations (13.9) and (13.10)

$$P_{as} = R_{s*}I*C_{vs}/(C_{as} + C_{vs}) + P_0 \tag{13.11}$$

From equation (13.11), it is clearly shown that the physiologically required pressure can be maintained in the hydraulically closed system if the product of flow and peripheral resistance and ratio of vascular distensibilities have certain fixed values. Hence if flow should step up as with exercise, resistance must be decremented proportionally. Heart rate is precisely controlled by the data obtained from the vasomotor. These vasomotors are carried to the heart by the vagus nerves and by the sympathetic nervous system. Step up the vagal activity, decrement the heart rate, and vice versa. As sympathetic activity increases, heart rate also increases and enhances contractility of the heart muscle as well. Thus high pressure, large pulse amplitude in the aorta and in the carotid arteries reflexly brings about reduction in pulse rate and contraction, and vice versa, so that pressure is brought back almost to its normal, required value.

13.3.2 Input–output relationship of the baroreceptor

Mean arterial pressure and pressure changes alter the firing rate in the afferent fiber from the pressure sensor. 3Bidirectional rate sensitivity [11] can be expressed as dynamic response of the baroreceptors. The relationship for the discharge rate $a(t)$ observed in an individual afferent fiber is given by equation (13.12):

$$a(t) = c_1\left\{(c_2 - \bar{p})\frac{dp^+}{dt} + (\bar{p} - c_3)\frac{dp^-}{dt} + c_4\frac{p(t) - (c_5\bar{p} + c_6)}{c_7 + p(t) - (c_5\bar{p} + c_6)}\right\} \tag{13.12}$$

where

$\frac{dp^+}{dt}$ and $\frac{dp^-}{dt}$ are the positive and negative pressure changes, respectively;

$\bar{p}$ is pressure average over the past few beats;

$P(t)$ is instantaneous pressure;

C_1–C_7 = constant.

The dependence of firing rate $n(t)$ on positive and negative pressure change and instantaneous pressure is expressed by equation (13.4). It has been observed that $(c_2 - \bar{p})$ is greater than $(c_3 - \bar{p})$, that is, $n(t)$ increases more with a $\frac{dp^+}{dt}$ than it is decreases with a $\frac{dp^-}{dt}$. The third term in equation (13.12) shows saturation. In the steady state ($p(t) = \bar{p}$), the discharge rate n (in steady state) is given by equation (13.13)

$$n = c_4\frac{\bar{p}(1 - c_5) - c_6}{c_7 + \bar{p}(1 - c_5) - c_6} \tag{13.13}$$

Hence, $\frac{c_6}{1 - c_5}$ must be steady state threshold at the baroreceptor

The derivative of n with respect to p (at steady state) is given by equation (13.14).

$$\frac{dn}{d\bar{p}} = c_4\frac{(1 - c_5)c_7}{c_7 + \bar{p}(1 - c_5) - c_6} \tag{13.14}$$

The square wave is used to represent the arterial pressure and heart, by using of peak systolic pressure (P_{sk}) and end diastolic pressure (P_{dk}), we can express the number of discharge for the kth beat (equation (13.15)):

$$N_k = c_{1\left[\{c_2(p_{sk} - p_{dk})\} + \frac{p_{sk} + p_{dk}}{2} - c_3\right]} \tag{13.15}$$

where

$p_{sk} - p_{dk}$ is the pulse amplitude of the kth beat;

$\frac{p_{sk}+p_{dk}}{2}$ is the estimate of mean pressure during the kth beat.

Equation (13.14) neglects the contribution of negative pressure change $(\frac{dp^-}{dt})$ within the kth beat.

13.4 Transformation of afferent baroreceptor signals into heart rate

This section represents a model simulating the observed transfer function between afferent nerve activity and heart rate. It was necessary to incorporate into these models two paths for signal processing. Once could be associated with the parasympathetic, the other with the sympathetic system. Heart rate changes induced by aortic pressure perturbations were small and slow if, initially heart rate was high (3 Hz) and pressure low. In this response the sympathetic system seemed to dominate. Second, at slower heart rate (2 Hz) and higher pressure, the heart was influenced quite quickly and to a larger degree by pressure fluctuations, here the parasympathetic is the most influential.

These two ways of heart rate control could be simulated class by the model presented in figure 13.5. The input to the model is the estimated number of afferent receptor nerve discharges per beat as computed by equation (13.14). This is thought to represent the signal from the sensor in both the carotid sinus and the aortic architecture. The lower path archive for small signal contains a linear, over-damped second order system. The second mode of regulation, occurring when vagal activity dominates, is described by the upper path of the model. Response is governed by a nonlinear, first order system. Capacitance C charges rapidly through the diode but discharges slowly through resistor R_2, causing asymmetry in the response. This path, however, is exclusively engaged only if mean pressure is greater than 80 mmHg.

13.4.1 Control of cardiac contractility

The autonomic nervous control system of ventricular contraction using the change in the systolic peak pressure in an iso-volumetrically contracting heart [3, 4] is

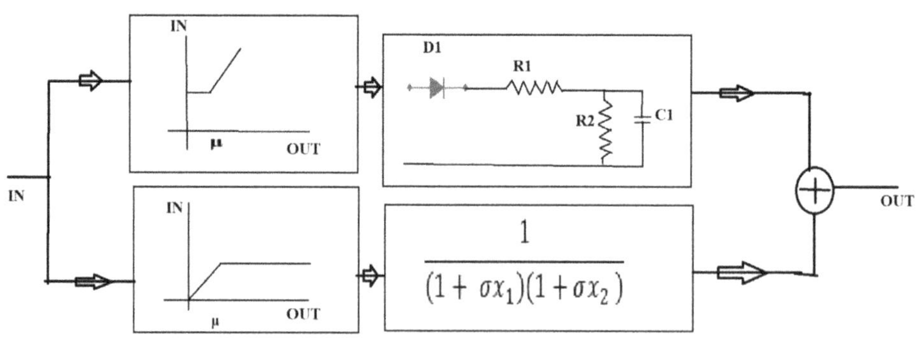

Figure 13.5. Overall relation between the measured model, indicative of afferent firing rate of the baroreceptor system, and the heart period T.

described in this section. Carotid sinus pressure (P_{lcs} and P_{rcs}) left and right is treated as an input for transfer function of receptor, after passing through the transfer function of receptor it is mixed and converted into sympathetic and vagal nerve signals (s′ and v′) inside the central nerve system. After that it passes through a number of transfer functions to change the effect of changes in the left ventricle systolic pressure (LVSP). The sympathetic nerve signal is operated on by different transfer functions depending on conditions such as $\frac{ds'}{dt}>0$, $\frac{ds'}{dt}=0$ or $\frac{ds'}{dt}<0$.

The steady-state control of LVSP is expressed by the following multifunctional equation (13.16)

$$\text{LVSP} = b_0 + b_1 s + b_2 v + b_3 S^2 + b_4 V^2 + b_5 \text{SV} \tag{13.16}$$

where b_0 represents LVSP under no automatic drive and b_1–b_5 are weight coefficient, b_2 and b_5 being negative. S and V are stimulations of sympathetic and vegal nerve.

13.4.2 Sympathetic control of fluid resistance of systematic arterioles

Resistance to study flow of blood is a function of the geometry of a vessel tree. The pressure drops in arterioles, normally the vessel accounts for approximately 75% of the total pressure drop between the aorta and capillaries. Action of the sympathatic nervous system is mediated by norepinephrine released at some nerve endings within the vascular walls, causing smooth muscle to contract, it appears that smooth muscle is most sensitive to concentration of O_2, CO_2 and nutrients in the blood. This locally induced reflex is more important for local flow regulation and thus for distribution of flow in the entire system. It may override the center, sympathetic control of peripheral resistance. The firing rate n_k from the pressure sensor is assumed to be transferred by the vasometer area to influence vascular resistance R as given by equation (13.17).

$$\frac{1}{R} = \frac{1}{R_0}\left\{1 + K\left(\frac{n_k}{n_{k0}} - 1\right)\left(1 - e^{-t/\tau_s}\right)\right\} \tag{13.17}$$

where

R_0 normal resting vascular resistance;

n_{ko} = normal reflecting afferent firing ratio;

τ_s = time constant of response, of the order of 12 s.

According to the above equation, increased nervous influx caused by high aortic pressure decreases the partial resistance R. Approximately 25 s elapse before the new level of R is approached.

13.4.3 Control of systemic venous volume

We know the significant influence of carotid pressure on venous volume [2]. A positive pressure step in the carotid, which decreased vascular resistance by an average of 40%, increased total venous volume by approximately 250 ml. This pooling effect, beginning after a delay of about 40 s, appeared to follow an experimental law with a time constant of approximately 3 min, thus venous return decreased transiently. After reaching steady state the venous distensibility was approximately 70% of its original value.

13.4.4 Model of cardiovascular control loop

The model of cardiovascular control loops has five mechanisms which are simulated as:

1. Heart rate based baroreceptor reflux action;
2. Cardiac contractility based baroreceptor reflux action;
3. Baroreceptor reflux action on peripheral vascular resistance;
4. Control of vascular resistance by side-line chemo-receptors;
5. Baroreceptor reflux action on venous compliance.

In the above system (given by figure 13.6), dead zone, a first order time lag system and a saturation element represent the nerve center system. The dead zone and saturation element show the required onset and saturation effect of the carotid/sinus receptors and the dead zone is 2 mmHg. The first order time lag arrangement in the heart rate control loop is determined based on response of heart rate stepwise change in the carotid sinus pressure [10]. The constants in the first order time lag system in the cardiac contractility control loop were chosen based on the response of LV pressure to change in intrasinus pressure [5]. The constants in the venous compliance control loop were obtained based on measurement of blood volume shift in the systematic vascular system when the pressure in carotid was lowered and raised in steps. The constants in the peripheral resistance control loops were chosen based on change in vascular resistance when transmural pressure in the carotid sinus was changed suddenly by applying sub atmospheric pressure to an air tight box enclosing the neck [1].

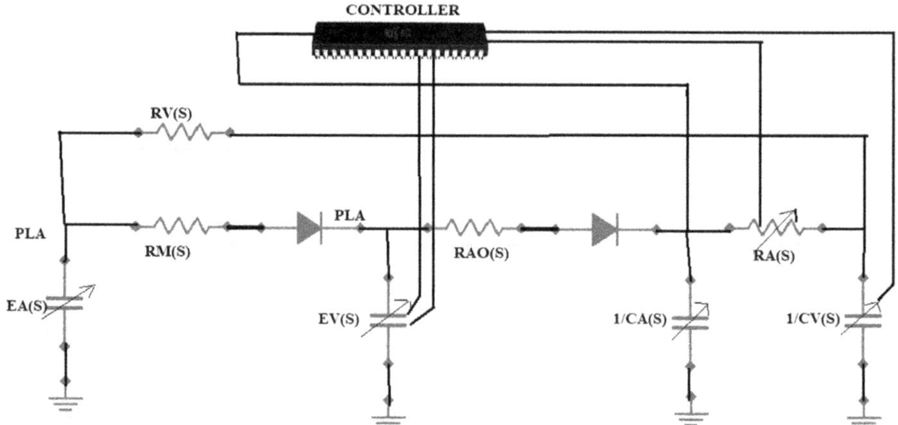

Figure 13.6. The electrical equivalent model of the cardiovascular system.

13.5 Behaviour of the controlled cardiovascular system

Output of the cardiovascular system with LV is very attractive because an assist pump gives unusual hemodynamic conditions. The peripheral blood flow can be controlled with the help of an assist pump which is self-reliant of blood ejection from the natural heart.

13.5.1 Cardiac rhythm caused by the LV assist pump

A self-oscillatory system with external periodic power will oscillate at the same frequency as the external force when the characteristics oscillating period of external force differs somewhat from that of the self-oscillatory system. It is known that the chance of entrainment depends on the power of external force [3].

13.5.2 Animal experiment

The practical was performed on a dog having weight between 10 to 15 kg and also performed on a sheep having weight 28 to 35 kg with the help of chronic experiments we can neglect the domination of anesthesia, intermittent positive pressure and artificial respiration. All these provide side effects on the autonomic nervous system control. Anesthesia was performed with intravenous 25 mg kg^{-1}. after intubation, ventilation was maintained by receptors and a left thoracotomy was performed through the fifth intercostal space. In a dog we can insert the cannula of the assist pump through the LA. And in a sheep we insert the cannula of the assist pump through the apex of LV. The outflow cannula of the assist pump was anastomosed to the downward aorta. In dogs, P-wave was obtained from RA with the help of an electrode and this electrode was injected via the jugular vein into the RA. The electrode was inserted through the epicedium and inter-costal muscle in the sheep. The assist pump operated in several modes according to the obtained ECG. We can calculate the P–P interval and the time interval between the P-waves of ECG to start the assist pump.

With the help of the LV assist pump we can show the cardiac rhythm. When the distance or ratio of the pumping interval to the R–R interval approached 1, then the R–R interval altered and finally the R–R interval synchronized with the pumping interval. The experiment is performed where the arterial pressure is maintained constant and self-reliant of the phase of LVAD driving. The arterial pressure is responsible for changing the interval between P–P. The phase dependency of the mean arterial pressure was acquired in the experiment. The mean arterial pressure exhibited no significant change for different LVAD driving phases.

A macroscopic studies model of an 'external periodic force to self-oscillatory system' shows that the entrainment phenomenon is revealed at the time of LVAD fixed rate pumping. It is capable of being assigned to the sensual habitual heart rate management system. The phase addiction of the sinoatrial (SA) node plays a crucial role in the model and the phasic modification of the SA node by aortic nerve simulation is explained by a simple model. In the model only the vegal effect is included and the sympathetic effect is neglected for simplicity. Parasympathetic fibers, parts of the vegas nerve, originate in the medullary vasomotor area and innervate the atria of the heart. An increased discharge rate results in an increase of acetylcholine release at the nerve endings, its direct effect is to slow down the rate of discharge of SA node. Acetylcholine is hydrolyzed by cholinesterase. Formulation of this depressing action of the parasympathetic system is given below.

The change of x_1 (w.r.t. to time) of acetylcholine at the SA node is proportional to the rate of acetylcholine release minus the rate at which acetylcholine is hydrolyzed, which is proportional to the concentration x_1 at its site of action, and represented by equation (13.18).

$$\frac{dx_1}{dt} = k_2 f_a - k_1 x_1 \tag{13.18}$$

where

k_1 is rate constant of acetylcholine destruction;

f_a is the vegal firing frequency; and

k_2 is a constant of proportionality.

The effects the neural transmitter, acetylcholine, has on the nodal cell is given by equation (13.19):

$$f_a(t) = \exp(-(t - T)^2/K_3) \tag{13.19}$$

where T is the transit time of the impulse obtained from the carotid sinus, $f_a(t)$ is

$$f_a(t) = \exp(-(t - (T + \phi)))^2/K_3 \tag{13.20}$$

The pacemaker potential $x_2(t)$ is given by equation (13.21)

$$\frac{dx_2}{dt} = \frac{x_2}{c + x_1} + \frac{k_4 - v_{hyp}}{c + x_1} \tag{13.21}$$

where

c is basal acetylcholine concentration;

k_4 is the gain of the positive feedback process of the membrane potential depolarization;

v_{hyp} is the hyperpolarization potential in the absolute refractory period.

Equations (13.19), (13.20), and (13.21) were solved by a digital computer, the constant used in these equations are summarized in the below table.

Constant	Value
K_1	0.05 s
K_2	1
K_3	0.02
k	1
p	0.1 s
c	0.122
Tr	0.18 s
V_{hyp}	−90 mV
V_{th}	−60 mV
Control P–P interval	0.6 s

13.5.3 Role of circulatory system in LV by-pass surgery

To maintain the circulation we use a circulatory assist device. It also provides backup for recovery of the failing heart. At the same time we observe the effect of the assist pump. For analyzing we use the blood ejection period (T_S) and delay time.

13.6 Cardiovascular circulatory system and assist pump model

For the simplicity study of the model pulmonary circulation is neglected. A variable capacitance and resistance are connected in series. Both these are coupled with a pneumatically operated diaphragm type assist pump and the diaphragm has elastic properties so that it tenses and is compressed according to oxygen passes. A geometric structure form helps for calculation of inertance of tubes. By using constant pressure (200 mmHg), various delay times and ejection periods, a computer simulation of hemodynamics during the sidestep from LV to aorta is performed. Equations (13.22a), (13.22b), and (13.22c) define the performance index of the assist pump as:

$$TTI = \int_0^{T_c} PLV(t)dt \qquad (13.22a)$$

$$EVR = \int_{T_{sys}}^{T_c} PAO(t)dt/TTI \qquad (13.22b)$$

$$PAOM = \frac{1}{T_c}\int_0^{T_c} PA0(t)dt \qquad (13.22c)$$

where

TTI denotes tension time index;
PLV(*t*) means pressure in *LV*;
EVR means endocardial viability ratio;
PAO(*t*) means the aortic pressure; and
PAOM denotes mean aortic pressure.

EVR is used for myocardial O_2 supply demand ratio. Simulation for both the normal and the failing heart with and without cardiovascular control loop were carried out.

13.7 Conclusion

In physiology, a primary interest lies in regulation and control. With the help of a quantitative description, we can convert the control system into the cardiovascular system. The joint effect of a physiologist and system engineer gives powerful and useful integration of physiological supervision and governance of the circulatory system. Here we are considering baroreceptor reflex control, because well-defined studies have accumulated around this reflex. The interesting response of the baroreceptor reflex to the artificial pump was demonstrated together with the mathematical model. From the above observation we can say that the mathematical structure adds to a quantitative understanding of governance of the cardiovascular system and with the help of these models we can do a quantitative assessment of the circulatory system. Two important points should be considered fom a system engineer's point of view. Although it is difficult to eliminate the central nervous system and the endocrine controller, it is essential for the controller to discriminate the plant achievement and its modification. The other noticeable feature of the approach of this system lies in formulating principles in organized complexity. With respect to its ultimate goal, the focus is placed on figuring out the complete structure output; adding new data, dimensions or constraints may force a huge change in the model.

References

[1] Bevegard B S and shepherd J T 1966 Circulatory effects of stimulating the carotid arterial stretch receptors in man at rest and during exercise *J. Clin. Invest.* **45** 132–42

[2] Braunwald E J R Jr, Ross J Jr, Kahler R L, Gaffney T E, Goldblatt A and Mason D T 1963 Reflex control of the systemic venous bed: effect on venous tone of vasoactive drugs,and of baroreceptor and chemoreceptor stimulation *Cir. Res.* **12** 539–52

[3] Levy M N, Ng M, martin P and Zieske H 1966 Sympathetic and parasympathetic interactions upon the left ventricle of the dog *Cir. Res.* **19** 5–10

[4] Levy M N and Zieske H 1970 Mechanism of synchronization in isorythmic dissociation. I. Experiment on dog *Cir. Res.* **27** 429–43

[5] Martin P J, Levy M N and Zieske H 1969 Bilateral carotid sinus control of ventricular performance in the dog *Cir. Res* **24** 321–37

[6] Noordergraaf A, Vardouw P D and Boom H B K 1963 The use of an analog computer in a circulation model *Progr. Cardiovasc* **5** 419–39

[7] Sawaga K 1973 The circulation and its control II: neural and humoral control of the heart and vessel ed J H U Brown and D S Gann *Engineering Principle in Physiology* vol 2 (New York: Acadmic Press) pp 73–100

[8] Shoukas A A and sagawa K 1973 Control of total systemic vascular capacity by the carotid sinus baroreceptor reflex *Cir. Res.* **33** 22–33

[9] Suga H K, Sagawa and Shoukas A 1973 Load independence of the intentaneous pressure volume ratio of the canine left ventricle and effect of epinephrine and heart rate on the ratio *Cir. Res* **32** 314–22

[10] Topham W S 1967 The control of cardiac output during exercise *Physical Basic of Circulatory Transport* (Philadelphia: W.B. Saunders) pp 77–90

[11] Warner H R 1964 Simulation as a tool for biological research *Simulation* **1** 57–63

Chapter 14

Control aspects of a heart assistive device

Arindam Bit, Jasjit S Suri and Khemraj Deshmukh

The main focus for design of a circulatory assistance technique is to achieve rapid and stable recovery of the cardiac pumping function. It is a little difficult to get reproducible results and to evaluate the work of an assistive pump (AP) over a wide range of internal and external conditions in the experimental studies of processes associated with AP/organism interaction. Therefore, experimental studies have been performed simultaneously with the application of physical and mathematical modeling.

14.1 Introduction

The basic concept of an AP control system is to receive current information from the field (body), and according to this information and set point values, we find an error signal, and that error signal provides tight regulation in the system. If we follow this step then the requirement is for a biologically informative signal and the AP control system is defined to provide effective, strong and safe pump operation (figure 14.1).

14.2 AP phase control

For all the methods of pulsatile assistive circulation, the main criteria for safe operation is orthogonal of the AP, and the cardiac output (CO) is represented by equation 14.1:

$$\int_0^T Q(t)Q_P(t)dt = 0 \tag{14.1}$$

where

$Q(t)$ is the blood flow from the left ventricle (LV) into the aorta;
$Q_P(t)$ is the blood flow from the AP into the aorta;
T is the cardiac cycle.

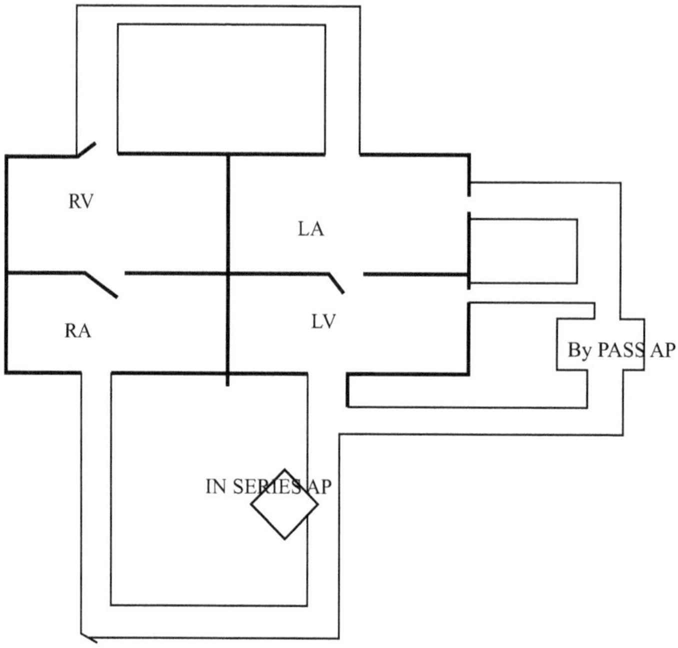

Figure 14.1. Two way mode of connection of an assist pump (AP) to the cardiovascular system.

If the condition (equation (14.1)) does not match, coincidence in output phase from the LV and the AP raises hydraulic cardiac load which may result in major complications, even heart arrest. From equation (14.1), the AP safety operation which is connected in series with LV should be defined by equation (14.2).

$$K_S = 1 - \int_0^{T_S} Q_P(t)dt / \int_O^{T_S} Q(t)dt = 0 \tag{14.2}$$

where T_s is the LV systolic duration.

Some amount of blood from the LV is bypassed if we connect the AP to the LV-aorta or left atrium (LA)-aorta. It is mandatory to calculate volume of bypassing for the calculation of K_s with the equation (14.3).

$$\alpha = \int_0^{T} \{Q(t) + Q_P(t)dt\} / \int_0^{T} Q(t)dt \tag{14.3}$$

The safety coefficient for the AP can be defined by equation (14.4).

$$K_S = 1 - \int_0^{T} Q_P(t)dt / \alpha \int_O^{T} Q(t)dt \tag{14.4}$$

Equation (14.4) shows that if α increases, the value of K_s also increases and orthogonally can be neglected for the lager value of α [1, 2]. In order to evaluate the efficiency of an AP connected in series, endocardium vitality ratio EVR is realized in the Data-scope-80 apparatus [3], designed as followed by equation (14.5):

$$EVR = DPTI/TTI \tag{14.5}$$

where

TTI is the tension-time index,
DPTI is the diastolic pressure-time index.

The conditions to appraise the productiveness of intra-aortic balloon pumping are:

$$E = C_1\bar{P}_d + C_2\bar{P}_S + C_3(P^0_{ed} - P_{ed})^2 \tag{14.6}$$

where

$\bar{P}_d$ = is mean diastolic aortal pressure;
$\bar{P}_S$ = is mean systolic aortal pressure;
P^0_{ed} is the desired end diastolic pressure;
P_{ed} is the end diastolic aortal pressure;
C_1, C_2, C_3 are empiric coefficients.

The automatic control system maximizes a functional *E* by regulating the delay and duration of the intraaortic balloon pump operation. To meet condition of AP and LV matching, information on phase structure of cardiac activity should be available. For this aim, most investigators use the QRS complex of ECG to define the beginning of an electric systole of the heart, which precedes a mechanical contraction [4].

14.3 Amplitude AP control

The problem of amplitude control is not a big issue for in-series AP, but it is critical for bypass. It is possible to establish/show that the fruitful off-loading of myocardium with a series AP can be achieved at equal LV and AP stroke volume (SV). Considering a dependence of the aortal volume and arterial pressure linear [6], the following can be written by equation (14.7)

$$P_{a = 1/c_a(v_a - v_{ao})} \tag{14.7}$$

where

$1/c_a$ is the volume aortal elasticity;
v_a is the aortal volume;
v_{ao} is the unstressed aortal volume.

Equation (14.7) gives the information about the aortal volume step-up due to a respective pressure step-up proportional to the volume elasticity [5].

The most serious issue of the control system for assisted circulation with the help of series APs is safe and precise cardiovascularization. In the case of a bypassed AP control system the major issue is matching the amplitude of AP and LV in the range of change in the pumping function of the heart and the venous return. But the bypass AP control is a complicated issue because there is no theoretical concept available in previous studies. Most authors give the solution for manual AP control with the help of a change in pressure throughout injection and in vacuum during inflation.

To estimate the method for the bypassed AP control we can carry out an analysis of the variation in the starting inflection reflecting a dependence of CO on atrial pressure under AP. any step-up in VR under continuous AP output can be realized only by a step-up in CO [7].

The absolute cardiac AP output curve $\sum Q$ is the heart curve removed upwards by a value of Q_P. an increase in *VR* is associated with the increase in atrium pressure which restricts *VR*. At the time of hard heart failure, the negative feedback plays an important role. On the basis of error signal the control system provides the concept of stabilization of heart atrial pressure, of a given and measured atrial pressure controlling AP output [8].

This is a vertical line crossing an atrial pressure axis in the point of given value of P_0. Any increase in *VR* is a response to an increase in output. The habilitation is an interpretation of the method based on SV control, proportionally to atrial pressure as shown in equation (14.8) [11]

$$Q = \beta P_{at} + Q_0 \tag{14.8}$$

where

β is gain;

Q_0 is the pump output at $P_{at} = 0$.

With the help of gain β we can find the slope of output characteristics of the AR. The response of APs is to change atrial pressure. All these methods of bypassed AP control can be applied in practice. The development of the mathematical circulatory model capable of reproducing heart failure would allow an optimal option of the algorithm for bypassed AP control. Before the model is selected and its main units and links are defined, a brief mutative diagnostic of hemodynamic shifts at heart failure should be considered using the bioengineering analysis of a dependence of CO on atrial pressure.

14.4 Bioengineering analysis of heart failure

LV and right ventricle (RV) and bilateral heart failure can occurr due to structural and functional disorder. This decrease in the pumping function of the LV, which is related with the reality that the RV is endured with the initial output, results in a step-up in the left atrial (LA) pressure. As a result, the pressure is transferred to the pulmonary artery augmentation of the RV load and reduces its output. A step-up in pulmonary circulation pressure may be to antecedent pulmonary edema. A disturbance in pulmonary function, under this condition, results in fall off in oxygen saturation of the arterial blood involving further disturbance in the state of the organism and the myocardium (second hard cycle). In addiction, the boosted RV load leads gradually to degradation of its pumping function (third hard cycle). At the RV ventricular failure secondary to temporary operation of LV with output higher than that of the right one the blood is propelling out of the pulmonary circulation of the systematic circulation. However, the vascular capacity of systematic circulation is 6–7 times as much as that of the pulmonary circulation. If a step-up in the mean pressure and RA pressure would be relatively small, we observe a degradation in the left atrial pressure due to degradation in blood volume pulmonary circulation. The LV output would degrade to a value equal to the right ventricle failure, total blood flow degrades at the minimum change in the left and right atrial pressure. The bilateral heart failure encourages the pressure step-up in both atrial pressures, but in the LA to a considerably higher level. Consider hemodynamic shift originating from heart failure when a bypassed AP is continued. Degradation in the LA pressure reduces the RV load of the pulmonary edema due to AP switching on.

The two important facts, however, should be considered:

1. degradation in LV output is only when the operating point leaves the area of saturation of the CO curve when we start the bypassed AP;
2. secondly, the step up in total flow results in a step-up in arterial pressure and therefore in an additional LV load.

Thus the bioengineering analysis allows, if AP is on, the qualitative development of hemodynamic modification to take place under heart failure condition. Fuller analysis can be performed with the mathematical heart failure model intended for the evaluation of effectiveness of various control algorithms.

14.5 Automatic bypassed AP control system

The servo system utilizes a pressure sensor in the atrium in the feedback circuit to realize a selected AP control algorithm [8, 9]. If the systems fail to provide long term and safe AP work, a pressure sensor must be inserted in the blood flow. The pulsatile pressure located in the atrium demands the use of an averaging filter, giving additional delay. The system is in unstable mode onto the dynamic load [10]. We consider a separate phase of bypassed AP operation in detail assuming that when the AP is completely filled then injection always happens. In order to get the orthogonal condition, the inflation phase should begin after closing the aortal valve and the effect from the delay of R-wave of the ECG. Inflation duration is traced by an air pressure pulse time AP immediately after the injection; the inflation phase is run until the AP fills up to a fixed level. Then, the AP is waiting for the next air pressure pulse. This period in which change in AP volume equals zero, can be called a 'zero-th phase'.

Define a desired inflation duration τ_i equal to a period between the end of an anterior inflation phase and the onset of posterior injection phase through equation (14.9)

$$\tau_i = T - \tau_o \tag{14.9}$$

where

T is cardiac circle;

τ_o is inflation duration.

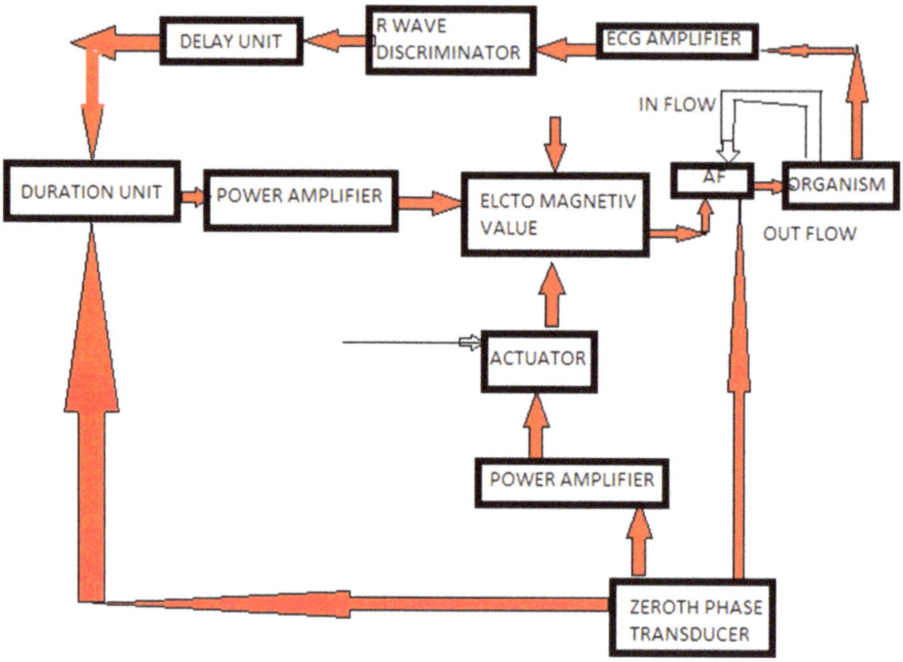

Figure 14.2. Principle of AP control.

True inflation duration, τ_i, corresponds with a period between the end of an injection phase and the moment of fixed inflation of the AP. Thereby a period of the zero-th phase is defined by equation (14.10).

$$\Delta\tau = \tau_i^* - \tau_i \tag{14.10}$$

The value of $\Delta\tau$ may be defined with an analysis of diastolic phase of the air pressure curve in the AP pneumatic chamber on the air flow curve [12, 13]. The zero-th phase can be eliminated by SV control. Figure 14.2 shows the principle of AP control.

With the help of injection pressure $P_{s,}$ the SV of the system is regulated, according to an error signal $\Delta\tau$. In figure 14.2 SV is regulated proportionately to the atrial pressure. The SV is defined by equation (14.11):

$$q_{\text{out}} = \int_0^T Q_{P(t)dt} \tag{14.11}$$

where $Q_P(t)$ is the instantaneous blood flow from AP into the aorta.

Equation (14.11) is further simplified to equation (14.12) as:

$$q_{\text{out}} = \bar{Q}_P\tau_c \tag{14.12}$$

where $\bar{Q}_P$ is the average flow form AP into the aorta. The following equation (14.13) can be written neglecting delay losses

$$\bar{Q}_P = \frac{\bar{P}_0 - \bar{P}_a}{R_{\text{out}}} \tag{14.13}$$

where

$\bar{p}_0$ = is mean injection pressure;
$\bar{p}_a$ = is mean aortal pressure;
R_{out} is the AP output resistance.

From equations (14.12) and (14.13), the following equation (14.14) can be defined

$$q_{\text{out}} = (p_0 - p_a)\tau_o / R_{\text{out}} \tag{14.14}$$

so the value of q_{out} is defined by aortal pressure, AP output resistance and injection pressure at τ_o = const. Similarly, the inflation volume can be evaluated using equation (14.14).

$$\mathrm{q}_{\text{in}} = (\bar{p}_{\text{LA}} - \bar{p}_i)\tau_{\mathrm{i}}/\mathrm{R}_{\text{in}} \tag{14.15}$$

where

$\bar{p}_{\text{LA}}$ is mean left atrial pressure in inflation phase;
$\bar{p}_i$ is the inflatica vacuum;
R_{in} is the AP input resistance.

The following equation (14.16) can be written taking into account that the injection always follows the fixed inflation:

$$\tau_i = \frac{q_{\text{out}} R_{\text{in}}}{\bar{p}_{\text{LA}} - \bar{p}_i} \tag{14.16}$$

Zero-th phase can be defined by equation (14.17).

$$\Delta\tau = \tau_i^* - \frac{q_{\text{out}} R_{\text{in}}}{\bar{p}_{\text{LA}} - \bar{p}_i} \tag{14.17}$$

If $\bar{p}_i = 0$ and the inflow process occurs only on account of the atrial pressure and $\Delta\tau$ *tends to* 0, the following equation (14.18) is available for the static condition

$$q_{\text{out}} = \bar{p}_{\text{LA}} \tau_i^* / R_{\text{in}} \tag{14.18}$$

Thus any change in the atrial pressure results in the corresponding change in SV. Thereby the slope of AP pumping characteristics would be defined by the τ_i^*/R_{in} ratio. According to equation (14.18), invariance of SV to aortal pressure is an additional advantage of the system.

14.6 Fluid mechanical simulator to cardiovascular system

The cardiovascular system can be used as a fluid circulatory system. This system consists of two pumps connected in series with systematic and pulmonary circuits. In this section, designing a mechanical simulator of the total cardiovascular system has been represented. The objective of the chapter is to test the function of various artificial organs as well as determining their optimal drive control method. This section provides the knowledge of the system design and the basic properties of this simulator.

The basic conditions of our circulatory simulator are listed below.

1. Simulation of the cardiovascular system of a 20 kg dog: in designing the model, we combine similar elements of the cardiovascular system together, on the basis of physiological and anatomical data. Consequently, the simulator has two pulsatile pumps (corresponding to the right and left heart), an elastic pump (blood vessel) and peripheral resistance (with the help of this we can get suitable pressure drop).
2. Macroscopic cardiovascular system as a completely closed loop: alternately we use water (20 °C) in the place of blood, making its volume equal to the total blood volume of a 20 kg dog (1600 ml). If we fix standard pump flow equivalent to the cardiac output from the natural heart, the travelling time of the total circulation will be similar in both the simulator and the natural circulation. These characteristics make easy the process of examining the response of dynamic behavior of the total circulatory system.
3. Simulation of the living body feedback mechanism including automatic control: systemic circulation is affected by the automatic nervous system. However, we cannot simulate all the automatic functions. So we select only aortic pressure control which alters systemic circulation and the control of cardiac output from the left heart, which functions on the basis of the Starling law for our simulator.

Following the above conditions, we have designed a simulator that has a pressure–flow relationship in the main part of the cardiovascular system that performs the same as found in the living/human body. Figure 14.3 shows a simplified diagram of the designed system. It is divided into two parts: (1) a fluid circulatory system to simulate the hemodynamics of the cardiovascular system; and (2) a mini computer system that is used for data acquisition as well as control of the fluid circulatory system.

14.7 Design of a mechanical simulator

The simulator consists of two pulsatile pumps for the right and left heart [17, 18]. We designed the heart pump in a simple axial configuration to improve function instead of shape. The required blood volume for the heart (160 ml, which is 10% of the total circulating blood volume) is divided equally among the four parts. Each pump is made of three plug-ins: housing, elastic bladder and disc type heart valves based on a hydrodynamic study [19, 22]. In the pneumatically driven artificial pump, the elastic bladder of the ventricle is scattered by air pressure, causing the blood to be forced out. The flow of the blood is always in one direction by means of inlet and outlet artificial valves.

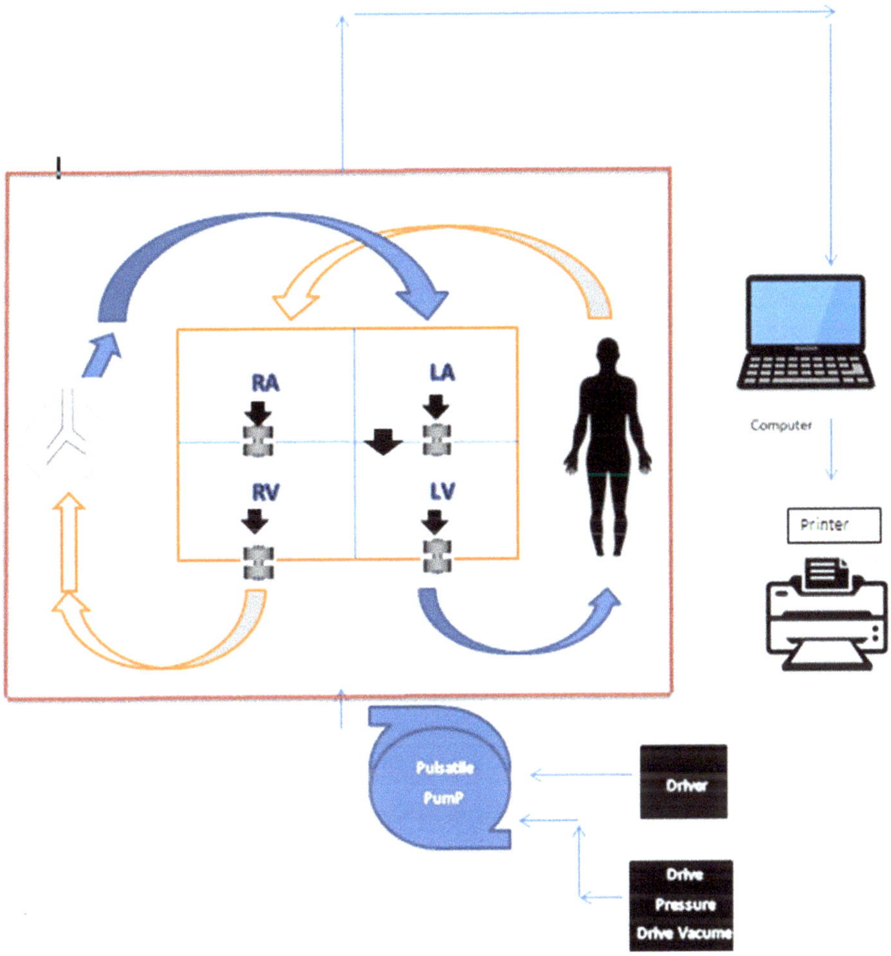

Figure 14.3. Schematic diagram for a mechanical simulator for total circulatory system.

Figure 14.4 indicates our control drive structure for our heart pump. It has pneumatic circuits which are used to adjust the contraction of right and left heart pumps and the systolic/diastolic phase is determined by the electronic circuits.

14.7.1 Design of a vascular system for the mechanical circulatory simulator

According to the physiological data, 10% of the total blood volume (1600 ml) of a 20 kg dog is located within the heart [14]. The remaining blood is distributed between the systematic circulation and the pulmonary circulation in a ratio of 7:1.

In the human body 'blood volume' harmonizes to that of the aorta, vena cava and pulmonary arteries and according to anatomical data we select the ratio of diameter/tube length in each element. In our simulator the vascular components are replaced by elastic rubber tubes. These tubes have the same volume and diameter-to-length ratio as those of the corresponding elements in the living body. To obtain value of

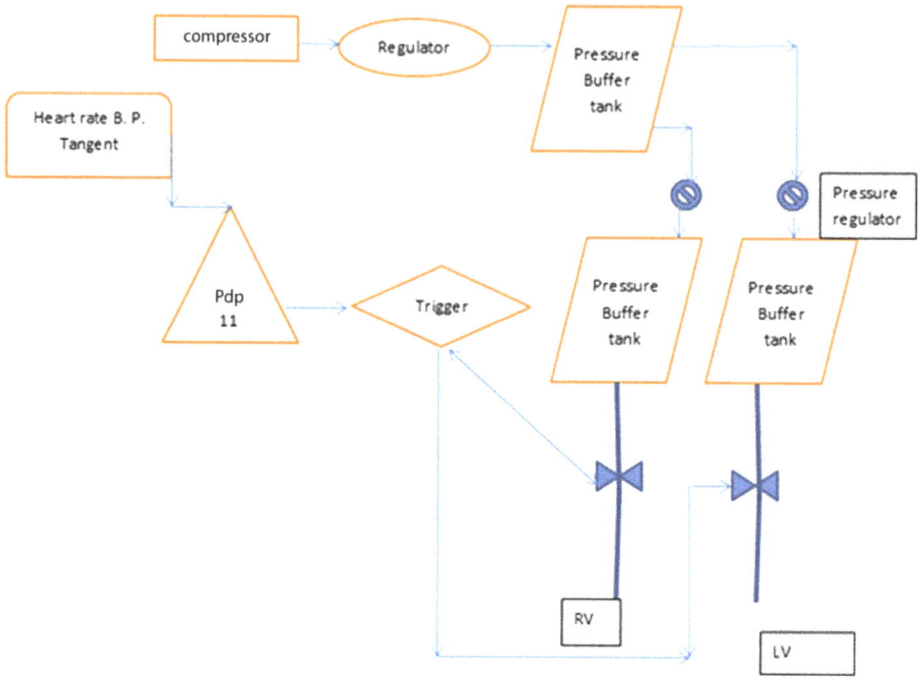

Figure 14.4. Pneumatic circuits for a heart driver.

compliance (dv/dp) and volume pressure component the same as in the human body [23], we must find out the thickness (t) of each elastic tube; we consider that each elastic tube should be thin enough to obtain a linear stress–strain relationship. The step-up in elastic tube diameter is given by following mathematical expression (14.19)

$$\Delta R = \frac{\left(\frac{D}{4}\right)^2 \Delta p}{2Et} X\left(2 - \frac{1}{m}\right) \tag{14.19}$$

where

p is the pressure difference, $\frac{1}{m}$ = poisson ratio = 0.475, E = Young's modulus = 2.45×10^{-4}, t = thickness (cm), R = radius (cm), D = inner diameter (cm). From the above, we can express the increment of the volume (Δv) by equation (14.20):

$$\Delta v = \pi(R - \Delta R)^2 L - \pi R^2 L \tag{14.20}$$

neglecting ΔR^2, we have $\Delta v = \pi LD\,\Delta R$. Thus the compliance ($c$) can be written by equation (14.21).

$$C = \frac{\Delta v}{\Delta p} = \pi l D^3 \frac{\left(2 - \frac{1}{m}\right)}{8Et} \tag{14.21}$$

Table 14.1. Dimension and compliance of elastic tubes used for vessels induced from numerical analysis.

Element	Diameter (mm)	Length (mm)	Thickness (mm)	Compliance (cm^3 dyn^{-1})
Aorta	17	300	0.36	0.99×10^{-4}
Vena cava	20	300	0.025	23.01×10^{-4}
Pulmonary arteries	20	230	0.34	1.34×10^{-4}

The numerical result is shown in table 14.1. We organize compliance test on the latex tubes for a vessel. When we inject water with the help of a syringe, the pressure of the elastic tube suddenly changes and that pressure change is measured.

The compliance of our aorta is close about to the theoretical value, provided that the pressure remains less than 140 mmHg. For our pulmonary arteries, we can calculate the compliance of both 0.3 and 0.4 mm thick tubes. Within the physiological pressure range, the curve of a 0.3 mm tube is closer to the theoretical one.

14.8 Design of control valve system

14.8.1 Aortic pressure control system

In a living body, the cardiovascular system has automatic nervous system control, a function of that automatic nervous system is maintaining the variation in blood pressure to a minimum throughout the system. In this simulator, a pre-capillary resistance was used with the help of which we can handle ‘aortic pressure’. It is very difficult to check out whether the dynamic behaviour of our designed control system closely matches the natural control system. The reason is that the control system of a living body can be considered as a multivariable/multi-parameter control system. Therefore, we have tentatively tried to make identical the characteristics of the designed system with the data calculated according to the mathematical circulatory model [15]. The aortic pressure control mechanism employed in the model senses aortic pressure by means of either a chemoreceptor which is use as pressure transducer or a baroreceptor coupled with a high-pass filter. Then, the signal moves via a low-pass filter for handling of the resistance which is used to control aortic pressure. In our simulator, consideration was given only to the signals from the baroreceptor (act as a pressure transducer), because they have a considerable effect on the variation of arterial pressure. The system consists of both high-pass filter and low-pass filter which is connected in series. From this situation, the system shown in figure 14.5 describes a continuous system for controlling aortic pressure.

In figure 14.5, Q_{in} = left ventricular output, Q_{out} = flow rate across the valve, m is the position of valve lift, p is the aortic pressure, and c is the aortic compliance. This system is characterized by the equation (14.22):

$$Q_{in} - Q_{out} = c\frac{dp}{dt} \tag{14.22}$$

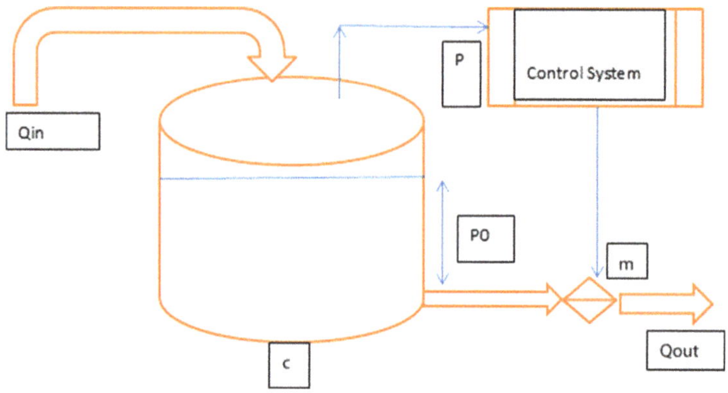

Figure 14.5. Schematic representation of 'aortic' compartment model.

$Q_{out} = Q_{out}(m, p)$. Considering the variation from the equilibrium point, a Laplace transform is made from the above equation in the form of equations (14.23) and (14.24).

$$\Delta Q_{in} - \Delta Q_{out} = sc\Delta p \tag{14.23}$$

$$\Delta Q_{out} = \frac{dQ}{dm} - \left| p_0 \Delta m + \frac{dQ}{dp} \right| m_o \Delta p \tag{14.24}$$

The first term on the right side of equation (14.24) represents the variation in flow rate across the valve relative to the variation in the valve lift position. The second term on the right side of the equation expresses the variation in flow rate across the valve relative to the deviation in pressure that occurs when the valve lift is fixed at the operating position.

The relationship between the baroreceptor and variation in the valve lift was established as follows. The baroreceptor calculates the aortic pressure as a deviation from the equilibrium point is converted to valve lift displacement by a certain function. When the variable range of the valve lift is adjusted to correspond to that of the aortic pressure, there is a change in the extent of manipulation of the system. If we consider the slope in figure 14.6 as $\frac{\Delta m}{\Delta p} = k$, it shows the variation in variable range of the valve lift shifting the gain (k) of the control system. The time constant of the high-pass filter in the aortic pressure control system is large enough to be omitted from the analysis of valve lift control, i.e., the variation in valve lift (Δm) may be expressed by equation (14.25).

$$m = \frac{k}{1 + sT}\Delta p \tag{14.25}$$

where,

$\frac{1}{1+sT}$: linear delay element;

T: time constant.

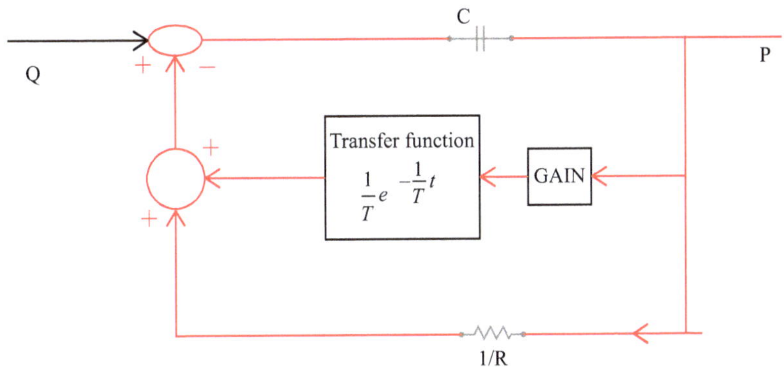

Figure 14.6. Diagram for an 'aortic pressure' feedback control system.

When equation (14.22) is substituted into equation (14.24)

$$\Delta Q_{\text{out}} = \frac{dQ}{dm} - \left| p_0 \frac{k}{1 + sT} \Delta p + \frac{dQ}{dp} m_1 \right| p \tag{14.26-a}$$

$$\Delta Q_{\text{out}} = \left(\frac{dQ}{dm} - | p_0 \frac{k}{1 + sT} \Delta p + \frac{1}{R} \right) \Delta p \, | \tag{14.26-b}$$

where

$$\frac{1}{R} = \frac{dQ}{dp} lm_0$$

With the help of equations (14.22) and (14.25) we made the diagram shown in figure 14.6. A program for the control of pre-capillary resistance in our simulator was then designed. In this program, we can vary the aortic pressure from 35 to 165 mmHg, with the maximum sensitivity of the corresponding range of the needle valve set at 1.0.

In a response characteristics test, the control gain and sensitivity were varied within the range from 0.4 to 1.0 in steps of 0.1. The *S*/*D* ratio of the LV was changed by step input. The test of positive steps inputs will be conducted in the future. The following observation can be drawn:

1. if we provide the variation in sensitivity cause variation in damping ratio and the final value;
2. better stability can be obtained from low sensitivity;
3. it might be taken long time to reach final value due to low sensitivity;
4. critical damping will happen at *S* by 0.5.

It is hardly difficult to calculate the control nature of the recently developed pre-capillary resistance apparatus by animal experiments. Therefore, we conclude that the dynamic response of aortic pressure in the system is illustrated in figure 14.7.

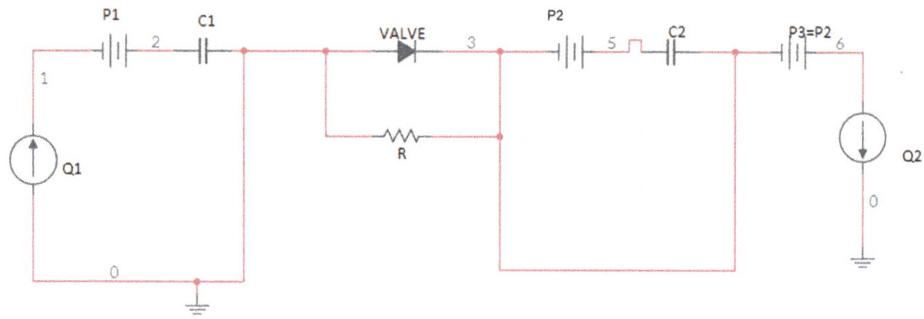

Figure 14.7. Representation of an 'aortic compartment model' which determines 'aortic pressure' under automatic control.

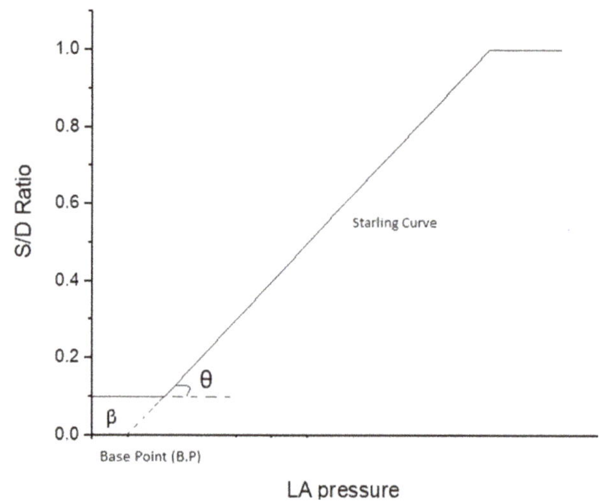

Figure 14.8. Determine of the 'Starling curve'.

The results of mathematical simulation and those of the mechanical simulation were compared in order to select the optimum control sensitivity for pre-capillary resistance.

Figure 14.7 is a basic model of systemic circulation with Q_1: left cardiac output (l min^{-1}), Q_2: flow rate across valve (l min^{-1}), C_1: arterial compliance (cm^3 mmHg^{-1}), C_2: venous compliance (cm^3 mmHg^{-1}), R: pre-capillary resistance (mmHg*min/l), P_1: aortic pressure (mmHg), P_2: venous pressure (mmHg), P_3: constant pressure (mmHg). To select the R value for the figure, autonomic nervous system model and capillary resistance of the systemic circulation were converted or drafted to mathematical formulas. These were then input into the computer finding the aortic pressure response relative to the change in cardiac output from the left heart. From figure 14.8 it is possible to obtain the derivation from the equilibrium point of the characteristics, as shown in equations (14.27) and (14.28).

$$Q_1 - Q_2 = C_1(\mathrm{d}p_1/\mathrm{d}t) \tag{14.27}$$

$$P_1 - P_2 = RQ_2 \tag{14.28}$$

Since P_2 is constant, $P = P_1 - P_2$; equations (14.27) and (14.28) then become:

$$Q_1 - Q_2 = C_1\{d(P_1 - P_2/dt)\} \tag{14.29}$$

$$P = RQ_2 \tag{14.30}$$

From equation (14.29),

$$Q_1 - Q_2 = C_1(dp_1/dt) + C_1(dp_2/dt) \tag{14.31}$$

$$Q_1 - Q_2 = C_1(dp_1/dt) \tag{14.32}$$

Laplace transformation of equations (14.31) and (14.32)

$$Q_1(s) - Q_2(s) = SC_1P(s) \tag{14.33}$$

$$P(s) = RQ_2(s) \tag{14.34}$$

Eliminating $Q_2(s)$ from equations (14.33) and (14.34)

$$\frac{P(s)}{Q(s)} = \frac{R}{1 + sC_1R} \tag{14.35}$$

If flow rate Q_1 is the step input :

$$DP = (DQ/S) \times \frac{R}{1 + sC_1R} \tag{14.36}$$

By inverse Laplace transformation we have $DP = DQ.\ R(1 - e^{\frac{-t}{R_1C_1}})$. Since the value of P_1 is the above plus the initial value, it may be expressed by equation (14.37)

$$P_1 = RQ_0 + P_{2+} + DQ.\ R.\left(1 - e^{\frac{-t}{R_1C_1}}\right) \tag{14.37}$$

The equation derived here involves linear delay. Since the time constant is negligible we can redefine equation (14.37) into (14.38) as:

$$e^{\frac{-t}{R_1C_1}} = 0 \tag{14.38}$$

Therefore, equation (14.37) can be written as

$$P_1 = RQ_0 + P_2 \tag{14.39}$$

This equation was utilized for finding deviation in aortic pressure. For selection of proper sensitivity the following conditions should be taken into account:

1. because it simulates a biological system, it must not produce any oscillation;
2. the error in the study state must be that of the Guyton model;
3. the time required to obtain the final value must be between 5 and 14 s [15].

14.8.2 Pump flow control system

The Starling curve represents the output performance curve of the natural heart. In the heart, the output is controlled according to change in the LA pressure. In order to realize this mechanism in our simulator, we need to include a system by which the pump output can be automatically controlled in accordance with changes in the LA pressure. The pump output can be dependent on the number of parameters. Of these parameters, the *S*/*D* ratio was selected because it is easy to control electrically, and it permits a wide range of output fluctuations to be produced. In the pump properties, we express them as LA pressure versus cardiac output using the *S*/*D* ratio as the parameter.

The functioning of natural heart pump phenomena is greatly different from that in our simulator. Unless some kind of control is provided for the pump, it is very difficult for the pump to perform with phenomena similar to those of the natural heart. In view of the above, with the help of a proper selected value of *S*/*D* ratio that produces the desired output curve, the preload is evaluated. LA pressure was refined through the computer to transmit signals to the control-drive console via the opening time of the solenoid valve.

In the computer we refined the Starling curve as follows:

1. pulse rate is fixed 80 bpm. Only *S*/*D* ratio is varied;
2. *S*/*D* ratio is varied in the range $10 < SD < 100$;
3. Select base point β and tangent θ of the Starling curve for input into the computer.

By means of the above procedure, the straight line shown in figure 14.8 is established within the computer. The *S*/*D* ratio is then calculated from the line. An output control test was done using our ordinary *in vitro*, test apparatus. In this test, pump drive pressure was 0.25 kg cm^{-2}, and four combinational values of β and θ lines are given in table 14.2.

If the graph is plotted between LAP–flow relationship for different Starling curves, then providing variations in parameters β and θ of the fluctuation in input pressure, one can obtain the desired value of the Starling curve. If we choose a large value of θ, then the flow rate is changed and the resulting pressure area will also change. When θ is small the flow characteristics are reversed. Intermediate values of θ produce uniform flow characteristics throughout the entire range of pressure.

Table 14.2. Dimensions of the Starling curve.

S. No.	Base point (β)	Tangent (θ)
1	3.33	0.00
2	0	0.02
3	−3.33	0.012
4	3.33	0.014

14.9 Basic characteristics of the mechanical simulator

The individual element was tested and inter-connected to form a total circulation system with a closed circuit loop. Table 14.3 shows the volume and dimensions of each simulator component. As for the measuring system, with the use of an electromagnetic blood flow meter we can calculate pump flow, the probe of which is integrated in the out flow portion of the left heart pump. Strain gauge type pressure transducers are used to measure the pressure but the output of the transducers is very low so we can connect pressure transducers to the amplifier in a polygraph. These hydrodynamic data are considered as an input to a computer.

14.9.1 Hydrodynamics under standard conditions

The presetting valve lifts of semi-fixed resistance used in the 'systemic and pulmonary' resistance was determined to be 5 and 7, providing pressure losses of 35 mmHg and 10 mmHg, respectively. Since the compliance value for these in the living body are not clear, the following compliance values were formulated from previous experimental data:

systemic capillary tanks : 3.0×10^{-7} cm^5 dyn^{-1};

pulmonary vein tank : $6.5v \times 10^{-4}$ cm^5 dyn^{-1}.

Table 14.3. Dimension and compliance of elastic tubes used for a 'vessel' induced from numerical analysis.

Component	Volume (V_{ml})	Length (L_{mm})	Diameter (mm)	Thickness (mm)
Heart	160			
LA	40	75	26	1.0
LV	40	75	26	1.0
RA	40	75	26	1.0
RV	40	75	26	1.0
Vessels	1440			
Systematic circulation	1240			
Aorta	70	300	17	0.4
Aortic resistance	25			
Sys. vein tank	850		110	
Cap. resist	25			
Vena cava	95	300	20	0.1
Connecting tube	175	990	14	
Pulmonary cir	200			
Pul. arteries	70	230	20	0.3
Pul. resist	25			
Pul. vein tank	80		38	
Connecting tubes	25	140	14	
Total	1600			

Cir.: Circulation; Pul.: Pulmonary; Cap.: Capillary.

Both the systemic capillary and pulmonary vein tanks are required to have the same volume of air of 60 ml. After these exercises, pressure wave forms were recorded for various parts. The blood pressure and mean blood pressure of both (simulator and goat) are the same at different locations. So the following points must be measured in evaluation of the pressure graph:

1. evaluation of reflux wave in a rigid pipe;
2. evaluation of overestimation of viscosity resistance in a windkessel model.

To express various vascular systems in terms of a windkessel model would be very essential for a mechanical model, i.e., in the analysis of its frequency properties. It would also be used for evaluating the mechanical model to understand the properties of elastic tubes. This may clarify the differences between the mechanical model and the living body and may provide useful information for the further improvement of such particular models.

14.9.2 Evaluation of Starling curve

Based on the result of the test on controlled output in LAP—flow relationship for different Starling curves, the basic performance test of the total circulatory simulator was conducted. The simulator was operated under the following conditions: left heart drive pressure 0.25 kg cm^{-2}; right heart drive pressure 0.08 kg cm^{-2}; control sensitivity for aortic pressure resistance 0.5.

After achieving a stable state for the system, to find the response of the system in the case of disturbance, a specific disturbance signal was given to the pressure transducer. Tests on different functions displayed in table 14.3, were performed. The test using control in accordance with function (2) resulted in the pressure stabilizing quickly at 100 mmHg, with the LA pressure and central venous pressure at the expected levels and the *S*/*D* ratio quite stable. It resulted in stable *S*/*D* ratio at 18% at which the basal flow was obtained. In contrast, function (4) took too much time for the atrial pressure to obtain the level of 100 mmHg showing that the system is not stable with these functions, but the LAP in this case was stable at 6 to 7 mmHg, the corresponding *S*/*D* ratio was constant at the minimum setting of 10%. For atrial resistance to be stabilized by the valve requires a minimum of 20 s. During this interval, the valve was closed to increase the atrial pressure. This means that the automatic flow control from the left heart had hardly any effect on the change in the specified range of the LAP. The ratio of *S*/*D* should be very small for the LAP to provide sufficient flow. From the above it can be seen that function (2) gives best performance and function (4) is unstable.

14.9.3 'Blood transfusion' experiment

Experiment on blood transfusion was performed as a test of the totally circulating system simulator. The normal simulator is a device that simulates the normal circulatory state of a living body. For this reason increase in total volume can be absorbed by the living body, but it cannot be possible in the simulator. In order to minimize this problem, we have designed a system with a volume assistant tank in the ‘caval vein’ and a volume storage tank simulating the peripheral vessel volume. In this light, the response to the increment of the circulating blood volume was explored, using an enhanced version of simulator.

The pump was driven not due to the control of the left ventricular output under Starling law, but rather to establish *S*/*D* ratio at the level of 18%. The test was conducted by the following methods on the proportional control sensitivity of the servo valve for the systematic resistance set at 0.0, 0.2, 0.4 and 0.6. Transfusion of 100 ml of water was performed five times at intervals of 2 min, totaling 500 ml. For determining of volume of blood transfusion in the animal for this we use the ratio of the body weight of the human subjects to that of the experimental animal considering the difference in the volume of blood in circulation. In terms of stability the following may noted:

1. the stability of aortic pressure becomes down as the sensitivity steps up, which was identified in the test on valve characteristics;
2. the circulatory system produces continuous oscillation at control sensitivity of 0.6 and above.

Comparisons were made of these data. It is evident from the figures that because there are no oscillations in its aortic pressure, the living body should belong to the area of control sensitivity 0.4 or below. The error in aortic pressure was figured out. In the living body, the error or aortic pressure is about 0 mmHg upon transfusion of 200 ml of the blood. In the simulator, if the sensitivity is very high, the aortic pressure or error is much smaller. Control sensitivity of 0.4 appeared to be optimal for simulation.

14.9.4 Comparative test on a single artificial heart (SAH)

Now we compare the test result obtain from the simulator and the result of the experiment conducted, a goat implanted with an SAH [16]. With the help of this test, it would be possible to justify the characteristics of the simulator in the systemic capillary area and to determine whether the device can actually be used in a practical situation. The test was carried out with the SAH connected as shown in figure 14.9. In this figure, the LV was replaced with the artificial heart. With this device, the responses were monitored upon blood transfusion in the goat with the ventricle stopped. Data were collected for the animal under two conditions: (1) normal volume; and (2) normal resistive condition.

A resistive element was added to the simulator by connecting it in parallel with the already installed pulmonary resistance [20]. The reason for the addition of this element is as follows:

1. the pulmonary vascular resistance characteristically increases initially, drops sharply at a certain point and then rises again. This makes it difficult for the convectional needle valve by itself to simulate the condition;
2. greater flow volume than before was desired.

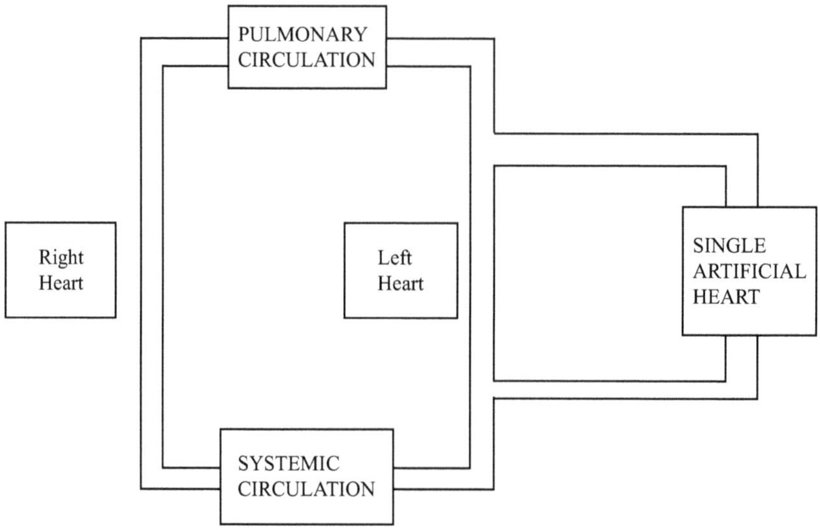

Figure 14.9. Schematic drawing of SAH mode in a goat.

Table 14.4. Driving pressure of the heart at different driving configurations.

Drive mode	Drive pressure kg cm^{-2}		*S*/*D* ratio
	Left	Right	
Drive condition fixed	0.30	0.0 (29.4)*	20
Automatic pressure level control	0.30	0.0 (29.4)*	Variable

Cir.: Circulation; Pul.: Pulmonary; Cap.: Capillary.

Data on the goat were obtained under two different circumstances: (1) the drive circumstances fixed; and (2) LAP of 5 mmHg by the use of our automatic control system designed [21]. The simulation was done with the circulating blood volume changed by the drive condition, as shown in table 14.4. The LAP was made constant by changing the *S*/*D* ratio.

Flow control based on Starling law was not occupied. First a steady state was obtained under the drive condition, as shown in table 14.4. A test was then performed by the following procedures to measure different pressure, flow, volumes and resistances.

1. From the venous area 50 ml of blood was removed twice.
2. 100 ml of blood was transformed to return the system to the normal state.
3. 200 ml blood was added to the venous assistant tank six times.
4. The *S*/*D* ratio was adjusted during the above to provide the LAP of 5 mmHg.
5. After removing 200 ml of blood the *S*/*D* ratio was again adjusted to provide the LAP of 5 mmHg.

The above steps are repeated five times.

The central venous pressure was almost identical. Because of the excessive gain in aortic pressure control, the aortic pressure produced by the simulator tended to be in the vicinity of 100 mmHg. This could be corrected by reducing the control sensitivity below 0.4.

At central venous pressure (CVP) greater than 12 mmHg, the flow volume produced by the simulator demonstrated properties that closely resembled those of the living body. At CVP less than 12 mmHg, differences were observed between these two systems. This seemed to be attributable to the narrow flow controllable range, which is a drawback of the pulsatile pump we used in the present experiment. It was found that this device can reproduce the function of the SAH with considerable accuracy:

1. requirement for design and assemble of a pulsatile pump with a larger range of variable flow;
2. arrangement of a pulmonary resistance control system that is convenient with the characteristics of the resistance.

14.10 Conclusion

From the above details, we have been able to achieve, at any part of the mechanical model, a pressure–flow relationship closely resembling those of the living body. It was confirmed in the SAH test that the device can readily be adjusted to produce a systemic capillary resistance value closely related to those of the living body. In the design of the mechanical simulator, we concentrated on the improvement of the performance and accuracy of the simulator. The following are problems that not have been described in this chapter, but should be determined for further improvement:

1. bronchial flow caused approximately 10% difference between the right and left heart output. The output of the right and left heart are identical to each other in this model [19];
2. it will mandatory to think about how to reproduce heart failure models and such abnormal models such as for a ventricular septal defect;
3. conflicting information on pressure metabolic response will be difficult to obtain with the simulator. This is because the input/output relationship is not understood very well at present. However, it may be possible to create a control algorithm through the accumulation of data from animal experiments.

Most of the time, the structure of the drive system is very large because optimal drive control is unknown. The present study was designed to eliminate such a problem.

References

[1] Kirsch U, Nassari M and Buchrl E 1966 Left ventricular bypass in experimental left heart failure *Trans. Am. Soc. Artif. Intern. Organs* **12** 53

[2] Akers w, Bannon W O, Hall G and Liotta D 1966 Design and operation of paracoporeal bypass pumping *Trans. Am. Soc. Artif. Intern. Organs* **12** 86

[3] Bregman D *et al* 1977 A pulsatile assist device (PAD) for use during cardiopulmonary bypass *Ann. Thorac. Surg.* 24 574–81
[4] Callaghan P, Watkins D H and Klink E J 1964 Orthopasic postsystolic myocardial augmentation: application of the technique to the treatment of shock *Archiv. Surg.* **89** 354
[5] Bregman D and Goetz R H 1972 A failsafe cardiac assist system for intraaortic balloon pumping *Trans. Am. Soc. Artif. Intern. Organs* **13** 505
[6] Rashvsky N 1964 *Some Medical Aspect of Mathematical Biology* (Springfield, IL: Charles C. Thomas Publisher)
[7] Oleon H I J F D, Chevrier J and Dubost C 1967 A time and volume controlled extracorporeal pump. Technical description, performance and experimental application to the left heart supporting *Trans. Am. Soc. Artif. Intern. Organs.* **13** 317
[8] Klopp E and Jacobs L *et al* 1969 Use of left atrial pressure as a control parameter for total left heart bypass *Trans. Am. Soc. Artif. Intern. Organs* **15** 391
[9] Bednaric B *et al* 1972 Experimental evalution of left atrial pressure controlled left ventricular bypass *Scriptamedidica* **45** 307
[10] Willdevuur C, Marva G and Crosby M 1968 An artificial heart sensitive to atrial volume *Trans. Am. Soc. Artif. Intern. Organs* **14** 276
[11] Klain M and Marva G *et al* 1971 Can we achieve over 100 hour's survival with total mechanical heart? *Trans. Am. Soc. Artif. Intern. Organs* **17** 437
[12] Rosenberg G and Landis D *et al* 1978 Determining arterial pressure, left atrial pressure and cardiac output form the left pneumatic drive line of the total artificial heart *Trans. Am. Soc. Artif. Intern. Organs* **24** 341
[13] Affeld K, Echt M and Mohmaupt A *et al* 1975 A new method for drift free measurement of venous pressure *Eur. Soc Artif Organs* **2** 142
[14] Green H 1944 *Medical Physics: The Year Book* 1 (Chicago, IL: Medical Publisher) pp 210
[15] Guyton A C 1956 Pressure–volume curve of the entire arterial and venous system in the living animal *Am. J. Physiol.* **184** 253–8
[16] Tanko H 1984 Circulatory maintenance with a single artificial heart (SAH) *Jap. J. Artif. Organs.* **13** 109–12
[17] Tsuciya K and Umezu M 1975 Mechanical simulator to cardiovascular system *Memoirs of the School of Science and Engineering, Waseda Univsity* 39 (Tokyo: Waseda University) 1–14 p
[18] Umezu M 1979 Engineering analysis of cardiovascular system for developing a mock circulator *J. Tokyo Womens Medical College* **49** 626–39
[19] Umezu M 1983 Design of pulse duplicator system for prosthetic heart valves *Proc. of the Multi-national Instrumentation Conf. (Shanghai)* Part II, pp 751–8
[20] Umezu M 1984-1 Mechanical simulator for testing artificial organs *Progress in Artificial Organs—1983 (Proc. ISAO)* **2** 907–12
[21] Umezu M 1984-2 Microcomputer based automatic level control system for left ventricular assist device *Life Support System (Proc ESAO)* **2** 258–61
[22] Umezu M 1984-3 Comparable hydrodynamic study of 26 commercially available prosthetic heart valve *Life Support System (Proc. ESAO)* **2** 70–3
[23] Wada A 1984 Mechanical simulation model of blood circulatory system with automatic control *Master Thesis* Department of Mechanical Engineering, Waseda University

www.ingramcontent.com/pod-product-compliance
Ingram Content Group UK Ltd.
Pitfield, Milton Keynes, MK11 3LW, UK
UKHW051902100926
8556IPUK00005B/735

9 780750 320863